Artificial Intelligence and Communication Techniques in Industry 5.0

The book highlights the role of artificial intelligence in driving innovation, productivity, and efficiency. It further covers applications of artificial intelligence for digital marketing in Industry 5.0 and discusses data security and privacy issues in artificial intelligence, risk assessments, and identification strategies.

This book:

- Discusses the role of artificial intelligence applications for digital manufacturing in Industry 5.0.
- Presents blockchain methods and data-driven decision-making with autonomous transportation.
- Covers reinforcement learning algorithm and highly predicted models for accurate data analysis in industry automation.
- Highlights the importance of robust authentication mechanisms and access control policies to protect sensitive information, prevent unauthorized access, and enable secure interactions between humans and machines.
- Explains attack pattern detection and prediction which play a crucial role in ensuring the security of business systems and networks.

It is primarily written for senior undergraduates, graduate students, and academic researchers in the fields of electrical engineering, electronics and communication engineering, computer engineering, industrial engineering, manufacturing engineering, and production engineering.

Advances in Manufacturing, Design and Computational Intelligence Techniques

Series Editor – Ashwani Kumar – *Senior Lecturer, Mechanical Engineering, at Technical Education Department, Uttar Pradesh, Kanpur, India*

The book series editor is inviting edited, reference and textbook proposal submission in the book series. The main objective of this book series is to provide researchers a platform to present state-of-the-art innovations, research related to advanced materials applications, cutting-edge manufacturing techniques, innovative design and computational intelligence methods used for solving nonlinear problems of engineering. The series includes a comprehensive range of topics and its application in engineering areas such as additive manufacturing, nanomanufacturing, micromachining, biodegradable composites, material synthesis and processing, energy materials, polymers and soft matter, nonlinear dynamics, dynamics of complex systems, MEMS, green and sustainable technologies, vibration control, AI in power station, analog-digital hybrid modulation, advancement in inverter technology, adaptive piezoelectric energy harvesting circuit, contactless energy transfer system, energy efficient motors, bioinformatics, computer aided inspection planning, hybrid electrical vehicle, autonomous vehicle, object identification, machine intelligence, deep learning, control-robotics-automation, knowledge based simulation, biomedical imaging, image processing and visualization. This book series compiled all aspects of manufacturing, design and computational intelligence techniques from fundamental principles to current advanced concepts. Books in the series include,

Hybrid Metal Additive Manufacturing: Technology and Applications
Edited by Parnika Shrivastava, Anil Dhanola, and Kishor Kumar Gajrani

Thermal Energy Systems: Design, Computational Techniques, and Applications
Edited by Ashwani Kumar, Varun Pratap Singh, Chandan Swaroop Meena, and Nitesh Dutt

Artificial Intelligence and Communication Techniques in Industry 5.0
Edited by Payal Bansal, Rajeev Kumar, Ashwani Kumar, and Daniel D. Dasig, Jr.

For more information about the series, please visit www.routledge.com/Advances-in-Manufacturing-Design-and-Computational-Intelligence-Techniques/book-series/CRCAIMDCIT?publishedFilter=alltitles&pd=published,forthcoming&pg=1&pp=12&so=pub&view=list?publishedFilter=alltitles&pd=published,forthcoming&pg=1&pp=12&so=pub&view=list

Artificial Intelligence and Communication Techniques in Industry 5.0

Edited by
Payal Bansal, Rajeev Kumar, Ashwani Kumar,
and Daniel D. Dasig, Jr.

CRC Press is an imprint of the
Taylor & Francis Group, an **informa** business

Designed cover image: NMStudio789/Shutterstock

First edition published 2025
by CRC Press
2385 NW Executive Center Drive, Suite 320, Boca Raton FL 33431

and by CRC Press
4 Park Square, Milton Park, Abingdon, Oxon, OX14 4RN

CRC Press is an imprint of Taylor & Francis Group, LLC

© 2025 selection and editorial matter, Payal Bansal, Rajeev Kumar, Ashwani Kumar and Daniel D. Dasig, Jr.; individual chapters, the contributors

Reasonable efforts have been made to publish reliable data and information, but the author and publisher cannot assume responsibility for the validity of all materials or the consequences of their use. The authors and publishers have attempted to trace the copyright holders of all material reproduced in this publication and apologize to copyright holders if permission to publish in this form has not been obtained. If any copyright material has not been acknowledged please write and let us know so we may rectify in any future reprint.

Except as permitted under U.S. Copyright Law, no part of this book may be reprinted, reproduced, transmitted, or utilized in any form by any electronic, mechanical, or other means, now known or hereafter invented, including photocopying, microfilming, and recording, or in any information storage or retrieval system, without written permission from the publishers.

For permission to photocopy or use material electronically from this work, access www.copyright.com or contact the Copyright Clearance Center, Inc. (CCC), 222 Rosewood Drive, Danvers, MA 01923, 978-750-8400. For works that are not available on CCC please contact mpkbookspermissions@tandf.co.uk

Trademark notice: Product or corporate names may be trademarks or registered trademarks and are used only for identification and explanation without intent to infringe.

ISBN: 9781032798202 (hbk)
ISBN: 9781032798196 (pbk)
ISBN: 9781003494027 (ebk)

DOI: 10.1201/9781003494027

Typeset in Sabon
by Newgen Publishing UK

Contents

Aim and Scope ix
Preface xi
About the Editors xv
Contributors xix
Acknowledgment xxv

1 Digital Manufacturing: Artificial Intelligence in Industry 5.0   1
ASHWANI KUMAR, PAYAL BANSAL, ABHISHEK KUMAR,
ASHISH KUMAR GUPTA, AND AMIT CHOUDHARI

2 The Rise of Industry 5.0: How Artificial Intelligence Is Shaping the Future of Manufacturing   26
VEERENDRA SINGH AND RAJENDRA KUMAR

3 Industry 5.0 with Artificial Intelligence: A Data-Driven Approach   47
NAMITA KATHPAL, PRATIMA MANHAS, JYOTI VERMA, AND SEEMA JOGAD

4 An Explorative Study on the Use of Artificial Intelligence in Oman High School Education   55
POOJA CHHABRA AND SAMEER BABU M

5 Evolving Industries: A Journey from Industry 1.0 to 5.0   73
SANJEEV KUMAR, GEETA TIWARI, AND NEERAJ TIWARI

6 The Industrial Revolution: From Mechanisation (1.0) to Smart Automation (5.0)   87
BABITA JHA, SARTHAK GARG, AND SARTHAK DHINGRA

7 Enhancing the Digital Economy in the Context of the
   Fourth Industrial Revolution: The Case of Vietnam   101
   NGUYET THAO HUYNH AND CHIEN-VAN NGUYEN

8 Machinery to Mind: Navigating the Transformation from
   Industry 1.0 to Industry 5.0   116
   M. C. SHANMUKHA, K. C. SHILPA, AND A. USHA

9 The Role of Cutting-Edge Technologies in Revolutionary
   Industry 5.0   128
   ABHAY BHATIA

10 Managing Industry 5.0: The Next Frontier for Artificial
   Intelligence and Machine Learning Algorithms   154
   VARSHA SAHNI AND EKTA BHAGGI

11 Industry 5.0: Security Based on Optimal Thresholding   167
   ARCHIKA JAIN AND DEVENDRA SOMWANSHI

12 Industry 5.0: Revolutionizing Energy Management
   through Smart Grid Integration and Sustainable Solutions   185
   SONAL I. SHIRKE, PAYAL BANSAL, AND SUSHIL JAIN

13 Dynamic Human–Artificial Intelligence Collaboration
   Framework for Adaptive Work Environments in
   Industry 5.0   210
   NISHA BANERJEE AND ANIKET BHATTACHARYEA

14 An Overview of Artificial Intelligence Algorithms for
   Future Generation   229
   RITU AND BEBESH TRIPATHY

15 Artificial Intelligence Algorithms: Conspectus and Vision   239
   ARPITA TEWARI

16 A Historical, Present, and Prospective Review of
   Artificial Intelligence's Role in Securing Personal
   Information and Private Data   257
   SHRUTI GUPTA, NAVIN KUMAR GOYAL, AND AJAY KUMAR

17 Blockchain-Enabled Sentiment Analysis with OpenID
   Authentication for Data Security and Integrity   273
   RAKTIM KUMAR DEY, DEBABRATA SARDDAR,
   RAJESH BOSE, SHRABANI SUTRADHAR, AND SANDIP ROY

18	Introduction to Application of Artificial Intelligence in Agricultural Industry: Industry 5.0 Use Case	289
	MRUNALINI BHANDARKAR, PAYAL BANSAL, AND BASUDHA DEWAN	
19	Artificial Intelligence and Computer Vision in Sustainable Vertical Farming: A Key Component of Industry 5.0 Revolution	306
	ARCHANA BHAMARE AND PAYAL BANSAL	
20	Technological Solutions for Waste Management in Indian Hotels: Exploring Feasibility and Environmental Implications	318
	MONIKA, KIRAN SHASHWAT, AND UMANG BHARTWAL	
21	A Study on the Use of AI in Academic Curriculum and Its Role in the Industry 5.0 Revolution	333
	POOJA CHHABRA AND SAMEER BABU M	
22	Analyze the Performance of Multileg Thermocouples with Different Thermoelectric Materials and Device Optimization Using Multiphysics Simulation	346
	MAYANK MEWARA, SHOBI BAGGA, SAPNA GUPTA, AND RAHUL PRAJESH	
23	Multi-Objective Optimization Strategy for Enhanced Accuracy and Productivity in AWJ Machining	360
	ABHIJIT SAHA, RAJEEV RANJAN, ASHOK KUMAR YADAV, ASHWANI KUMAR, AND PAYAL BANSAL	
24	Enhancing Hospital Management Through Robotics: Developing a Custom API	373
	S. JAVEED HUSSAIN, SIVASAKTHIVEL THANGAVEL, AND ASHWANI KUMAR	

Index 393

Aim and Scope

Artificial intelligence (AI) plays a crucial role in enhancing the supply chain and logistics management processes. With AI-powered analytics and optimization algorithms, companies can optimize inventory levels, predict demand patterns, and streamline transportation routes. This leads to improved efficiency, reduced costs, and faster delivery times. Adaptive and adversarial attacks in Industry 5.0 requires a comprehensive and proactive approach, combining robust defense mechanisms, adversarial training, input validation, regular updates, and responsible AI practices. By adopting these measures, organizations can enhance the security and trustworthiness of their AI systems in the face of evolving cyber security threats.

The book *Artificial Intelligence and Communication Techniques in Industry 5.0* has 24 chapters which lays the groundwork for reader understanding of Industry 5.0 by first providing a solid foundation in the core concepts of AI. This equips them with the essential knowledge to navigate the exciting world of human–machine collaboration in manufacturing. Next, the book embarks on a fascinating journey, tracing the evolution of manufacturing from its humble beginnings in Industry 1.0 to the cutting-edge world of Industry 5.0. This historical perspective allows reader to appreciate the dramatic advancements that have transformed the manufacturing landscape. With this foundation set, the book explores deep into the heart of Industry 5.0: the harmonious collaboration between humans and machines. It explores how AI empowers humans by taking over repetitive tasks, allowing us to focus on our strengths—creativity, problem-solving, and strategic decision-making. The book then highlights the transformative power of AI for manufacturing processes. It explains how AI can be harnessed to optimize production lines, streamline workflows, and ensure consistent quality.

The book *Artificial Intelligence and Communication Techniques in Industry 5.0* discovers how AI facilitates seamless collaboration between humans and machines, utilizes intelligent systems to optimize processes, and leverages predictive capabilities to anticipate and address potential issues before they arise. Furthermore, the book examines how Industry 5.0 revolutionizes supply chain optimization. It details how AI can be used to streamline logistics, optimize inventory management, and ensure efficient delivery of materials throughout the production process. Finally, the book showcases the power of AI in action by exploring success stories in AI-powered manufacturing. Real-world examples demonstrate the tangible benefits of AI, providing a glimpse into the exciting future of intelligent manufacturing.

The book explains various concepts of AI, communication techniques, and Industry 5.0. It explores that Industry 5.0 is the next generation of manufacturing, characterized by the seamless integration of AI, robotics, and automation. AI and automation can handle repetitive tasks, freeing up human workers to focus on higher-level cognitive tasks, such as problem-solving, creativity, and decision-making. For instance, AI-powered robots can take over tasks like welding on an assembly line, allowing human workers to focus on design and quality control. Industry 5.0 promotes collaborative workspaces where humans and machines work side-by side. This requires safety protocols to ensure a harmonious relationship. Sensors can be embedded in machinery to detect human presence and prevent accidents. By automating tasks and improving efficiency, Industry 5.0 can lead to significant increases in productivity. Additionally, AI can be used to analyze data and identify new opportunities for innovation. For example, AI can be used to optimize factory layouts or predict equipment failures before they happen.

This book, *Artificial Intelligence and Communication Techniques in Industry 5.0*, caters to a wide audience at the forefront of technological advancement. Postgraduate students, PhD researchers, and aspiring researchers will find the content invaluable as they explore into various engineering fields. The book also serves as a comprehensive reference for professionals like data scientists, mechanical, electrical, and software engineers, as well as experts in heat transfer and specialists in digital manufacturing. The book's reach extends beyond specific disciplines. Its content is highly relevant to research and development in a vast array of industries, including heat transfer, manufacturing, design engineering, automotive manufacturing, aviation, electronics, and civil engineering. This makes it a crucial resource for anyone engaged in research and innovation across these diverse sectors.

Editor
Payal Bansal
Rajeev Kumar
Ashwani Kumar
Daniel D. Dasig, Jr.

Preface

The book presents 24 chapters dealing with the applications areas of artificial intelligence (AI) and communication techniques in Industry 5.0. **Chapter 1** explores the applications of AI in various aspects of manufacturing, including intelligent automation, predictive maintenance, and quality control. By leveraging AI's capabilities for data analysis, machine learning, and decision-making, manufacturers can achieve significant gains in efficiency, productivity, and product quality. **Chapter 2** explores this transformation, detailing the key features of Industry 5.0, human–machine collaboration, sustainability, and advancements like AI. **Chapter 3** describes the various emerging applications of AI in the era of Industry 5.0 revolution. **Chapter 4** investigates the high school writing curriculum in the Sultanate of Oman and anticipates challenges stemming from the integration of AI. In recent years, Oman has emphasized educational reforms to enhance student competencies in writing, aligning with global trends towards technological advancement.

Chapter 5–6 elaborate on the historical evolution of industrial revolutions and their far-reaching implications on society and the economy. It has a flow that starts with the advent of Industry 1.0 characterized by steam power and mechanization, flows through Mass Production (Industry 2.0), the Digital Revolution (Industry 3.0) to the era of Smart Manufacturing (Industry 4.0). The focus now moves towards Industry 5.0, defining the concept of the human-centric future where technology blends with human capabilities. **Chapter 7** highlights the case study of Fourth Industrial Revolution in Vietnam. Research results show that payment efficiency, technology factors, convenience, service factors and risks affect decisions of non-cash payments. The research results also discuss solutions for developing the digital economy in the coming time. **Chapter 8–9** highlight the machinery to mind journal and shed light on Industry 5.0 and its remarkable applications. **Chapter 10** covers significant advances in machine learning, deep learning, robotics, and natural language processing. The research provides detailed information about AI and machine learning algorithms that will revolutionize Industry 5.0.

Chapter 11 solves the problem of intellectual property theft that can be resolved via watermarking. It is a method of concealing data or retrieving information in real pictures, real audio–video, and paper work so that it canot be easily retrieved by an unauthorized individual. DCT, DWT, LSB, and SSM are some of the algorithms used to insert watermarks in digital photographs. Chapter 12 explores that energy management systems are improved when Smart Grid technology and Industry 5.0 principles are combined, making them more sustainable and effective. Chapter 13 aims to implement a framework that focuses on AI and the communication technologies to create collaboration between human workers and intelligent systems in Industry 5.0 dynamically, for a user-friendly integration within the industrial work environment. Chapter 14 also explores new AI paradigms that have the potential to expand the field of AI research and deployment in the future including explainable AI, federated learning, and algorithms based on quantum computing. In addition, the chapter looks at the moral and societal ramifications of AI algorithms becoming more widely used.

Chapter 15–24 are application-based chapters. Chapter 15 has a comprehensive objective of acting as a valuable resource for academia, industrial professionals, and decision-makers in diverse real-world scenarios and application domains, with a specific emphasis on the technical perspective. It centers on the utilization of AI algorithms in various domains such as Internet of Things (IoT) data, cyber-security data, mobile data, business data, health data, and social media data. These applications are derived from extensive research in addition to the implementation of machine learning, deep learning, and transformer models. Chapter 16 analyzes the last decade of cybercrime data and checks existing trends, and potential future approaches to explore the use of AI in identifying and preventing cybercrime. In continuation, Chapter 17 introduces a novel sentiment analysis system that utilizes Ethereum blockchain and OpenID authentication for improved data security and integrity. Traditional sentiment analysis approaches vary based on data type and processing methods, leading to diverse technical procedures and potential cultural biases in results. This study addresses these issues by proposing a system that harnesses blockchain technology and OpenID authentication, ensuring enhanced privacy and protection against data tampering.

AI in agriculture is highlighted in Chapters 18–19. In Chapter 18, an attempt is made to explore the use of AI in the agriculture sector and the challenges faced in its deployment. In India, agriculture is vital for economic advancement. One of the significant challenges faced by Indian farmers is selecting the appropriate crop based on the specific region. Chapter 19 emphasizes how the significance of AI in hydroponics and vertical gardening is useful to the agricultural industry. This study also emphasizes the creation of a computer vision-based AI plant growth monitoring system. Real-time and automated plant growth and health evaluation will be made possible by the system using CV algorithms and AI techniques.

Chapter 20 endeavors to shed light on the efficacy, challenges, and outcomes associated with the utilization of technology in the reduction of food waste. To achieve this, a comprehensive analysis of technological interventions that have been applied by various hotels in India is being employed. **Chapter 21** looks at how AI is improving academic writing and how that helps Industry 5.0. In universities, AI tools are making academic writing easier. **Chapter 22** presents a comparative study of multileg thermocouple with different thermoelectric materials from lower bandgap toward higher bandgap and different conducting plate materials at low temperature ranges and investigates influence on voltage sensitivity. **Chapter 23** deals with multi-objective optimization strategy for enhanced accuracy and productivity in AWJ machining. The results provide useful information for optimizing the AWJ machining procedure for graphite-filled glass fiber reinforced epoxy composites, leading to increased accuracy and productivity. **Chapter 24** deals with the integration of robotics and automation into healthcare systems. Using a cutting-edge robotic platform like the TEMI robot, hospital managers can solve many tasks related to hospital management with ease. It is presented in this chapter that an Application Programming Interface (API) has been developed that can be customized for the TEMI robot and used to streamline the hospital management operations.

Editor
Payal Bansal
Rajeev Kumar
Ashwani Kumar
Daniel D. Dasig, Jr.

About the Editors

Payal Bansal has received a Diploma, BE, MTech, and PhD degree in the area of Electronics and Communication Engineering. She has vast teaching experience of more than 15 years in reputed organizations. In her professional career, she has been associated with reputed organizations. In her research, she contributed remarkable research in the field of radio frequency communication and developed a filter that can work as a protected sheet to prevent high-frequency radio signals. Currently, she is working as Dean R& D & Associate Professor in the Department of Electronics and Communication Engineering, Poornima College of Engineering, Jaipur, Rajasthan. She has published more than 20 research articles in peer-reviewed journals including IEEE Explore, SCOPUS, and SCI indexed, in which eight journal papers, nine conference papers, seven book chapters published, five national and two international books, five patents and one copy right is there. Her main area of research is RF communication, Internet of Things, wide band antenna designs, and machine learning. She has organized several national and international conferences, symposia, and webinars. Bansal has also published three patents in the field of antenna designs, Internet of Things-based applications under the Intellectual Property Rights of Government of India. She is actively associated with NAAC SSR preparations at universities interface promotional activities. She is a professional member of several national and international bodies.

Rajeev Kumar is a proficient academician and academic administrator with more than 14 years of experience in developing the strategy toward excellence in professional education. Kumar is currently serving as Professor of Computer Science and Engineering Department, Moradabad Institute of Technology, Moradabad, Uttar Pradesh, India. He earned the intellect in distinction as PhD (Computer Science), DSc (Post-Doctoral Degree) in Computer Science, and Postdoctoral Fellowship (Malaysia). He has done certification in Data Science and Machine Learning using python and R Programming from IIM Raipur and certification from IBM, Google, etc. Senior member of IEEE and core team member of IEEE young professional committee and he is having membership of Computer Society of India, and AIEEE. He has participated in many training programs in leadership and also delivers expert talks on how to develop or enrich the curriculum, development of PO, PSO, PEOs, and how to design the vision and mission and the implementation. His academic areas of interest and specialization include artificial intelligence, cloud computing, e-governance and networking. Kumar is actively associated with NAAC SSR preparations at universities interface. He has visited international countries—London (United Kingdom), Mauritius, Dubai—for academic promotional activities. Under his supervision, nine scholars were awarded their PhD and four scholars' research work is ongoing in the field of computer science. He is awarded three times best project supervisor award. Kumar guides institutions toward academic excellence; promotes a culture of research, innovation, and leads new curricula and program development efforts. He is an expert in designing and deployment academic curriculum, and program inception. He organizes the seminar, workshops for the students and faculty to promote the research. Kumar also nurtures the ecosystem for incubation for the students as well as mentors. Under his leadership, many ideas are under incubation, and few are incubated and developed prototypes. Kumar developed a short-term course on artificial intelligence, machine learning, deep learning and its application and published 11 patents (National/International) mostly in Computer Science and Engineering to serve the nation in the field of research. Kumar has authored and coauthored more than 115 research papers in refereed international journals (SCI and non-SCI) and conferences like *IEEE, Springer, American Journal of Physics* (USA), and many international and national conferences, like IIT (International Conference), etc. and serves as editor of different international journals and is a reviewer in SCI and Non-SCI journal, serves as a committee member in many IEEE Society, Springer international conferences. He has delivered guest lectures and keynote speeches, and has chaired sessions in many national and international conferences, webinars, faculty development programs and workshops. Kumar is associated with many high-impact society memberships like IEEE, ACM, CSI, IAENG, and ISC. His key contributions include introducing the activity-based learning system and skill-based learning system through project-based learning instead of classroom teaching in all programs across the university and the development and maintenance of a student-centric ambiance on the campus.

Ashwani Kumar holds a PhD in Mechanical Engineering, specializing in Mechanical Vibration and Design. Currently serving as Senior Lecturer in Mechanical Engineering (Gazetted Officer Group B) at the Technical Education Department of Uttar Pradesh, India, under the Government of Uttar Pradesh, he assumed this role in December 2013. Prior to this, Kumar worked as Assistant Professor in the Department of Mechanical

Engineering at Graphic Era University, Dehradun, India, with an NIRF Ranking of 55, from July 2010 to November 2013. With over 13 years of combined experience in research, academia, and administration, Kumar has undertaken various roles, including Coordinator for AICTE-Extension of Approval, Nodal Officer for PMKVY-TI Scheme (Government of India), Internal Coordinator for the CDTP scheme (Government of Uttar Pradesh), Industry Academia Relation Officer, Assistant Centre Superintendent (ACS) in the Institute Examination Cell, and Zonal Officer for conducting the Joint Entrance Examination (JEE-Diploma), and Sector Magistrate for State Assembly Election held in Uttar Pradesh. As an academician and researcher, Kumar serves as the Series Editor for five book series published by CRC Press (Taylor & Francis, USA) and Wiley Scrivener Publishing, titled as Advances in Manufacturing, Design and Computational Intelligence Techniques, Renewable and Sustainable Energy Developments, Smart Innovations and Technological Advancements in Mechanical and Materials Engineering, Solar Thermal Energy Systems, and Computational Intelligence and Biomedical Engineering. Furthermore, he is the Guest Editor of a special issue titled "Sustainable Buildings, Resilient Cities, and Infrastructure Systems" in Buildings (ISSN: 2075-5309, I.F. 3.8). Kumar holds the position of Editor-in-Chief for the *International Journal of Materials, Manufacturing and Sustainable Technologies* (IJMMST, ISSN: 2583-6625) and serves as the Editor of the *International Journal of Energy Resources Applications* (IJERA, ISSN: 2583-6617). Additionally, he acts as a guest editor and editorial board member for eight international journals and is a Review Board Member for 20 prestigious international journals indexed in SCI/SCIE/Scopus, including *Applied Acoustics, Measurement, JESTEC, AJSE, SV-JME,* and *LAJSS*. Kumar has a prolific publication record, with more than 100 articles in journals, book chapters, and conferences. He has authored, co-authored, and edited more than 40 books in Mechanical, Materials, and Renewable Energy Engineering. Holding three patents, he is recognized for his contributions to academia and research, earning the title of Best Teacher for excellence in both fields. Actively engaged in academic pursuits, Kumar has successfully supervised 15 BTech, MTech, and PhD theses and serves as an external doctoral committee member of SRM University, New Delhi. His current research interests encompass AI and ML in Mechanical Engineering, Smart Materials and Manufacturing Techniques, Thermal Energy Storage, Building Efficiency, Renewable Energy Harvesting, and Sustainable Mobility. Kumar's Orcid is 0000-0003-4099-935X, and Google Scholar web link is https://scholar.google.com/citations?hl=en&user=KOILpEkAAAAJ.

Daniel D. Dasig, Jr. is currently an Associate Professorial Lecturer 6 of Graduate Studies of Science and Computing and De La Salle University Dasmarinas, Philippines. Dasig has PhD in Educational Management, and Doctor of Business Administration degree. He also holds a MS Engineering major in Computer Engineering, and Bachelor of Science in Computer Engineering, and currently pursuing a degree in Doctor of Public Administration. He has published a copious of journal articles indexed in ESCI, SCI and Scopus, and co-edited books for engineering and technology research domains. Currently, he is the Dean—Institute of Engineering, and Program Chair—Computer Engineering of Mandaluyong College of Science and Technology.

Contributors

Shobi Bagga
Electronics Engineering Department Rajasthan Technical University, Kota, Rajasthan, India

Nisha Banerjee
NSHM Knowledge Campus, Kolkata, West Bengal, India

Payal Bansal
Associate Professor, Department of Electronics, Poornima Institute of Engineering and Technology, Jaipur, Rajasthan, India

Ekta Bhaggi
Department of Computer Science and Engineering, Lovely Professional University, Phagwara, Punjab, India

Archana Bhamare
Research Scholar, Poornima University, Jaipur, Rajasthan, India and Assistant Professor, PCCOE, Pune, Maharashtra, India

Mrunalini Bhandarkar
Research Scholar, Poornima University, Jaipur, India and Assistant Professor, Pimpri Chinchwad College of Engineering, Pune, Maharashtra, India

Umang Bhartwal
Assistant Professor, Suresh Gyan Vihar University, Jaipur, Rajasthan, India

Abhay Bhatia
Department of Computer Science & Engineering, Roorkee Institute of Technology, Roorkee, Uttarakhand, India

Aniket Bhattacharyea
Developer Advocate, Draft.dev India

Rajesh Bose
Professor, Department of Computer Science & Engineering, JIS University, West Bengal, India

Pooja Chhabra
Lincoln University College, Malaysia

Amit Choudhari
Mechanical Engineering Department, Cleveland State University, Cleveland, OH 44115, USA

Basudha Dewan
Assistant Professor, Poornima University, Jaipur, Rajasthan, India

Raktim Kumar Dey
Assistant Professor, Department of Computational Sciences, Brainware University, West Bengal, India

Sarthak Dhingra
Christ (Deemed to be University), Delhi, NCR, India

Sarthak Garg
Christ (Deemed to be University), Delhi, NCR, India

Navin Kumar Goyal
Associate Professor, Department of Computer Engineering, Poornima Institute of Engineering & Technology, Jaipur, Rajasthan, India

Ashish Kumar Gupta
School of Mechanical and Aerospace Engineering, Oklahoma State University, Stillwater, OK 74078, USA

Sapna Gupta
Electronics Engineering Department Rajasthan Technical University, Kota, Rajasthan, India

Shruti Gupta
Assistant Professor, Department of Computer Engineering, Poornima Institute of Engineering & Technology, Jaipur, Rajasthan, India

S Javeed Hussain
Global College of Engineering and Technology, Muscat, Oman

Nguyet Thao Huynh
Thu Dau Mot University, Binh Duong, Vietnam

Archika Jain
Swami Keshvanand Institute of Technology, Management & Gramothan (SKIT), Jaipur, Rajasthan, India

Sushil Jain
Poornima University, Jaipur, Rajasthan, India

Babita Jha
Christ (Deemed to be University), Delhi, NCR, India

Seema Jogad
Department of Computer Science & Engineering, Bennett University, Greater Noida, Uttar Pradesh, India

Namita Kathpal
Department of Electronics & Communication Engineering, Manav Rachna International Institute of Research & Studies, Faridabad, Haryana, India

Abhishek Kumar
J. Mike Walker '66 Department of Mechanical Engineering, Texas A&M University, 100 Mechanical Engineering Office Building, College Station, TX 77843-3123, USA

Ajay Kumar
Assistant Professor, School of Engineering & Technology, Raffles University, Neemrana, Rajasthan, India

Ashwani Kumar
Senior Lecturer, Department of Mechanical Engineering, Technical Education Department, Kanpur, Uttar Pradesh, India

Rajendra Kumar
Department of Physics, Faculty of Engineering and Technology, Rama University, Kanpur, Uttar Pradesh, India

Sanjeev Kumar
Department of Electrical Engineering, Swami Keshvanand Institute of Technology, Management & Gramothan, Jaipur, Rajasthan, India

Sameer Babu M
Lincoln University College, Malaysia

Pratima Manhas
Department of Electronics & Communication Engineering, Manav Rachna International Institute of Research & Studies, Faridabad, Haryana, India

Mayank Mewara
Electronics Engineering Department Rajasthan Technical University, Kota, Rajasthan, India

Monika
Associate Professor, UITHM, Chandigarh University, Mohali, Punjab, India

Chien-Van Nguyen
Thu Dau Mot University, Binh Duong, Vietnam

Rahul Prajesh
CSIR-CEERI, Pilani, Rajasthan, India

Rajeev Ranjan
Department of Mechanical Engineering, Dr. B.C. Roy Engineering College, Durgapur, West Bengal, India

Ritu
Department of Computer Science & Engineering, Chandigarh College of Engineering, Chandigarh Group of Colleges, Jhanjeri-140307, Punjab, India

Sandip Roy
Professor, Department of Computer Science & Engineering, JIS University, West Bengal, India

Abhijit Saha
Department of Mechanical Engineering, Haldia Institute of Technology, Haldia, West Bengal, India

Varsha Sahni
Department of Computer Science and Engineering, Lovely Professional University, Phagwara, Punjab, India

Debabrata Sarddar
Assistant Professor, Department of Computer Science & Engineering, University of Kalyani, West Bengal, India

M. C. Shanmukha
Department of Mathematics, PES Institute of Technology and Management, Shivamogga, Karnataka, India

Kiran Shashwat
Assistant Professor, UITHM, Chandigarh University, Mohali, Punjab, India

K. C. Shilpa
Department of AI & ML, PES Institute of Technology and Management, Shivamogga, Karnataka, India

Sonal I. Shirke
Poornima University, Jaipur, Rajasthan, India and
Pimpri Chinchwad College of Engineering, Pune, Maharashtra, India

Veerendra Singh
Department of Mechanical Engineering, Faculty of Engineering and Technology, Rama University, Kanpur, Uttar Pradesh, India

Devendra Somwanshi
Poornima College of Engineering, Jaipur, Rajasthan, India

Shrabani Sutradhar
Assistant Professor, Department of Computational Sciences, Brainware University, West Bengal, India

Arpita Tewari
Department of Information Technology, Manipal University, Jaipur, Rajasthan, India

Sivasakthivel Thangavel
Global College of Engineering and Technology, Muscat, Oman

Geeta Tiwari
Department of Computer Engineering, Poornima College of Engineering, Jaipur, Rajasthan, India

Neeraj Tiwari
Professor, Poornima University, Jaipur, Rajasthan, India

Bebesh Tripathy
Department of Computer Science & Engineering, Chandigarh University, Mohali, Punjab, India

A. Usha
Alliance School of Applied Mathematics, Alliance University, Anekal-Chandapura Road, Bengaluru, Karnataka, India

Jyoti Verma
Department of Electronics & Communication Engineering, Manav Rachna International Institute of Research & Studies, Faridabad, Haryana, India

Ashok Kumar Yadav
Department of Mechanical Engineering, Raj Kumar Goel Institute of Technology, Ghaziabad, Uttar Pradesh, India

Acknowledgment

The successful completion of this book, *Artificial Intelligence and Communication Techniques in Industry 5.0,* would not have been possible without the invaluable contributions of several key parties. First and foremost, we extend our deepest gratitude to CRC Press (Taylor & Francis) and their dedicated editorial team. Their unwavering support and insightful guidance throughout the entire process were instrumental in shaping this book into a valuable resource. Their commitment to excellence significantly enhanced the overall quality of the work, from initial concept to final publication.

We also express our sincere appreciation to the numerous contributors and reviewers who generously shared their expertise. Their insightful perspectives enriched every chapter and undeniably contributed to the depth and breadth of the content. Their illuminating feedback ensured the accuracy and comprehensiveness of this book, making it a valuable reference for the scholarly community.

Furthermore, we acknowledge the crucial role of artificial intelligence (AI) in enhancing various aspects of Industry 5.0. AI-powered analytics and optimization algorithms play a pivotal role in optimizing supply chain and logistics management, leading to improved efficiency, reduced costs, and faster delivery times. However, with the rise of AI comes the challenge of adaptive and adversarial attacks. Addressing these challenges requires a comprehensive and proactive approach, combining robust defense mechanisms, adversarial training, input validation, regular updates, and responsible AI practices.

Finally, we dedicate this book to the passionate and dedicated individuals whose tireless efforts are driving advancements in Industry 5.0 using AI. Their commitment inspires us, and it is our hope that this work serves as a source of inspiration and a valuable reference for the scholarly community. We recognize and honor the collaborative spirit that unites us all in the pursuit of excellence and progress within the dynamic intersection of engineering and technology.

Chapter 1

Digital Manufacturing

Artificial Intelligence in Industry 5.0

Ashwani Kumar, Payal Bansal, Abhishek Kumar, Ashish Kumar Gupta, and Amit Choudhari

1.1 INTRODUCTION

Artificial intelligence (AI) serves as the central nervous system of Industry 5.0, propelling manufacturing toward a future of unprecedented efficiency, customization, and optimization. Unlike previous revolutions that focused primarily on automation, AI empowers a more collaborative and intelligent production environment. AI excels in data analysis, a crucial element in Industry 5.0's focus on real-time decision-making. By investigating vast amounts of data from sensors embedded in machines, production lines, and throughout the supply chain, AI can identify patterns, predict equipment failures, and optimize resource allocation. This translates into proactive maintenance, reduced downtime, and a more streamlined production flow. Furthermore, AI's machine learning (ML) capabilities empower intelligent automation. AI algorithms can continuously learn and improve based on data, enabling robots to perform complex tasks with greater precision and adaptability. This fosters a more flexible production environment capable of handling variations in product design and shorter production runs, a hallmark of mass personalization in Industry 5.0. Beyond automation, AI plays a vital role in quality control. By analyzing product data and real-time sensor readings, AI can detect anomalies and potential defects early in the production process. This not only minimizes waste but also ensures consistent product quality, a key factor in customer satisfaction and brand reputation [1–3].

AI's impact extends beyond the factory floor. By analyzing customer data and market trends, AI can provide valuable insights into consumer preferences. This enables manufacturers to anticipate demand fluctuations, personalize product offerings, and optimize pricing strategies. This data-driven approach fosters agility and responsiveness, allowing manufacturers to stay ahead of the curve in a dynamic market. AI is not merely a tool in Industry 5.0; it is the driving force behind its transformative potential. From optimizing production processes to personalizing products and ensuring quality, AI empowers manufacturers to achieve new heights of efficiency, flexibility, and customer focus. As AI continues to evolve, the possibilities for supercharging manufacturing in Industry 5.0 are truly boundless [4].

1.2 TRANSFORMATIVE JOURNEY: INDUSTRY 1.0 TO 5.0

Humanity's relationship with production has undergone a remarkable evolution, marked by distinct eras of innovation. This journey, often referred to as the *Industrial Revolutions*, has fundamentally reshaped economies, societies, and the very way we live. Figure 1.1 embarks on a brief exploration of these revolutions, highlighting the key advancements that propelled each era [5–8].

- **Industry 1.0: The Age of Steam (1760s–1840s)**—The Industrial Revolution began in the 18th century, fueled by the power of steam. The invention of the steam engine ushered in an era of mechanization, replacing manual labor with machine power. Factories emerged, powered by water and steam, enabling mass production for the first time. This era witnessed advancements in textiles, iron, and coal production, laying the foundation for future industrial growth.
- **Industry 2.0: The Rise of Mass Production (1870s–1914)**—The Second Industrial Revolution, characterized by electrification and the assembly line, further revolutionized manufacturing. The invention of the electric motor and light bulb provided a cleaner and more efficient power source. Pioneered by Henry Ford, the assembly line concept dramatically increased production efficiency and affordability, leading to the mass production of consumer goods such as automobiles.

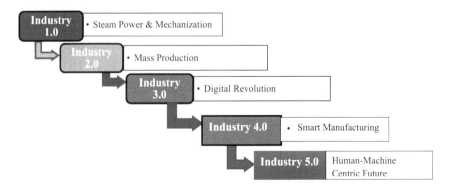

Figure 1.1 Key highlights of transformative journey Industry 1.0 to 5.0.

- **Industry 3.0: The Digital Revolution (1960s–Present)**—The Third Industrial Revolution, marked by the rise of computers and automation, brought a wave of digitalization to manufacturing. Electronics, information technology, and automation played a key role. Programmable logic controllers (PLCs) and industrial robots revolutionized production processes, while computers facilitated data collection, analysis, and control.
- **Industry 4.0: The Smart Factory (2010s–Present)**—The Fourth Industrial Revolution, also known as Industry 4.0, is characterized by the rise of the "smart factory." This era witnessed the convergence of cyber–physical systems, the Internet of Things (IoT), and Big Data analytics. Machines became interconnected, collecting and exchanging data in real time, leading to intelligent automation, predictive maintenance, and highly optimized production processes.
- **Industry 5.0: Human-Centric, Personalized, and Sustainable (Present–Future)**—Industry 5.0 represents the emerging paradigm in manufacturing. It builds upon the foundation of Industry 4.0, but with a renewed focus on human centricity, personalization, and sustainability. While automation remains crucial, the emphasis shifts toward collaboration between humans and intelligent machines.

1.3 ARTIFICIAL INTELLIGENCE IN INDUSTRY 5.0

AI, ML, and Big Data play a critical role in Industry 5.0. AI facilitates intelligent decision-making, while ML algorithms optimize processes and predict maintenance needs. Big Data analysis provides valuable insights into customer preferences, production bottlenecks, and resource utilization. Figure 1.2 shows that Industry 5.0 envisions a future where humans and machines work together seamlessly. Collaborative robots (cobots) take over repetitive and potentially dangerous tasks, while humans contribute their problem-solving skills, creativity, and adaptability. This fosters a more flexible and responsive production environment. Industry 5.0 represents a paradigm shift toward a more harmonious relationship between humans and machines. It is not about replacing humans with machines, but rather about empowering a workforce equipped with the skills to leverage these powerful tools [9–10]. With advanced technologies, manufacturers can cater to individual needs. Imagine a scenario where a customer can design a customized pair of shoes online, specifying the color, material, and fit. Industry 5.0, with its focus on agile manufacturing, can make this a reality through technologies like three-dimensional (3D) printing and AI-powered design tools. Human expertise remains crucial in areas like product design, customer interaction, and quality control [11]. Collaborative robots, designed to work alongside humans, take over the physically demanding or hazardous tasks. This allows humans to focus on higher-level activities that require creativity,

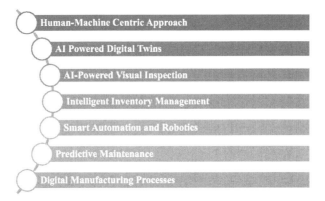

Figure 1.2 Role of AI in Industry 5.0.

problem-solving, and decision-making. The human touch remains vital in areas like process innovation, product development, and ensuring ethical considerations that are factored into AI-powered decisions [12].

1.3.1 AI-Powered Digital Manufacturing Processes

The landscape of manufacturing is undergoing a significant transformation driven by the convergence of automation and robotics with cutting-edge technologies like AI. This fusion, known as smart automation and robotics, is paving the way for a future where machines could not only perform tasks efficiently, but also collaborate with human workers and make intelligent decisions (Figure 1.2). Three key concepts that define this exciting new era—collaborative robots (cobots), AI-powered machine control, and predictive maintenance—have been explained as follows:

- **Collaborative Robots:** Gone are the days of industrial robots confined to caged workspaces. Collaborative robots, or cobots, are a new breed of robots designed to work safely alongside humans. Equipped with integrated safety features, cobots can operate in close proximity to human workers without posing a threat. This fosters a collaborative environment where humans and robots complement each other's strengths. Cobots excel at repetitive and potentially dangerous tasks such as welding, material handling, and assembly line work. This frees up human workers to focus on higher-level cognitive tasks such as quality control, process optimization, and supervision. The collaborative nature of cobots also leads to greater flexibility on the production floor, allowing for faster product changeovers and adaptation to fluctuating demands [13–14].

- **AI-Powered Machine Control:** AI is rapidly transforming the way in which machines operate and make decisions. In the context of smart automation and robotics, AI-powered machine control plays a crucial role in optimizing performance, improving efficiency, and enhancing production quality. One key application of AI is in ML. By analyzing vast amounts of data collected from sensors and process parameters, ML algorithms can "learn" optimal operating conditions for various machines. This allows for adjustments to be made in real time, ensuring machines function at peak efficiency and minimize energy consumption. AI can also be used for predictive maintenance, which we will explore further in the next section. Another critical role of AI is in intelligent control systems. These systems use AI algorithms to analyze data from sensors, cameras, and other sources to make real-time decisions about robot movements and actions. This enables robots to adapt to changes in the environment, such as variations in product size or unexpected obstacles [15].
- **Predictive Maintenance:** Traditionally, maintenance of industrial equipment has been reactive, with repairs being carried out only after a breakdown occurs. This approach can lead to significant downtime and lost productivity. Predictive maintenance, a cornerstone of smart automation and robotics, takes a proactive approach by utilizing AI and data analysis to predict potential equipment failures before they happen. Sensors embedded in robots and machines continuously collect data on factors like vibration, temperature, and power consumption. AI algorithms analyze this data and identify patterns that could indicate an impending failure. This allows for preventive maintenance to be scheduled, minimizing downtime and ensuring smooth operation of the production line. Additionally, by predicting failures early, businesses can avoid costly repairs and replacements [16].

1.3.2 AI-Powered Smart Automation and Robotics

The integration of cobots, AI-powered machine control, and predictive maintenance within smart automation and robotics creates a powerful synergy (Figure 1.2). Collaborative robots enable humans and machines to work together seamlessly, while AI-powered control systems optimize performance and facilitate predictive maintenance, thereby ensuring smooth operation. This combination leads to several advantages [17–18], which are discussed as follows:

- **Increased Productivity:** By automating repetitive tasks and minimizing downtime, smart automation and robotics significantly improve production output.

- **Enhanced Quality:** AI-powered control systems and real-time adjustments ensure consistent product quality.
- **Reduced Costs:** Predictive maintenance prevents equipment failures and minimizes the need for emergency repairs, leading to cost savings.
- **Improved Safety:** Cobots create a safer work environment by taking over potentially dangerous tasks.
- **Greater Flexibility:** Smart automation and robotics allows for faster adaptation to changing production demands.

Smart automation and robotics, empowered by collaborative robots, AI-powered machine control, and predictive maintenance, signifies a significant leap forward in manufacturing. This intelligent and collaborative approach paves the way for a future where humans and machines could work together to achieve greater efficiency, productivity, and innovation [19].

1.3.3 AI-Powered Supply Chain Optimization

Industry 5.0 ushers in a new era of manufacturing, characterized by human–machine collaboration, mass personalization, and sustainable practices. This revolution extends to the realm of supply chains, where three key areas—demand forecasting, intelligent inventory management, and automated logistics—become intertwined with advanced technologies to achieve unprecedented levels of optimization [20–22].

- **Demand Forecasting**: Gone are the days of relying solely on historical data and gut instinct. Industry 5.0 leverages AI and ML to create highly accurate demand forecasts. These algorithms analyze vast amounts of data, including social media trends, weather patterns, and real-time customer behavior, to predict future demand with remarkable precision. This allows manufacturers to tailor production schedules and inventory levels to meet fluctuating customer needs, minimizing the risk of stockouts or excess inventory.
- **Intelligent Inventory Management:** Inventory management becomes **highly dependent** on real-time data and automationSmart warehousing solutions powered by the IoT allow for constant monitoring of stock levels. AI algorithms analyze this data, coupled with the demand forecasts, to automatically trigger reordering when necessary. This ensures manufacturers have the right amount of raw materials on hand to meet production demands without excessive storage costs.
- **Automated Logistics:** The physical movement of goods within the supply chain is transformed by automation. Autonomous vehicles like self-driving trucks and drones handle long-distance and last-mile deliveries, optimizing routes and minimizing transportation times. Robotics plays a crucial role in warehouses, with automated guided

vehicles (AGVs) and robotic arms streamlining picking, packing, and shipping processes. This automation reduces human error and streamlines operations, leading to faster delivery times and lower transportation costs.
- **The Synergy of Optimization:** The three areas, demand forecasting, intelligent inventory management, and automated logistics, work in a symphony of optimization within Industry 5.0. Accurate demand forecasts inform intelligent inventory management, ensuring the right materials are available. This, in turn, allows for efficient production and timely deliveries, facilitated by automated logistics. The entire supply chain becomes a well-oiled machine, with every step informed by data and powered by intelligent automation.

The ripple effects of a well-optimized supply chain in Industry 5.0 are significant. Manufacturers experience reduced costs, improved efficiency, and increased responsiveness to customer demands. Additionally, sustainability is enhanced through optimized resource utilization and reduced transportation emissions. Ultimately, Industry 5.0 paves the way for a more responsive, efficient, and sustainable supply chain ecosystem [23–28].

1.3.4 AI-Powered Product Quality and Inspection

Traditionally, product quality inspection has relied on human visual checks. However, this approach can be prone to errors due to fatigue, inconsistency, and limitations of human vision. Here is how AI and real-time defect detection are transforming product quality [29–33]:

- **AI-Powered Visual Inspection:** This involves using intelligent algorithms trained on vast amounts of image data. These algorithms can analyze product images with incredible precision, identifying even subtle defects that might escape the human eye. For example, AI can detect minuscule scratches on a phone screen, misprints on packaging, or inconsistencies in the size and shape of components.
- **Real-Time Defect Detection:** AI algorithms can process images and identify defects at high speeds. This allows for real-time quality control, where products are inspected as they move down the production line. Defective products can be flagged immediately, preventing them from being further processed or shipped to customers. This significantly reduces waste and ensures consistent product quality.

The benefits of AI-powered product quality and inspection are as follows:

- AI can detect a wider range of defects with greater consistency than human inspectors, which results in improved accuracy.

- Real-time defect detection allows for faster inspection and quicker identification of problems, which results in increased efficiency.
- By catching defects early, manufacturers can minimize waste and rework, leading to cost savings.
- AI systems can collect and analyze vast amounts of data on defects, providing valuable insights for process improvement.

AI-powered visual inspection and real-time defect detection are revolutionizing product quality control. By leveraging these technologies, manufacturers can ensure consistent quality, improve efficiency, and ultimately deliver a superior product to their customers. AI-powered visual inspection has been used across various industries, from automotive and electronics to food and pharmaceuticals. It also ensures ensuring proper labeling, printing accuracy, and damage detection on packaging lines. AI-powered medical device inspection can detect critical medical equipment with high precision [34–38].

1.3.5 Human–Machine Collaboration in Industry 5.0

Industry 5.0 marks a significant shift in the relationship between humans and machines within the manufacturing landscape. While advancements in AI and automation might seem like a threat to human workers, this new era is paving the way for a more collaborative and empowering future. This study explores into the evolving role of human workers in Industry 5.0, emphasizing the rise of human–AI collaboration and the emergence of new skillsets, while acknowledging potential challenges like job displacement due to automation [39–44].

- **From Oversight to Collaboration:** Industry 5.0 moves away from the human-as-overseer model prevalent in earlier industrial revolutions. AI is no longer a stand-alone tool to be monitored. Instead, humans and AI become collaborative partners, each contributing their unique strengths. AI excels at tasks requiring data analysis, pattern recognition, and repetitive execution. Humans, on the other hand, bring invaluable problem-solving skills, creativity, adaptability, and the ability to handle unforeseen situations. This collaborative approach fosters a more flexible and responsive production environment, capable of adapting to dynamic market demands and rapid innovation cycles [45–49].
- **Evolving Skillsets for a New Era:** The rise of AI necessitates a shift in the skillsets required for success in Industry 5.0. While technical skills remain important, the emphasis is now on human-centric abilities. Workers will need strong communication, collaboration, and critical thinking skills to effectively work alongside AI partners. Data

literacy will be crucial, enabling workers to interpret and utilize data generated by AI systems to optimize processes and make informed decisions. Additionally, creativity and design thinking will be paramount for developing innovative products and solutions that cater to a personalized market.
- **The Upskilling Imperative:** The transition to Industry 5.0 requires a focus on continuous upskilling and reskilling of the workforce. Educational institutions and training programs need to adapt to equip individuals with the necessary skills to thrive in this new paradigm. Upskilling programs can focus on areas like data analysis, AI fundamentals, design thinking, and human–machine collaboration. By investing in workforce development, companies can ensure a smooth transition and leverage the full potential of human–AI collaboration.
- **Addressing Job Displacement Concerns:** It is undeniable that automation within Industry 5.0 may lead to job displacement in certain sectors, particularly those involving highly repetitive tasks. However, the focus should not be on job replacement, but rather on job transformation. New roles will emerge that require a different blend of skills. For instance, the need for AI specialists, data analysts, and human–machine interface designers will increase significantly. Additionally, the rise of customized production will require skilled workers for tasks like 3D printing, programming robots, and managing complex supply chains.
- **A Human-Centric Approach:** Industry 5.0 represents a promising future for manufacturing, one where humans and AI could work together to achieve remarkable feats. By embracing a human-centric approach, investing in upskilling initiatives, and focusing on collaboration, we can navigate potential challenges and unlock the true potential of this transformative era. The key lies in recognizing that humans and AI are not competitors, but rather complementary forces that can propel the industrial world toward a more sustainable, efficient, and innovative future [50–54].

1.3.6 AI-Powered Digital Twins

Imagine a virtual replica of a physical object, system, or process—a dynamic, data-driven model that mirrors its real-world counterpart in real time. This, in essence, is the concept of a digital twin. Figure 1.3 shows that digital twins act as a bridge between the physical and digital worlds, continuously updated with sensor data, historical records, and operational information. Digital twins are revolutionizing how we interact with physical assets. By creating a virtual replica that mirrors its real-world counterpart, organizations can gain unprecedented insights, optimize performance, and

10 Artificial Intelligence and Communication Techniques

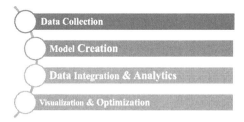

Figure 1.3 Steps involved in creating digital twins.

predict potential issues. Creating a digital twin involves several key steps as follows [28]:

- **Data Collection:** The foundation of any digital twin lies in the data it ingests (Figure 1.3). Sensors play a critical role here, acting as the eyes and ears of the physical asset. These sensors are strategically embedded to capture real-time data on various performance metrics. This data can encompass anything from temperature and vibration levels in a machine to pressure and flow rates in a pipeline. But the data does not stop there. Environmental conditions, like humidity and ambient temperature, are also crucial for understanding how the asset interacts with its surroundings. Additionally, operational parameters, such as energy consumption and throughput, provide valuable insights into the asset's overall efficiency. By collecting this comprehensive data stream, we paint a detailed picture of the asset's health and performance [29–30].
- **Model Creation:** With the data flowing in, it is time to construct the digital twin itself. This can be visualized as a 3D model or a software representation that mirrors the physical entity. There are two main approaches to model creation. One method leverages existing engineering design data, such as computer-aided design (CAD) models, to form the base of the digital twin. This data provides a precise geometrical representation of the asset, including its components and their interrelationships. The other approach involves creating a new model from scratch based on sensor data and real-world measurements. Here, advanced techniques like photogrammetry or laser scanning can be used to capture the asset's physical dimensions and translate them into a digital model. However, the model goes beyond just a geometric replica. To truly represent the physical asset, historical performance records are integrated. This historical data provides valuable context, allowing the digital twin to learn from past behavior and predict future trends. Additionally, any

other relevant information, such as maintenance logs and operating manuals, can be incorporated to enrich the digital model. By combining these elements, we create a comprehensive digital representation that captures the essence of the physical asset.
- **Data Integration and Analytics:** Data collection and model creation are just the first steps. The true power of digital twins lies in their ability to analyze and interpret the vast amount of data they gather. Here is where advanced analytics tools come into play. These tools process the combined data stream from sensors, historical records, and other sources. This data fusion allows for a holistic understanding of the asset's behavior. Powerful algorithms can then be applied to perform simulations, predict performance under different operating conditions, and identify potential issues before they escalate into real-world problems. Imagine simulating the impact of increased production demand on a machine or predicting when a critical component might require maintenance. This proactive approach enables organizations to make data-driven decisions and optimize their operations.
- **Visualization and Optimization:** The insights gained from data analysis are visualized through dashboards, simulations, or augmented reality applications. This empowers informed decision-making and process optimization.

1.3.6.1 Digital Twins Applications

Digital twins represent a paradigm shift in how we approach product development, manufacturing, and operations. By creating a virtual reflection of the physical world, they empower us to make data-driven decisions, optimize processes, and achieve significant improvements in efficiency, cost reduction, and overall performance. As digital twins evolve, they hold the potential to revolutionize industries and create a more sustainable and intelligent future. In Industry 5.0, digital twins are finding applications in various sectors as follows [31–35] (Figure 1.4):

- **Aerospace Industry:** In the high-stake world of aerospace, digital twins are proving their worth. Imagine virtually simulating the performance of a new aircraft design before a single metal sheet is bent. This allows engineers to optimize aerodynamics, test engine performance under various conditions, and identify potential issues early in the development process. Digital twins further empower airlines by predicting maintenance needs for their fleet. By analyzing sensor data from engines and other onboard systems, the digital twin can anticipate when a component might require servicing, enabling proactive maintenance and preventing costly breakdowns. Additionally,

12 Artificial Intelligence and Communication Techniques

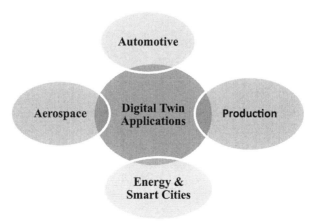

Figure 1.4 Industrial applications of digital twins.

  optimizing flight routes based on real-time weather data and air traffic control information becomes possible. This results in fuel savings, reduced emissions, and a smoother flying experience for passengers.
- **Automotive Industry:** The automotive industry is experiencing a digital transformation, and digital twins are at the forefront. Imagine designing and testing a new car virtually before building a physical prototype. This not only reduces development time and costs but also allows for a wider exploration of design possibilities. Digital twins can be used to optimize factory production lines by simulating assembly processes and identifying bottlenecks. This ensures smoother production flow, minimizes waste, and optimizes resource allocation. Furthermore, by analyzing sensor data from connected vehicles, digital twins can predict when a car might require maintenance, enabling proactive service and preventing breakdowns on the road. This not only enhances customer satisfaction but also reduces downtime and associated costs.
- **Manufacturing Industry:** Manufacturing is all about streamlining processes, maximizing efficiency, and delivering high-quality products. Digital twins are becoming a game-changer in this arena. Imagine a virtual replica of a factory floor that allows for real-time monitoring and optimization of production processes. By analyzing data from sensors on machines and production lines, the digital twin can identify inefficiencies, predict equipment failures, and suggest adjustments to optimize production flow. This translates into reduced waste, minimized downtime, and increased overall efficiency. Digital twins can further empower manufacturers to continuously improve

product quality. By simulating the impact of different manufacturing parameters on the final product, engineers can identify and rectify potential quality issues before they reach the production line. This leads to a consistent and superior product for the end customer.
- **Energy and Utilities:** The energy and utilities sector is facing the dual challenge of meeting growing demand while ensuring sustainability. Digital twins are emerging as a powerful tool to navigate these challenges. Imagine a virtual replica of the power grid that allows for real-time monitoring and optimization of energy distribution. By analyzing historical data and weather patterns, the digital twin can predict energy demand fluctuations. This enables utilities to proactively adjust power generation and distribution to meet demand efficiently, minimizing wasted energy and optimizing resource allocation. Additionally, digital twins can be used for predictive maintenance of critical infrastructure, such as power lines and transformers. By analyzing sensor data, the digital twin can identify potential equipment failures before they occur, allowing for timely maintenance and preventing costly disruptions. This translates into a more reliable and sustainable energy supply for consumers.
- **Building Smarter Cities: Optimizing Urban Systems with Digital Twins:** As cities grow, managing traffic flow, emergency response, and resource allocation becomes increasingly complex. Digital twins are offering a glimpse into a future of smarter cities. Imagine a virtual replica of a city that allows for real-time traffic monitoring and optimization. By analyzing data from traffic sensors and cameras, the digital twin can predict congestion hotspots and suggest dynamic route adjustments. This translates into reduced traffic congestion, improved air quality, and a smoother commute for citizens. Furthermore, digital twins can be used to simulate emergency response scenarios, such as natural disasters or fires. This allows city planners to identify potential bottlenecks and optimize emergency response routes, leading to faster and more effective response times. Digital twins can also be used to manage resource allocation for public services, such as waste collection and public transportation. By analyzing data on service demand and resource availability, the digital twin can optimize routes and schedules, leading to a more efficient and cost-effective public service system.

As technology advancements continue, digital twins are poised to become even more sophisticated. The increasing adoption of the IoT will lead to a proliferation of data feeding into these virtual models, further enhancing their accuracy and predictive capabilities. Integration with AI will enable them to learn from data, identify patterns, and make autonomous recommendations for optimization [55–61].

1.4 SUCCESS STORIES IN AI-POWERED MANUFACTURING

The integration of AI is revolutionizing manufacturing processes across various industries. Here, we explore five compelling case studies showcasing how companies are leveraging AI to achieve significant improvements in efficiency, cost reduction, and product quality.

1.4.1 Case 1: Ford Motors—Predictive Maintenance with AI

> **Challenge:** Unscheduled equipment downtime can significantly disrupt production schedules and lead to substantial financial losses.
> **Solution:** Ford Motors implemented an AI-powered predictive maintenance system in their engine plants. Sensor data from machinery is fed into AI algorithms that analyze historical trends and predict potential equipment failures.

Benefits:

- **Reduced Downtime:** By anticipating equipment issues before they occur, Ford has minimized unplanned downtime by a significant margin, leading to smoother production flow.
- **Cost Savings:** Early detection of potential failures allows for proactive maintenance, reducing the need for expensive repairs and part replacements.
- **Improved Equipment Lifespan:** By addressing maintenance needs promptly, Ford extends the lifespan of their machinery, optimizing equipment utilization.

1.4.2 Case 2: Boeing—AI-Powered Defect Detection in Aircraft Manufacturing

> **Challenge:** Maintaining the highest quality standards in aircraft manufacturing is crucial for safety. Traditional visual inspections can be time-consuming and prone to human error.
> **Solution:** Boeing employs AI-powered visual inspection systems equipped with high-resolution cameras. These systems analyze images of aircraft components, identifying even minute defects that might escape human detection.

Benefits:

- **Enhanced Quality Control:** AI ensures a more thorough and consistent inspection process, minimizing the risk of defective parts making it into final assemblies.
- **Improved Safety:** By identifying potential issues early, Boeing strengthens the overall safety and reliability of their aircraft.
- **Increased Efficiency:** AI-powered inspections are significantly faster than manual processes, allowing for quicker turnaround times and increased production throughput.

1.4.3 Case 3: Adidas—AI-Driven Demand Forecasting and Customization

Challenge: Traditional forecasting methods often struggle to keep pace with rapidly changing consumer preferences in the fashion industry.

Solution: Adidas leverages AI algorithms to analyze vast amounts of data, including sales trends, social media sentiment, and weather patterns. This data is used to generate highly accurate demand forecasts for various product lines. Additionally, Adidas utilizes AI to personalize the customer experience by offering custom shoe designs and on-demand manufacturing options.

Benefits:
- **Reduced Inventory Costs:** Accurate demand forecasting minimizes the risk of overproduction or understocking, leading to optimized inventory management.
- **Increased Sales:** By anticipating customer preferences, Adidas can produce the right products in the right quantities, leading to higher sales and customer satisfaction.
- **Enhanced Customer Experience:** AI-powered customization caters to individual needs, fostering a more engaging and personalized shopping experience.

1.4.4 Case 4: Siemens—AI-Powered Optimization of Production Lines

Challenge: Optimizing complex production lines to maximize efficiency and minimize waste can be a major challenge for manufacturers.

Solution: Siemens employs AI-powered production line optimization software. This software analyzes real-time data from sensors throughout the production line, identifying bottlenecks and suggesting adjustments to optimize workflow and resource allocation.

Benefits:
- **Increased Efficiency:** AI helps to identify and eliminate inefficiencies in the production process, leading to faster production pace and higher output.
- **Reduced Waste:** By optimizing resource allocation, Siemens minimizes material waste and energy consumption throughout the manufacturing process.
- **Improved Decision-Making:** Real-time data analysis by AI empowers production managers to make informed decisions that can improve overall production line performance.

1.4.5 Case 5: Schneider Electric—AI-Powered Supply Chain Management

Challenge: Managing a complex global supply chain can be challenging, especially when dealing with fluctuating demand and potential disruptions.

Solution: Schneider Electric employs AI-powered supply chain management solutions that analyze historical data, market trends, and real-time logistics information. This allows them to predict potential disruptions and proactively adjust their supply chain accordingly.

Benefits:
- **Enhanced Supply Chain Resilience:** AI helps anticipate and mitigate potential disruptions, ensuring a more resilient and responsive supply chain.
- **Reduced Transportation Costs:** By optimizing logistics based on real-time data, Schneider Electric minimizes transportation costs and delivery times.
- **Improved Customer Satisfaction:** Consistent product availability and timely deliveries contribute to enhanced customer satisfaction and loyalty.

These five cases demonstrate the transformative power of AI in manufacturing. By leveraging AI solutions, companies are achieving significant improvements in efficiency, cost reduction, and product quality, paving the way for a more intelligent and sustainable future for the manufacturing industry [62–70].

1.5 CHALLENGES AND OPPORTUNITIES

While AI holds immense potential for revolutionizing manufacturing, its path to widespread adoption is riddled with obstacles. Figure 1.5 shows that challenges can be broadly categorized into three areas: data security, integration costs, and talent gaps. Addressing these concerns is crucial for manufacturers to unlock the true potential of AI and gain a competitive edge. The cornerstone of AI is data. Manufacturing facilities generate vast amounts of data from sensors, machines, and production processes. This data becomes the fuel for AI algorithms, enabling them to learn, identify patterns, and optimize operations. However, this data-driven approach raises significant security concerns [71–75], as follows:

- **Data Breaches and Industrial Espionage:** Manufacturing data is a goldmine for competitors. Information about production processes,

Digital Manufacturing: Artificial Intelligence in Industry 5.0   17

Figure 1.5 Challenges associated with AI for revolutionizing manufacturing.

product designs, and supply chains can be immensely valuable. A security breach involving AI systems could leave a manufacturer vulnerable to industrial espionage, causing significant financial losses and a potential erosion of intellectual property.
- **Data Privacy and Regulations:** Manufacturing often involves sensitive information, such as trade secrets and customer data. As regulations like the General Data Protection Regulation (GDPR) become more stringent, manufacturers must ensure compliance when collecting, storing, and processing data for AI applications. Failure to do so can lead to hefty fines and reputational damage.
- **Securing the AI System Itself:** AI models are complex software programs susceptible to hacking. Malicious actors could potentially manipulate training data or exploit vulnerabilities in the AI system to disrupt production processes or steal sensitive information.

Integrating AI into existing manufacturing infrastructure can be a costly endeavor. Running AI algorithms often requires significant computing power. Manufacturers may need to invest in new hardware, such as high-performance servers and specialized AI chips. Additionally, the software licenses for AI tools and platforms can be expensive. To leverage AI effectively, manufacturers need a robust data infrastructure. This may involve centralizing data from various sources, implementing data cleaning and labeling processes, and building secure storage solutions. These upgrades can be resource-intensive and require ongoing maintenance. Integrating AI systems with existing manufacturing software and equipment can be complex. Legacy systems may not be compatible with new AI tools, requiring additional engineering effort and potentially causing production disruptions during the integration process [32–33].

Extracting value from AI requires a skilled workforce. Developing, deploying, and maintaining AI systems requires expertise in data science, ML, and AI engineering. These specialists are in high demand across various industries, making them difficult and expensive to recruit. For AI to be successful in manufacturing, it needs to be tailored to specific industry processes and equipment. This necessitates a workforce that possesses both strong technical skills and a deep understanding of manufacturing operations. The introduction of AI can lead to significant changes in how work is done on the factory floor. Manufacturers need to invest in training programs and change management initiatives to ensure employee buy-in and address potential anxieties about job displacement due to automation.

These challenges are not insurmountable. However, by implementing robust data security protocols, carefully planning integration projects, and investing in workforce development, manufacturers can pave the way for a successful AI-powered future [76–80].

1.6 CONCLUSION

Industry 5.0 marks a shift toward human-centered manufacturing, where AI plays a crucial role in driving personalization, sustainability, and economic growth. A key transformative power of AI lies in its ability to analyze vast amounts of data to achieve mass customization. This allows manufacturers to cater to individual customer needs and preferences, creating a new paradigm of highly tailored products and services. Furthermore, AI empowers manufacturers to optimize resource utilization, minimizing waste and environmental impact. By analyzing data on energy consumption and materials usage, AI algorithms can identify inefficiencies and suggest adjustments to production processes, promoting sustainable practices. Additionally, AI-powered decision-making based on real-time data analysis fosters greater efficiency and productivity, leading to economic growth for manufacturers. However, the transformative potential of AI hinges on responsible development and ethical considerations. Bias in training data can lead to discriminatory outcomes, and a lack of transparency in AI algorithms can raise concerns about accountability. Therefore, it is crucial to ensure fairness, transparency, privacy, and accountability throughout the AI development and deployment process. By prioritizing ethical considerations, manufacturers can harness the transformative power of AI while building trust and ensuring a positive impact on society.

REFERENCES

1. Chander, B., Pal, S., De, D., & Buyya, R. (2022). Artificial intelligence-based Internet of Things for Industry 5.0. In: S. Pal, D. De, & R. Buyya (eds.), *Artificial Intelligence-based Internet of Things Systems*. Springer, Cham. https://doi.org/10.1007/978-3-030-87059-1_1
2. Alarifi, A., & Yoo, P. D. (2020). Industry 4.0 and digital twins for enterprise industrial automation: a review and future directions. *Enterprise Information Systems*, 14(12), 1843–1878.
3. Barreto, L., Amaral, C. A., & Cunha, M. (2020). Digital twins enabling Industry 5.0: a review of the relevant capabilities. *International Journal of Mechanical and Industrial Engineering*, 14(8), 1051–1062. https://doi.org/10.1080/15432757.2020.1783432
4. Berrahoum, L., Zitouni, M., & Ziadi, T. (2023). A review of security aspects for digital twins in the context of Industry 5.0. *Journal of Network and Computer Applications*, 222, 103824. https://doi.org/10.1016/j.jnca.2023.103824
5. Chen, J., Cheng, X., & Liu, M. (2022). Industry 5.0: a conceptual model for intelligent manufacturing systems with human–machine co-creation. *Chinese Journal of Mechanical Engineering*, 35(1), 1–11. https://doi.org/10.1186/s10013-022-01329-w
6. Cheng, Y., Li, F., Zhang, Y., & Liu, Z. (2022). Digital twin-driven human–AI collaborative decision-making for sustainable manufacturing in Industry 5.0. *Journal of Cleaner Production*, 363, 132404. https://doi.org/10.1016/j.jclepro.2022.132404
7. Da Xu, L., Xu, E. L., & Li, L. (2021). Industry 5.0: a human-centric era from AI to ecological civilization. *Engineering*, 7(1), 1141–1148. https://doi.org/10.1016/j.eng.2020.12.032
8. Garcia, C. G., Nunez-Valdez, E. R., Garcia-Diaz, V., Pelayo G-Bustelo, B. C., & Lovelle, J. M. C. (2018). A review of artificial intelligence in the Internet of Things. *International Journal of Interactive Multimedia and Artificial Intelligence*, 5(4).
9. Ding, K., Zhang, X., Cheng, Y., & Liu, M. (2023). Human–AI co-creation for service design in Industry 5.0. *Computers & Industrial Engineering*, 183, 106742. https://doi.org/10.1016/j.cie.2023.106742
10. Lee, S. K., Bae, M., & Kim, H. (2017). Future of IoT networks: a survey. *Applied Sciences*, 2017(7), 1072. https://doi.org/10.3390/app7101072.
11. Dubey, R., Gunasekaran, A., & Papadopoulos, T. (2020). Industry 4.0 and the circular economy: a state-of-the-art review and future directions.

Journal of Cleaner Production, 277, 123201. https://doi.org/10.1016/j.jclepro.2020.123201
12. Eltayeb, O. M., Wu, Q., Li, F., & Sun, Y. (2022). Security and privacy challenges in Industry 5.0: a survey. *IEEE Access*, 10, 130504–130524. https://doi.org/10.1109/ACCESS.2022.3211191
13. Ahn, H.-J., Park, S. H., & Ryu, K. (2020). Human–AI hybrid intelligence for real-time defect detection in semiconductor manufacturing. *Computers & Industrial Engineering*, 140, 106228.
14. Berggren, C., & Matheson, D. (2021). A human-centric approach to artificial intelligence in advanced manufacturing. *Technovation*, 112, 102223.
15. Chanias, S., Habib, M. N., & Nassif, A. I. (2022). Intelligent inventory management: a review. *International Journal of Production Research*, 60(8), 2422–2453.
16. Griepentrog, B., & Jie, L. (2019). Mass personalization: the key to Industry 5.0. *Procedia CIRP*, 83, 310–315.
17. Lee, J., Bagheri, B., & Kao, H.-A. (2015). A cyber-physical systems architecture for Industry 4.0 manufacturing systems. *Manufacturing Letters*, 3, 18–23.
18. Ungureanu, A. V. (2020). The transition from Industry 4.0 to Industry 5.0. The 4Cs of the global economic change. In C. Nastase (ed.), *Lumen Proceedings: Vol. 13. 16th Economic International Conference NCOE 4.0 2020.* pp. 70–81.
19. Li, Z., Wang, Y., & Liu, C. (2020). Review of application of machine learning in predictive maintenance. *Reliability Engineering & System Safety*, 199, 107050.
20. Qin, J., Yu, R., Liu, Z., Zhou, D., Fang, X., & Li, Y. (2023). A survey on human–machine collaboration in intelligent manufacturing. *IEEE Transactions on Industrial Informatics*, 19(8), 3802–3810.
21. Tao, F., Cheng, J., Qi, Q., Zhang, M., Zhang, Z., & Sui, F. (2018). Digital twin-driven product design, manufacturing and service with big data. *International Journal of Advanced Manufacturing Technology*, 99(1–4), 949–960.
22. Wang, S., Wu, J., & Duan, Y. (2020). AI-powered predictive maintenance for industrial equipment: a review. *IEEE Access*, 8, 141821–141844.
23. Zheng, P., Yang, Y., & Liu, C. (2022). Survey on smart manufacturing systems communication and networking technologies. *IEEE Transactions on Industrial Informatics*, 18(1), 367–376.
24. Grieves, M. (2019). From Industry 4.0 to Industry 5.0: moving from automation to autonomy & human–technology collaboration. *Procedia Manufacturing*, 38, 12–17.
25. Lee, J., Bagheri, B., & Kang, H. J. (2019). Cyber–physical systems for industry 4.0 manufacturing. *Manufacturing Letters*, 3(2), 40–43.
26. Manyika, M., Chui, M., & Osborne, M. (2017). *A Human World: Jobs in the Age of Algorithms* (2nd ed.). McKinsey Global Institute: New York.
27. Miner, L., & Lee, J. (2019). Uncertainty in supply chain logistics: a call for research that reflects practice. *Journal of Business Logistics*, 40(2), 382–394.
28. Qin, J., Liu, Y., & Zhong, R. Y. (2023). AI-based intelligent manufacturing: a survey. *Engineering Science*, 18(2), 1001–1012.

29. Xu, L., Xu, E., & Li, L. (2018). Industry 4.0: state of the art and future trends. *International Journal of Computer Technologies and Applications*, 9(5), 485–491.
30. Yang, Q., Yang, L., Chen, T., & Tong, Y. (2019). Federated machine learning: concept and applications. *ACM Transactions on Intelligent Systems and Technology*, 10(2), 19 pages. https://doi.org/10.1145/3298981
31. Davenport, T. H. (2018). From analytics to artificial Intelligence. *Journal of Business Analytics*, 1, 2,73–2,80. https://doi.org/10.1080/2573234X.2018.1543535
32. Meruje, M., GwaniSamaila, M., Virginia, N. L., Franqueira, M. M. F., & MoraisInâcio, P. R. (2018). A tutorial introduction to IoT design and prototyping with examples, Internet of Things A to Z: technologies and applications. In Q. F. Hassan (ed.), *The Institute of Electrical and Electronics Engineers, Inc* (1st ed.). Wiley & Sons, Inc.
33. Patel, K. K., & Patel, S. M. (2016). Internet of Things—IOT: definition, characteristics, architecture, enabling technologies, application & future challenges. *International Journal of Engineering Science and Computing*, 6(5).
34. Özdemir, V., & Hekim, N. (2018). Birth of Industry 5.0: making sense of Big Data with artificial intelligence, 'The Internet of Things' and next-generation technology policy. *OMICS A Journal of Integrative Biology*, 22(1). https://doi.org/10.1089/omi.2017.0194
35. Sharma, I., Garg, I., & Kiran, D. (2020). Industry 5.0 and smart cities: a futuristic approach. *European Journal of Molecular & Clinical Medicine*, 07(08), 2515–8260.
36. Kumar, A., Singla, Y., & Namboodri, T. (2024). *Globalization and International Issues in Sustainable Manufacturing*. CRC Press: Boca Raton, FL, USA; Chapter 01; pp 1–18. http://dx.doi.org/10.1201/9781003467496-1
37. Gajrani, K. K., Prasad, A., & Kumar, A. (2022). *Advances in Sustainable Machining and Manufacturing Processes*. CRC Press: Boca Raton, FL, USA. https://doi.org/10.1201/9781003284574
38. Awasthi, M. K., Kumar, A., & Dutt, N. (2024). *Recent Innovation, Technological Development and Futuristic Applications of Nanofluids*. CRC Press: Boca Raton, FL, USA; Chapter 1.
39. Singh, S., Behura, S. K., Kumar, A., & Verma, K. (2022). *Nanomanufacturing and Nanomaterials Design: Principles and Applications*. CRC Press: Boca Raton, FL, USA. https://doi.org/10.1201/9781003220602
40. Awasthi, M. K., Dutt, N., & Kumar, A. (eds.) (2024). *Nanofluids Technology for Thermal Sciences and Engineering: Research, Development, and Applications*. Taylor & Francis (CRC Press): Boca Raton, FL, USA. ISBN 9781032799117.
41. Kumar, A., Thangavel, S., Awasthi, M. K., Singh, V. P., & Dutt, N. (2024). *Safety and Environmental Considerations in Nano Fluid-Based Systems*. CRC Press: Boca Raton, FL, USA; Chapter 06.
42. Awasthi, M. K., Kumar, A., Dutt, N., & Singh, S. (eds.) (2024). *Computational Fluid Flow and Heat Transfer: Advances, Design, Control and Applications*. CRC Press: Boca Raton, FL, USA. https://doi.org/10.1201/9781003465171
43. Sharma, V. K., Kumar, A., Gupta, M., Kumar, V., Sharma, D. K., Sharma, S. (2022). *Additive Manufacturing in Industry 4.0: Methods, Techniques,*

Modeling, and Nano Aspects. Taylor & Francis (CRC Press): Boca Raton FL, USA. https://doi.org/10.1201/9781003360001

44. Awasthi, M. K., Dutt, N., Kumar, A., & Hedau, A. (2024). *Introduction to Mathematical and Computational Methods*. CRC Press: Boca Raton, FL, USA; Chapter 1. https://doi.org/10.1201/9781003465171-1
45. Kumar, A., Dutt, N., & Awasthi, M. K. (eds.) (2024). *Heat Transfer Enhancement Techniques: Thermal Performance, Optimization and Applications*. John Wiley & Sons: USA.
46. Kumar, A., Awasthi, M. K., Dutt, N., & Singh, V. P. (2024). *Recent Innovation in Heat Transfer Enhancement Techniques*. John Wiley & Sons: USA; Chapter 01.
47. Kumar, A., Kumar, A., Singh, V. P., Dutt, N., & Kutar, S. P. (2022). Study of material removal rate process parameters in WEDM. *International Journal of Materials, Manufacturing and Sustainable Technologies*, 1(2), 31–42. https://doi.org/10.56896/IJMMST.2022.1.2.010
48. King M, F. L., Kumar, A., Kumar, A., & Kumar, A. (2023). Introduction to optics and laser-based manufacturing technologies. In *Laser-based Technologies for Sustainable Manufacturing*. CRC Press: Boca Raton, FL, USA; Chapter 1; pp. 1–43. https://doi.org/10.1201/9781003402398-1
49. Verma, V., Thangavel, S., Dutt, N., Kumar, A., & Weerasinghe, R. (eds.) (2024). *Highly Efficient Thermal Renewable Energy Systems: Design, Design, Optimization and Applications*. CRC Press: Boca Raton, FL, USA; pp. 1–341. https://doi.org/10.1201/9781003472629
50. Dutt, N., Hedau, A., Kumar, A., Awasthi, M. K., Hedau, S., & Meena, C. S. Thermohydraulic performance investigation of solar air heater duct having staggered D-shaped ribs: numerical approach. *Heat Transfer*, 53(3), 1501–1531. https://onlinelibrary.wiley.com/doi/10.1002/htj.22998
51. Dewangan, A. K., Moinuddin, S. Q., Cheepu, M., Sajjan, S. K., & Kumar, A. (2023). *Thermal Energy Storage: Opportunities, Challenges and Future Scope*. CRC Press: Boca Raton, FL, USA; Chapter 2; pp. 17–28. http://dx.doi.org/10.1201/9781003395768-2
52. Kumar, A., Pathak, A., Kumar, A., & Kumar, A. (2023). Physics of laser-matter interaction in laser-based manufacturing. In *Laser-based Technologies for Sustainable Manufacturing*. CRC Press: Boca Raton, FL, USA; Chapter 2; pp. 45–54. https://doi.org/10.1201/9781003402398-2
53. Pathak, A., Kumar, A., Kumar, A., Kumar, A., & King M, F. L. (2023). Application of laser technology in the mechanical and machine manufacturing industry. In *Laser-based Technologies for Sustainable Manufacturing*. CRC Press: Boca Raton, FL, USA; Chapter 6; pp. 107–155. https://doi.org/10.1201/9781003402398-6
54. Kumar, A., Byadwal, M., Kumar, A., Kumar, A., & King M, F. L. (2023). Laser micromachining in biomedical industry. In *Laser-based Technologies for Sustainable Manufacturing*. CRC Press: Boca Raton, FL, USA; Chapter 8; pp. 169–206. https://doi.org/10.1201/9781003402398-8
55. Ahmadizadeh, M., Heidari, M., Thangavel, S., Naamani, E. A., Khashehchi, M., Verma, V., & Kumar, A. Technological advancements in sustainable and renewable solar energy systems. In *Highly Efficient Thermal Renewable*

Energy Systems: Design, Optimization and Applications. CRC Press: Boca Raton, FL, USA; Chapter 2. https://doi.org/10.1201/9781003472629-2.

56. Verma, V., Thangavel, S., Dutt, N., Kumar, A., & Weerasinghe, R. Recent development of thermal energy storage: solar, geothermal and hydrogen energy. In *Highly Efficient Thermal Renewable Energy Systems: Design, Optimization and Applications*. CRC Press: Boca Raton, FL, USA; Chapter 2. https://doi.org/10.1201/9781003472629-1
57. Chitt, M., Thangavel, S., Verma, V., & Kumar, A. Green hydrogen productions: methods, designs and smart applications. In *Highly Efficient Thermal Renewable Energy Systems: Design, Optimization and Applications*. CRC Press: Boca Raton, FL, USA; Chapter 16; pp. 261–276. https://doi.org/10.1201/9781003472629-16
58. Kumar, A., Kumar, A., & Kumar, A. (2023). *Laser Based Technologies for Sustainable Manufacturing*. Taylor & Francis (CRC Press): Boca Raton FL, USA. https://doi.org/10.1201/9781003402398
59. Kumar, R., Prasad, A., & Kumar, A. (2023). *Sustainable Smart Manufacturing Processes in Industry 4.0*. Taylor & Francis (CRC Press): Boca Raton FL, USA. https://doi.org/10.1201/9781003436072
60. Kumar, A., Gori, Y., Meena, C. S., Singh, V. P., & Sharma, V. K. *Design and Analysis of Heavy Vehicle Medium Duty Transmission Gearbox System*. CRC Press: Boca Raton, FL, USA, 2022; Chapter 4; pp. 67–86. https://doi.org/10.1201/9781003360001-4
61. Bansal, A., Bhardwaj, H. K., Sharma, V. K., & Kumar, A. (2022). *Static and Dynamic Behavior Analysis of Al-6063 Alloy Using Modified Hopkinson Bar*. CRC Press: Boca Raton, FL, USA; Chapter 6; pp. 107–124. https://doi.org/10.1201/9781003360001-6
62. Reddy, P. N., Verma, V., Kumar, A., & Awasthi, M. (2023). CFD simulation and thermal performance optimization of flow in a channel with multiple baffles. *JHMTR*, 10(2), 257–268. https://doi.org/10.22075/JHMTR.2023.31108.1458
63. Prasad, A., Kumar, A., & Gupta, M. (2023). *Advanced Materials and Manufacturing Techniques in Biomedical Applications*. John Wiley & Sons: USA, ISBN: 9781394166190. https://doi.org/10.1002/9781394166985
64. Saumya, S., Hemant, N., Gaurav, G., Ashwani, K., Tanuj, N., & Singla, Y. K. (2024). *Sustainability in Smart Manufacturing: Trends, Scope, and Challenges*. CRC Press: Boca Raton, FL, USA; ISBN 9781032740713.
65. Singh, B., Kaur, P., Yadav, A. K., Awasthi, M. K., & Kumar, A. (2024). *Design and Analysis of Solar Tracking System for PV Thermal Performance Enhancement*. In A. Kumar, N. Dutt, & M. K. Awasthi (eds.), *Heat Transfer Enhancement Techniques*. Scrivener Publishing LLC; Chapter 11; pp. 251–268.
66. Sharma, V. K., Kumar, V., Joshi, R. S., & Kumar, A. (2022). Effect of REOs on tribological behavior of aluminum hybrid composites using ANN. In V. K. Sharma, A. Kumar, M. Gupta, V. Kumar, D. K. Sharma, & S. K. Sharma (eds.), *Additive Manufacturing in Industry 4.0: Methods, Techniques, Modeling and Nano Aspects*. CRC Press: Boca Raton, FL, USA; Chapter 9; pp. 153–168. https://doi.org/10.1201/9781003360001-9

67. Verma, V., Meena, C. S., Thangavel, S., Kumar, A., Choudhary, T., & Dwivedi, G. (2023). Ground and solar assisted heat pump systems for space heating and cooling applications in the northern region of India—a study on energy and CO_2 saving potential. *SETA*, 59, 103405, ISSN 2213-1388. https://doi.org/10.1016/j.seta.2023.103405
68. Singh, S., Kumar, A., Behura, S. K., & Verma, K. (2022). Challenges and opportunities in nanomanufacturing. In *Nanomanufacturing and Nanomaterials Design: Principles and Applications*. CRC Press: Boca Raton, FL, USA; Chapter 2. https://doi.org/10.1201/9781003220602-2
69. Ahmed, A., Yadav, A. K., & Singh, A. (2023). Application of machine learning and genetic algorithm for prediction and optimization of biodiesel yield from waste cooking oil. *Korean Journal of Chemical Engineering*. 40, 2941–2956.
70. Khashehchi, M., Thangavel, S., Rahmanivahid, P., Heidari, M., Moazzeni, T., Verma, V., & Kumar, A. (2024). Solar desalination techniques: challenges and opportunities. In *Highly Efficient Thermal Renewable Energy Systems: Design, Optimization and Applications*. CRC Press: Boca Raton, FL, USA; Chapter 19; pp. 305–329. https://doi.org/10.1201/9781003472629-19
71. Dutt, N., Hedau, A. J., Kumar, A., Awasthi, M. K., Singh, V. P., & Dwivedi, G. (2023). Thermo-hydraulic performance of solar air heater having discrete D-shaped ribs as artificial roughness. *ESPR*. https://doi.org/10.1007/s11356-023-28247-9
72. Singh, V. P., Kumar, A., & Awasthi, M. K. (2024). Augmentation of Solar, Geothermal and Earth-Air Heat Exchanger in Sustainable Buildings. In A. Kumar, N. Dutt, & M. K. Awasthi (eds.), *Heat Transfer Enhancement Techniques*. Scrivener Publishing LLC; Chapter 6; pp. 119–158.
73. Kundu, A., Kumar, A., Awasthi, M. K., Dwivedi, G., & Shukla, A. K. An explorative study of earth-air heat exchanger for sustainable building applications. *International Journal of Energy for a Clean Environment* (in press).
74. Khan, O., Parvez, M., Kumari, P., Yadav, A. K., Akram, W., Ahmad, S., Parvez, S., & Idrisi, M. J. (2023) Modelling of compression ignition engine by soft computing techniques (ANFIS-NSGA-II and RSM) to enhance the performance characteristics for leachate blends with nano-additives. *Scientific Reports*, 13, 15429.
75. Heidari, M., Thangavel, S., Ghafri, K. A., & Kumar, A. (2024). Future trends and emerging research in nanofluids for aerospace applications. In *Nanofluids Technology for Thermal Sciences and Engineering: Research, Development, and Applications*. CRC Press: Boca Raton, FL, USA; Chapter 15; pp. 272–291.
76. Dutt, N., Hedau, A., Kumar, A., Awasthi, M. K., & Singh, V. P. (2024). Effect of duct length variation on solar air heater performance for smooth and D-shaped roughened absorber plate. *Heat Transfer* (in press).

77. Jaideep, P. Tomar, & A. Kumar. (2024). CFD Numerical Investigation of Thermal Performance of Diamond Shape Micro Rectangular Heat Exchanger. In A. Kumar, N. Dutt, & M. K. Awasthi (eds.), *Heat Transfer Enhancement Techniques*. Scrivener Publishing LLC; Chapter 7; 159–192.
78. Ahmed, A., & Yadav, A. K. (2024). Enhanced production of methane enriched biogas through intensified co-digestion process and its effective utilization in a biodiesel/biohydrogen fueled engine with duel injection strategies: ML-RSM based an efficient optimization approach. *International Journal of Hydrogen Energy*, 65, 671–686.
79. Khanam, S., Khan, O., Ahmad, S., Sherwani, A. F., Khan, Z. A., Yadav, A. K., & Ağbulut, Ü. (2024). A Taguchi-based hybrid multi-criteria decision-making approach for optimization of performance characteristics of diesel engine fuelled with blends of biodiesel-diesel and cerium oxide nano-additive. *Journal of Thermal Analysis and Calorimetry*, 149, 3657–3676.
80. Ahmed A., Yadav, A. K., Singh A., Singh, D. K., & Ağbulut, Ü. (2024). A hybrid RSM-GA-PSO approach on optimization of process intensification of linseed biodiesel synthesis using an ultrasonication reactor: enhancing fuel properties and engine characteristics with ternary fuel blends. *Energy*, 288, 129077.

Chapter 2

The Rise of Industry 5.0
How Artificial Intelligence Is Shaping the Future of Manufacturing

Veerendra Singh and Rajendra Kumar

2.1 INTRODUCTION

Industry 5.0 represents next-level evolution in industrial manufacturing and production. Basically it is an advance level of Industry 4.0, which focused on automation, digital technologies, and data exchange operation. While Industry 4.0 emphasized the combination of cyber security operation, Internet of Things (IoT), on-demand computing, and artificial intelligence (AI), Industry 5.0 seeks to reintroduce the human element into the manufacturing process in a more integrated and harmonious way (Javaid M, et al. 2020). The Industrial Revolution refers to a series of significant shifts in human history marked by profound changes in technology, manufacturing processes, and socioeconomic structures. There have been several key industrial revolutions.

2.1.1 First Industrial Revolution (Eighteenth to Nineteenth Century)

The First Industrial Revolution began in England in the late eighteenth century and spread to Europe as well as c. It was indicated by the process of textile manufacturing, the introduction of engines operated by steam power, and production of iron ore as well as coal companies (Longo F, et al. 2020).

Key Innovations: These include steam engines, mechanized spinning and weaving machines (e.g., the spinning jenny, power loom), and innovations in transportation (e.g., the steam locomotive, canals) (Figure 2.1).

2.1.2 Second Industrial Revolution (After Nineteenth to before Twentieth Century)

The Second Industrial Revolution comes into the light after nineteenth to before twentieth century, primarily in Western Europe and the United States. It was marked by advancements in steel production, electricity, and the expansion of the telegraph and telephone networks (Pathak A, et al. 2021).

Figure 2.1 Technology in the First Industrial Revolution.

Key Innovations: These include electric power generation and distribution, manufacturing of internal combustion engines, and mass production techniques (e.g., assembly line) (Figure 2.2).

2.1.3 Third Industrial Revolution (Late Twentieth Century)

The Third Industrial Revolution, also known as Digital Revolution or Information Age, began in the late twentieth century with the widespread adoption of digital technologies and automation (He D, et al. 2017).

Key Innovations: These include personal computers, the internet, telecommunications, digitalization of information, and automation technologies (e.g., robotics, computer-aided design/manufacturing) (Figure 2.3).

2.1.4 Fourth Industrial Revolution (Twenty-First Century)

The Fourth Industrial Revolution makes upon the digital revolution of the Third Industrial Revolution, defined by the mixing of technologies of automation and data exchange (Leone LA, et al. 2020).

Key Innovations: These include robotics, three-dimensional (3D) printing, nanotechnology, and biotechnology (Figure 2.4).

These industrial revolutions have had profound effects on society, economies, and the global landscape. They have transformed the way goods are produced, led to urbanization, created new industries and job opportunities, and fundamentally altered the nature of human interaction and

Figure 2.2 Technology in Second Industrial Revolution assembly line.

communication. The ongoing evolution of technology continues to shape the world in ways that were once unimaginable.

2.2 CHARACTERISTICS OF INDUSTRY 5.0

Industry 5.0 is a concept of advancements and principles of Industry 4.0, focusing more on the integration of human intelligence and capabilities into the manufacturing processes. While Industry 4.0 emphasized automation, data exchange, and smart technologies, Industry 5.0 seeks to maintain the level of technological innovation with human beings. Some characteristics of Industry 5.0 include the following (Majumdar A, et al. 2021):

- **Human–Machine Collaboration:** Industry 5.0 focuses on the collaboration between humans and machines. Despite automation of machines and robotics, which play a significant role, humans are integrated with analytical skills and adaptability.

The Rise of Industry 5.0   29

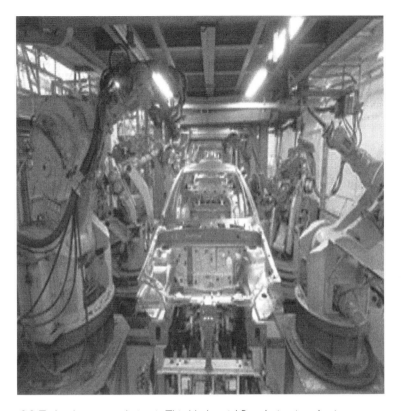

Figure 2.3 Technology upgradation in Third Industrial Revolution in robotics.

- **Customization and Flexibility:** Industry 5.0 prioritizes customization and flexibility in production processes. With human involvement, it becomes easier to adjust customer requirements and preferences, allowing more personalized products and services.
- **Resilience and Adaptability:** Industry 5.0 promotes resilience and adaptability in manufacturing systems. Human workers can quickly adapt to unexpected changes and challenges in the production environment, making the manufacturing process more robust and responsive.
- **Empowerment of Workers:** Industry 5.0 empowers workers by providing them with opportunities for skill development, training, and creativity. Human workers are encouraged to contribute ideas, innovate, and improve processes.
- **Ethical and Sustainable Practices:** Industry 5.0 focuses on ethical as well as sustainable practices in manufacturing. Human involvement enables a more ethical approach to decision-making, ensuring

Figure 2.4 Upgradation of technology in Fourth Industrial Revolution in robotics.

that the interests of workers, communities, and the environment are considered alongside business objectives.
- **Integration of New Technologies:** Industry 5.0 integrates new technologies such as AI, augmented reality (AR), and cloud computing to enhance human–machine collaboration and improve manufacturing processes.
- **Focus on Quality and Innovation:** Industry 5.0 focuses upon quality and innovation. Workers contribute to continuous improvement initiatives, driving innovation and ensuring that products meet high standards of quality and excellence.
- **Global Connectivity and Collaboration:** Industry 5.0 promotes global connectivity and collaboration among manufacturers, suppliers, and customers. Through digital platforms and networks, businesses

can collaborate across geographic boundaries to share knowledge, resources, and best practices.

Overall, Industry 5.0 represents a shift toward a more humanistic concept in manufacturing, where technology and people work intelligently for better innovation and productivity.

2.3 IMPORTANCE OF AI IN INDUSTRY 5.0

The importance of AI in Industry 5.0 signifies a paradigm toward humanistic manufacturing processes. In Industry 5.0, AI serves as a trigger for fostering combination of human and machine, driving innovation as well as enhancing productivity across diverse industrial domains. AI-driven predictive maintenance systems enable proactive equipment monitoring, minimizing downtime, and optimizing maintenance schedules. Furthermore, AI facilitates adaptive manufacturing, enabling real-time adjustments to production processes based on market demands and operational constraints. Leveraging computer vision and machine learning algorithms, AI enhances quality control processes, ensuring product quality and customer satisfaction. Moreover, AI-driven insights enable manufacturers to drive innovation, optimize supply chain management, and foster ethical AI adoption practices. In essence, AI enables manufacturers to achieve operational excellence, drive sustainable growth, and navigate the complexities of an interconnected digital landscape (Zhang C, Chen YA, 2020).

2.3.1 Techniques and Approaches

- **Machine Learning:** Machine learning is a part of AI which aims to develop algorithms and models in which computers are able to learn from data and improve their performance over time.
- **Deep Learning:** Deep learning is a subset of machine learning, which utilizes neural networks with many layers (therefore the term "deep") to learn representations of data. Deep learning performs well on specific tasks like picture identification and speech recognition.
- **Natural Language Processing (NLP):** NLP is a stream of AI that deals with communications between computer and human being by using natural language. It enables computer to interact human language and perform tasks like language conversation or translation and human sentimental analysis.

2.3.2 Challenges and Considerations

- **Ethical Concerns:** AI raises ethical considerations associated with bias in algorithms, privacy concerns, and automatic decision-making.

- **Data Privacy and Security:** AI has the potential to revolutionize industries, transform business processes, and enhance human capabilities. However, addressing ethical, privacy, and security considerations is crucial to ensuring the responsible evolution and growth of AI technologies.

2.4 IMPORTANCE OF AI IN INDUSTRIAL TRANSFORMATION

The importance of AI in industrial transformation is multifaceted and encompasses various aspects of modern manufacturing and production processes. The following points highlight the significance of AI in industrial transformation (Aslam F, et al. 2020):

- **Enhanced Efficiency and Productivity:** AI-driven automation streamlines production processes, reduces manual intervention, and optimizes resource utilization. This leads to increased efficiency and productivity in industrial operations.
- **Predictive Maintenance and Reduced Downtime:** This system examines equipment data in real time to identify anomalies and indicate reason for failures before it takes place. Manufacturers can reduce unplanned idle time, maintenance cost as well as prolong the lifespan of machinery and equipment.
- **Quality Assurance and Defect Identification:** Along with AI algorithm can automatically maintain quality assurance as well as identify defects in various industrial operations. By analyzing images and visual data, AI algorithms can detect inconsistencies by ensuring that only products meeting quality criteria are delivered to customers.
- **Optimized Supply Chain Management:** AI algorithms examine a large quantity of data from various sources to improve supply chain management processes such as inventory control and required amount of demand. This will not only improve supply chain efficiency but also improve by reducing inventory cost.
- **Customization and Personalization:** AI technologies enable mass customization and personalization of products to meet individual customer preferences and demands. With examining a customer's various data, AI algorithms produce tailored product configurations and enhance the overall customer experience.
- **Data-Driven Decision-Making:** AI allows informed decision-making by examining a large amount of data and identifying patterns. By leveraging AI-driven analytics, manufacturers can build informative decisions regarding strategic planning, leading to improved business outcomes and competitive advantage.

- **Innovation and New Product Development:** AI facilitates innovation and develops new products by enabling rapid prototyping, design optimization, and simulation-based testing. By leveraging AI-driven tools and techniques, manufacturers can accelerate the product development lifecycle, reduce time to market, and introduce innovative products that meet evolving customer needs and market trends.
- **Workforce Augmentation and Skill Enhancement:** AI complements human intelligence and capabilities by automating routine tasks, augmenting human decision-making, and enhancing workforce skills. Working alongside AI systems involves higher-value tasks, which require creative action, critical thought, as well as problem-solving abilities, leading to a more skilled and engaged workforce.

AI has a vital role for industrial transformation by improving efficiency, quality, and agility across manufacturing and production processes. By embracing AI technologies, manufacturers can make alterations to swap trade and maintain competitive in an increasingly digital and interconnected world.

2.5 KEY DRIVERS FOR AI ADOPTION IN INDUSTRY 5.0

The adoption of AI in Industry 5.0 is driven by several key factors that contribute to its transformative potential. Here are some of the key drivers for AI adoption in Industry 5.0 (Alhassan AB, et al. 2020):

- **Human–Machine Collaboration:** Industry 5.0 focuses attention on the association with human beings and machines. The basic purpose of this combination is to maintain a bridge with human workers and automated systems for improving productivity and making manufacturing processes efficient.
- **Advancements in AI Technologies:** The rapid advancements in AI technologies, particularly in machine learning, deep learning, and NLP, allow more sophisticated and intelligent automation solutions. AI-driven algorithms become more capable of analyzing complex data, recognizing patterns, and making predictions, empowering manufacturers to optimize operations and drive innovation.
- **Predictive Maintenance and Reliability:** AI-powered predictive maintenance systems enable manufacturers to check equipment's health in real time, predict possible downtimes, as well as schedule maintenance activities. By reducing unplanned idle time and optimizing maintenance schedules, AI adoption enhances equipment reliability, reduces repair costs, and improves overall operational efficiency.
- **Competitive Pressures and Market Dynamics:** In today's competitive business environment, manufacturers are under pressure for

minimizing product costs and improving quality of products to maintain a competitive edge. AI adoption enables manufacturers to streamline processes, optimize resources, and deliver high-quality products and services more efficiently, meeting market demands and customer expectations.
- **Regulatory Compliance and Quality Standards:** Compliance with regulatory requirements and quality standards is essential for manufacturers to ensure product safety, reliability, and compliance with industry regulations. AI-driven quality control systems enable manufacturers to detect defects, ensure product quality, and maintain compliance with regulatory standards, minimizing the risk of product recalls and liabilities.
- **Workforce Augmentation and Skill Enhancement:** AI technologies augment human intelligence and capabilities by automating routine tasks, providing decision support, and enhancing workforce skills. By leveraging AI-driven tools and training programs, manufacturers adapt to new technologies for performing well on higher-value tasks, which require high-value thinking as well as problem-solving abilities.

The adoption of AI in Industry 5.0 is driven by a combination of improved technology, regulatory requirements, as well as enhanced productivity, quality, and agility in manufacturing processes.

2.6 APPLICATIONS OF AI IN INDUSTRY 5.0

Industry 5.0 shows the convergence of traditional manufacturing practices with advanced digital technologies, including AI. AI has a pivotal role in conducting innovation and productivity across different sectors within Industry 5.0. In Industry 5.0, th following applications are available (Lima F, et al. 2019).

2.6.1 Smart Manufacturing and Robotics

Smart manufacturing and robotics play pivotal roles in the paradigm shift toward Industry 5.0, where the combination of advanced new technologies fosters greater connectivity, flexibility, and intelligence within manufacturing processes. Here is how smart manufacturing and robotics contribute to Industry 5.0 (Abdelmageed S, et al. 2020). Smart manufacturing emphasizes the interconnectivity of machines and devices through the IoT and Industrial Internet of Things (IIoT) (Matheus LE, et al. 2019). Robotics serve as an integral component in interconnected manufacturing environments, communicating data in real time to optimize production processes as well as improve decision-making.

Robotics enables advanced automation in manufacturing by performing tasks traditionally carried out by humans, such as assembly, welding, and material handling. Smart manufacturing platforms orchestrate the seamless integration of robotic systems with other manufacturing assets, optimizing workflow efficiency and reducing operational costs. Robotics equipped with sensors and diagnostic capabilities contribute to predictive maintenance strategies in smart manufacturing environments. Data collected from robotic systems enable predictive analytic algorithms to predict equipment failures and its schedule maintenance activities proactively, reducing unplanned idle time and enhances its performance.

2.6.1.1 Flexible Manufacturing Systems

Robotics and smart manufacturing technologies facilitate the development of flexible manufacturing systems, which adapt for changing production requirements and customer demands. Agile robotic systems can be reprogrammed and reconfigured quickly to perform different tasks and accommodate varying product specifications, enhancing manufacturing flexibility and responsiveness (Brown DG, et al. 2021).

In short, smart manufacturing and robotics are integral components of Industry 5.0, enabling manufacturer to achieve higher productivity as well as agility in the face of evolving market dynamics and customer expectations. The convergence of these technologies drives innovation and transformation across manufacturing industries, paving the way for the factories of the time ahead.

2.6.2 Supply Chain Improvement

Supply chain improvement within the context of Industry 5.0, driven by AI, entails the application of advanced algorithms and data analytics techniques to streamline and increase different facts of supply chain management. Here are specific applications of AI-based supply chain optimization in Industry 5.0 (Nguyen TN, et al. 2018).

Predictive analytics models anticipate transportation disruptions, optimize delivery routes, and improve overall logistics efficiency. AI-driven warehouse management systems optimize storage, picking, packing, and shipping processes within distribution centers. Robotics and automation technologies improve warehouse efficiency by automating repetitive tasks, reducing manual errors, and increasing throughput. AI-enabled visibility based on platforms provides on-time insights into inventory and shipment statuses. IoT sensors and connected devices track the movement of goods throughout the supply chain and risk management.

Figure 2.5 Smart supply chain management.

2.6.2.1 Supply Chain Risk Management

AI-based algorithms examine supply chain database to determine risks, vulnerabilities, and disruptions like supplier bankruptcies, geopolitical events, or natural disasters. Predictive analytics models assess the impact of potential disruptions and recommend risk mitigation strategies to improve supply chain flexibility. Dynamic pricing models maintain prices dynamically based on requirement elasticity, inventory levels as well as competitive positioning to increase revenue and profit. AI-based supply chain optimization in Industry 5.0 empowers organizations to create agile, responsive, and resilient supply chains capable of adapting to changing market conditions and customer expectations (Zambon I, et al. 2019) (Figure 2.5).

2.6.3 Personalized Production and Mass Customization

Personalized production and mass customization are key concepts within Industry 5.0, where AI plays an important role in enabling manufacturers to meet the growing demand for customized products at scale. Here is how personalized production and mass customization are facilitated by AI. This analysis helps manufacturers understand customer demands, allowing them to tailor products to specific preferences and anticipate evolving market trends. AI-powered design tools assist in creating customizable product configurations. Generative design algorithms use AI to inspect a large number of design possibilities, improving for both functionality and aesthetic appeal based on user-specified parameters.

2.6.3.1 Digital Twins for Personalization

Digital twins, virtual representations of physical products or systems, are used to create personalized product variations. AI-driven simulations and modeling help manufacturers visualize and customize digital twins to meet individual customer requirements.

AI algorithms optimize production schedules and resource allocation to accommodate personalized orders. Dynamic production planning considers on-time database based on customer demands, inventory and ensuring efficient and timely manufacturing of personalized products.

2.6.4 Autonomous Vehicles and Transportation Systems

Autonomous vehicles and transportation systems represent pivotal components of Industry 5.0, where the integration of AI drives transformative changes in mobility, logistics, and urban transportation. Here is how autonomous vehicles and transportation systems are evolving in the terms of Industry 5.0 based on AI (Tripathy HP, et al. 2020).

AI-powered fleet management systems optimize the operation of autonomous vehicle fleets for ride-sharing, delivery, and logistics applications. AI algorithms optimize route planning, dispatching, and vehicle allocation to minimize empty miles, reduce delivery times, and maximize resource utilization across the fleet. AI algorithms adjust traffic signals, reroute vehicles, and optimize traffic patterns based on current conditions, improving overall traffic efficiency and reducing travel times. Autonomous vehicles and drones equipped with AI capabilities enable efficient last-mile delivery solutions for e-commerce and logistics companies.

AI-driven autonomous shuttles and buses offer on-demand and shared mobility services in urban areas, complementing traditional public transportation systems. AI algorithms optimize passenger pickup and drop-off locations, dynamically adjust routes based on demand, and enhance the overall efficiency and accessibility of public transportation. Autonomous vehicles and transportation systems integrate with smart city infrastructure to create connected and sustainable transportation ecosystems. AI-powered traffic management, parking solutions, and public transit systems contribute to reducing emissions, improving air quality, and enhancing urban livability.

2.6.5 Energy Management and Sustainability

Energy management and sustainability are central themes for industries where integration of AI enables significant advancements in resource efficiency, environmental conservation, and sustainable practices across industrial sectors (Figure 2.6). Here is how energy management and sustainability initiatives are evolving. AI algorithms improve the operation of smart grids by examining on-time data on energy supply, demand, and distribution.

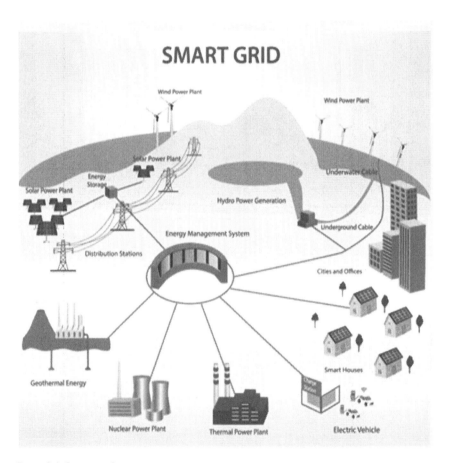

Figure 2.6 Smart grid management.

Predictive analytics and machine learning enable smart grids to anticipate fluctuations in energy demand, dynamically adjust power generation and distribution, and optimize energy flow to minimize wastage and maximize efficiency.

AI-driven energy monitoring systems detect anomalies and recommend energy-saving measures to reduce overall energy consumption and operational costs. This system examines the execution of energy infrastructure, such as turbines, generators, and Heating, Ventilation and Air Conditioning (HVAC) systems. Machine learning algorithms detect machine failures, identify required maintenance, and improve maintenance schedules to prevent costly breakdown and ensure reliability of energy systems.

AI algorithms predict renewable energy generation, maximize energy storage and distribution, and balance supply and demand to ensure grid stability and reliability while maximizing the use of clean energy sources. AI-powered Building Energy Management Systems (BEMS) analyze building energy usage patterns and optimize HVAC systems, lighting, and other energy-consuming devices to reduce energy waste. Machine learning algorithms adaptively adjust building controls based on occupancy patterns, weather forecasts, and energy tariffs to minimize energy consumption and maintain occupant comfort.

2.6.6 Healthcare Innovation

Healthcare innovation within the context of Industry 5.0, which emphasizes human–machine collaboration, can be greatly facilitated by AI (Figure 2.7). Here are several ways AI can drive innovation in healthcare within the framework of Industry 5.0 (Lutz J, et al. 2020):

- **Personalized Medicine:** AI-based algorithms can examine huge amount of datasets including genomics, patient records, and medical imaging to tailor treatments to individual patients. This approach, known as precision medicine, can improve treatment efficacy and reduce adverse effects.
- **Predictive Analytics:** AI can forecast patient outcomes and disease progression by analyzing historical patient data and identifying patterns. This enables healthcare providers to call earlier and prevent adverse events, leading to better patient outcomes and reduced healthcare cost.
- **Remote Monitoring and Telemedicine:** AI-based applications enable remote monitoring of patients' vital signs, symptoms, and medication adherence.
- **Diagnostic Support:** AI-based algorithms can assist healthcare providers in diagnosing diseases by analyzing medical images, pathology slides, and other diagnostic tests. Machine learning models

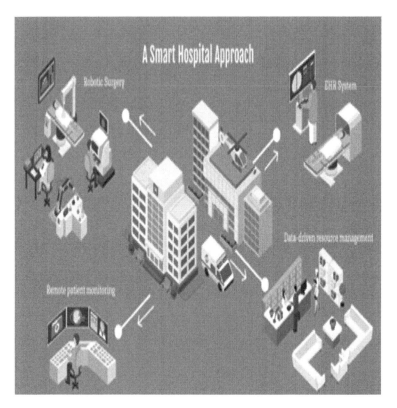

Figure 2.7 Smart hospital management.

can help identify patterns and anomalies that may be difficult for human practitioners to detect, leading to earlier and more accurate diagnoses.
- **Healthcare Operations Optimization:** AI algorithms can optimize hospital and healthcare system operations by analyzing patient flow, resource utilization, and staff scheduling. This can lead to improved efficiency and better allocation of resources.
- **Patient Engagement and Education:** AI-based virtual assistants can interact with patients, answer their healthcare-related questions, and provide personalized health education and coaching.

By leveraging advanced technologies and fostering combination between humans and machines, healthcare organizations can enhance efficiency, effectiveness, and accessibility across the continuum of care.

2.6.7 Customer Experience Enhancement

Customer experience enhancement within the context of Industry 5.0, which emphasizes human–machine collaboration, can be significantly augmented by AI. There are different ways AI can drive innovation in enhancing customer experiences:

- **Personalized Recommendations:** AI-based algorithms examine customer data and choices to provide customized product guidance and tailored experiences. This activity improves customer pleasure and improves relations for repeatedly purchasing.
- **Predictive Analytics:** This analysis predicts customer requirement and behaviors based on historical data and real-time interactions. By understanding customer preferences and anticipating future actions, businesses can proactively address customer issues and offer relevant solutions.
- **Customer Journey Optimization:** AI technologies enable businesses to map out and optimize the whole customer journey, from initial engagement to post-purchase interactions.
- **Visual Recognition:** AI-powered visual recognition technologies can enhance the customer experience by enabling features such as visual search and AR. Customers can use their smartphones to search for products using images or visualize how products will look in their environment, enhancing the shopping experience.
- **Dynamic Pricing and Offers:** AI-based algorithms can examine market fluctuations, competitor pricing, and customer behavior to dynamically adjust prices and customize offers in real time. This enables businesses to improve strategies in context with price and promotions to increase revenue.
- **Voice Commerce:** AI voice-activated-based assistants, such as Alexa and Google Assistant, enable seamless voice-driven shopping experiences. Customer uses voice commands to search for products, place orders, and track deliveries, simplifying the purchasing process and enhancing convenience.
- **Emotion Recognition:** AI-powered emotion recognition technologies can analyze facial expressions and voice tones to gauge customer emotions during interactions. Businesses can use this data to tailor their feedback and offers empathetic support, enhancing customer satisfaction and loyalty.

In short, AI-based innovation has the possibilities to transform customer experiences within the framework of Industry. By leveraging the new version of AI technologies and fostering combination between humans and machines, businesses can deliver highly personalized, seamless customer satisfaction.

2.7 CASE STUDIES AND EXAMPLES

There are some real-world examples of AI implementation within the context of Industry 5.0 across various industries, along with their impact on business processes and competitive advantage, detailed below.

2.7.1 Manufacturing Industry

- **Predictive Maintenance:** Companies like Siemens and General Electric use AI-based predictive maintenance systems to monitor machinery health and predict possible failures before they occur. This reduces downtime, optimizes maintenance schedules, and improves operational efficiency (Mahmood NH, et al. 2020).
- **Quality Control:** Automotive manufacturers like BMW and Ford employ AI-driven computer vision systems to inspect and detect defects in production lines. By automating quality control processes, they ensure product consistency and minimize defects so that customer satisfaction is kept high.

2.7.2 Healthcare Industry

- **Medical Imaging:** Companies like GE Healthcare and Philips Healthcare employ AI algorithms to examine medical images and help radiologists in diagnosing diseases such as cancer and neurological disorders. AI-powered image analysis improves diagnostic accuracy, reduces interpretation time, and enhances patient outcomes (Saxena A, et al. 2020).
- **Drug Discovery:** Pharmaceutical companies like Pfizer and Merck leverage AI-driven algorithms to step up the drug discovery mechanisms. By examining a huge amount of molecular database as well as simulating drug interactions, they identify potential drug candidates more efficiently, reducing time to market and research costs.

2.7.3 Retail Industry

- **Personalized Recommendations:** E-commerce platforms like Amazon and Alibaba employ AI-based algorithms that help toward customized product advice based on customer fondness and browsing history. By tailoring recommendations, they enhance customer engagement, increase sales, and foster brand loyalty.

- **Inventory Management:** Retailers such as Walmart and Target use AI-based demand prediction models to improve inventory as well as replenishment strategies. AI-driven insights help reduce stockouts, minimize excess inventory, and improve supply chain efficiency (Cioffi R, et al. 2020).

2.7.4 Financial Services Industry

- **Fraud Detection:** Banks and financial institutions deploy AI-powered fraud identifying systems for detecting skeptical activity and control fraudulent transactions. By analyzing transactional data and user behavior patterns, they mitigate financial risks and protect customer assets.
- **Algorithmic Trading:** Hedge funds and investment firms leverage AI algorithms to automate trading strategies and optimize investment decisions. AI-based trading system examines fluctuating market trends, detecting good time and carryout trades immediately, gaining a competitive edge in volatile markets.
- These examples demonstrate how AI technologies are transforming business processes and driving competitive advantage across diverse industries. By embracing AI-based innovations, organizations can enhance efficiency, productivity, and innovation while delivering greater value to customers and stakeholders.

2.8 FUTURE OUTLOOK

The future outlook for AI-driven industrial transformation within the context of Industry 5.0 is marked by several emerging trends, predictions, opportunities, and challenges.

2.8.1 Predictions for the Future of AI-Driven Industrial Transformation

- **Hyperautomation:** The convergence of AI, robotics, and automation will lead to hyperautomation, where end-to-end processes are fully automated. Hyperautomation will revolutionize industries by driving efficiency, agility, and innovation.
- **AI-Driven Personalization:** AI will enable hyper-personalized products, services, and experiences tailored to individual preferences and behaviors. Organizations will leverage AI to deliver customized offerings, anticipate customer needs, and enhance engagement.
- **Augmented Intelligence:** AI will augment human capabilities rather than replacing them entirely. Augmented intelligence will empower

workers with AI-driven insights, decision support, and automation tools, leading to more productive and fulfilling work experiences.
- **Industry-Specific AI Solutions:** AI solutions will be tailored to specific industry requirements and use cases. From manufacturing and logistics to healthcare and finance, industry-specific AI applications will drive innovation, efficiency, and competitiveness.
- **AI Ecosystem Collaboration:** Organizations will collaborate within the AI ecosystem to share data, expertise, and resources. Open innovation and collaboration will accelerate AI development, foster interoperability, and address complex challenges.

2.8.2 Opportunities and Challenges on the Horizon

- **Opportunities:** AI-driven industrial transformation offers opportunities for increased productivity, innovation, and competitiveness. Organizations can leverage AI to optimize operations, unlock new revenue streams, and deliver superior customer experiences.
- **Challenges:** Challenges include addressing ethical considerations, ensuring data privacy and security, managing workforce transition, and navigating regulatory complexities. Organizations must also address concerns related to algorithmic bias, accountability, and transparency.

The future of AI-driven industrial transformation is characterized by unprecedented opportunities for innovation and growth, along with complex challenges. By embracing emerging trends, fostering collaboration, and adopting responsible AI practices, organizations can navigate the evolving landscape as well as drive in this digital age successfully.

2.9 CONCLUSION

In conclusion, the incorporation of AI in driving the evolution represents a movement toward businesses operating and interacting with technology. Here is a summary of key points and reflections along with a call to action for industry leaders and decision-makers:

- Industry 5.0 emphasizes human–machine collaboration, where AI technologies augment human capabilities and enable greater efficiency, productivity, and innovation.
- AI-driven solutions such as predictive analytics, automation, and personalized experiences are revolutionizing industries across manufacturing, healthcare, finance, retail, and beyond.
- Challenges include ethical considerations, privacy concerns, workforce displacement, regulatory compliance, and sustainable AI deployment in Industry 5.0 environments.

- AI serves as a catalyst for driving Industry 5.0 evolution by unlocking new opportunities for optimization, automation, and human–machine collaboration.
- AI has the possible ways to revolutionize business models, disrupt traditional companies, and open new opportunities for growth and competitiveness in the global marketplace.

In conclusion, AI holds immense possibility to maintain structure of the future industry and drive unprecedented levels of innovation, efficiency, and value creation. By embracing AI-driven technologies and adopting a strategic approach to AI deployment, industry leaders and decision-makers can position their organizations for success in the digital era and unlock new opportunities for growth and competitiveness.

REFERENCES

1. Abdelmageed S, Zayed T (2020) A study of literature in modular integrated construction-critical review and future directions. J Clean Prod 277:124044.
2. Adadi A, Berrada M (2018) Peeking inside the black-box: a survey on explainable artificial intelligence (xai). IEEE Access 6:52138–52160.
3. Alhassan AB, Zhang X, Shen H, Xu H (2020) Power transmission line inspection robots: a review, trends and challenges for future research. Int J Electr Power Energy Syst 118:105862.
4. Aslam F, Aimin W, Li M, Ur Rehman K (2020) Innovation in the era of IoT and industry 5.0: absolute innovation management (AIM) framework. Information 11(2):124.
5. Brown DG, Wobst HJ (2021) A decade of FDA-approved drugs (2010–2019): trends and future directions. J Med Chem 64(5):2312–2338.
6. Chowdhury MZ, Shahjalal M, Ahmed S, Jang YM (2020) 6G wireless communication systems: applications, requirements, technologies, challenges, and research directions. IEEE Open J Commun Soc 1:957–975.
7. Cioffi R, Travaglioni M, Piscitelli G, Petrillo A, Parmentola A (2020) Smart manufacturing systems and applied industrial technologies for a sustainable industry: a systematic literature review. Appl Sci 10(8):2897.
8. Javaid M, Haleem A, Singh RP, Haq MI, Raina A, Suman R (2020) Industry 5.0: potential applications in COVID-19. J Industr Integr Manag 5(04):507–530.
9. He D, Ma M, Zeadally S, Kumar N, Liang K (2017) Certifcateless public key authenticated encryption with keyword search for industrial internet of things. IEEE Trans Ind Inf 14(8):3618–3627.
10. Leone LA, Fleischhacker S, Anderson-Steeves B, Harpe K, Winkler M, Racin E, Baquero B, Gittelsohn J (2020) Healthy food retail during the COVID-19 pandemic: challenges and future directions. Int J Environ Res Public Health 17(20):7397.
11. Lima F, De Carvalho CN, Acardi MB, Dos Santos EG, De Miranda GB, Maia RF, Massote AA (2019) Digital manufacturing tools in the simulation

of collaborative robots: towards industry 4.0. Braz J Oper Prod Manag 16(2):261–280.
12. Longo F, Padovano A, Umbrella S (2020) Value-oriented and ethical technology engineering in industry 5.0: a human-centric perspective for the design of the factory of the future. Appl Sci 10(12):4182.
13. Lutz J, Memmert D, Raabe D, Dornberger R, Donath L (2020) Wearables for integrative performance and tactic analyses: opportunities, challenges, and future directions. Int J Environ Res Public Health 17(1):59.
14. Majumdar A, Garg H, Jain R (2021) Managing the barriers of industry 4.0 adoption and implementation in textile and clothing industry: interpretive structural model and triple helix framework. Comput Ind 125:103372.
15. Mahmood NH, Böcker S, Munari A, Clazzer F, Moerman I, Mikhaylov K, Lopez O, Park OS, Mercier E, Bartz H, Jäntti R (2020) White paper on critical and massive machine type communication towards 6G. arXiv preprint arXiv:2004.14146.
16. Matheus LE, Vieira AB, Vieira LF, Vieira MA, Gnawali O (2019) Visible light communication: concepts, applications and challenges. IEEE Commun Surveys Tutor 21(4):3204–3237.
17. Nguyen TN, Ebrahim FM, Stylianou KC (2018) Photoluminescent, upconversion luminescent and nonlinear optical metal-organic frameworks: from fundamental photophysics to potential applications. Coord Chem Rev 377:259–306.
18. Pathak A, Kothari R, Vinoba M, Habibi N, Tyagi VV (2021) Fungal bioleaching of metals from refnery spent catalysts: a critical review of current research, challenges, and future directions. J Environ Manag 80:111789.
19. Tripathy HP, Pattanaik P (2020) Birth of industry 5.0: "the internet of things" and next-generation technology policy. Int J Adv Res Eng Technol 11(11):1904–1910.
20. Zambon I, Cecchini M, Egidi G, Saporito MG, Colantoni A (2019) Revolution 4.0: industry vs. agriculture in a future development for SMEs. Processes. 7(1):36.
21. Zhang C, Chen YA (2020) Areview of research relevant to the emerging industry trends: industry 4.0, IoT, blockchain, and business analytics. J Industr Integr Manag. 5(01):165–180 22.

Chapter 3

Industry 5.0 with Artificial Intelligence
A Data-Driven Approach

Namita Kathpal, Pratima Manhas, Jyoti Verma, and Seema Jogad

3.1 INTRODUCTION

The dawn of Industry 5.0 marks a significant evolution in the industrial landscape, characterized by the seamless integration of human expertise and cutting-edge technologies (Adel, 2022). At the heart of this revolution lies artificial intelligence (AI), poised to redefine the way industries operate, innovate, and interact with their environment (Olaizola, 2022). Unlike its predecessors, Industry 5.0 transcends automation to embrace a collaborative ecosystem where humans and machines work synergistically to drive productivity, efficiency, and sustainability (Johri et al., 2021). AI applications in Industry 5.0 are diverse and transformative, permeating every aspect of industrial operations; smart mining systems can be implemented by utilizing LLMs (Large Language models) in conjunction with Internet of Things (IoT) platforms (Chen et al., 2024). From predictive maintenance to intelligent supply chain management, AI empowers network automation, and businesses to optimize processes, anticipate market dynamics, and deliver personalized experiences to customers (Chi et al., 2023). Innovations, societal needs, legislative developments, and economic trends will all have an impact on Industry 5.0 as it develops. While 6G and mobile robots together with unmanned aerial vehicles (UAVs) or drones and next-generation mobile communication technology represent a potent combination that has the potential to revolutionize a number of industries with a wide range of applications and communication capabilities, by offering enhanced coverage, new opportunities for advanced automation, improved connectivity, new services, and efficient data collection are opened up. AI in conjunction with augmented intelligence (AuI) offers substantial advantages for Industry 5.0 production (Nguyen & Tran, 2023). The advent of Industry 5.0 signifies a transformative shift in the industrial landscape, characterized by the integration of advanced technologies such as AI with human expertise (Chandel & Sharma, 2023). The goal of the emerging field of human–computer interaction (HCI) is to create AI systems

DOI: 10.1201/9781003494027-3

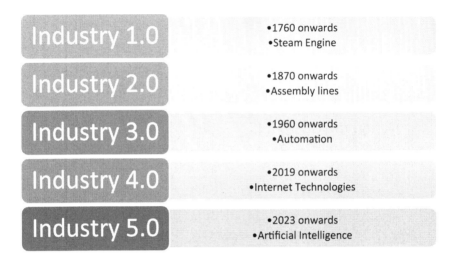

Figure 3.1 Different stages of Industrial Revolution.

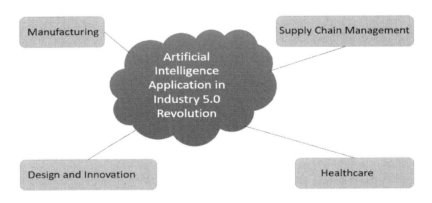

Figure 3.2 Diverse applications of AI in Industry 5.0.

that enhance rather than replace human abilities. The objective of Industry 5.0 is to improve how people and machines collaborate (Bigan, 2022).

This chapter presents different stages of the Industrial Revolution as depicted in Figure 3.1 followed by a comprehensive review of the diverse applications of AI in Industry 5.0 as shown in Figure 3.2, encompassing various sectors and domains. Through an extensive analysis of recent research, case studies, and industry practices, this chapter explores how

AI-driven solutions are revolutionizing industrial operations, and enhancing productivity, efficiency, and sustainability. Additionally, the chapter discusses the challenges and opportunities associated with the integration of AI in the Industry 5.0 revolution and offers insights into future directions and research avenues.

3.2 DIFFERENT STAGES OF THE INDUSTRIAL REVOLUTION

During early industrialization, i.e., Industry 1.0, industries began to mechanize, moving away from manual labor. Significant inventions such as the spinning jenny and the steam engine transformed textile production and powered early factories. Urbanization and the establishment of factory systems altered societal structures, marking the transition from agrarian to industrial economies (Mathur et al., 2022). The latter half of the 18th century witnessed a surge in industrial growth, i.e., Industry 2.0, innovations like the Bessemer process for steel production, and the widespread adoption of electricity revolutionized manufacturing. Transportation networks expanded with the proliferation of railroads and automobiles, facilitating trade and urbanization. The late 20th century saw the rise of digital technology, i.e., Industry 3.0, fundamentally altering industrial processes and communication. Computers, the internet, and telecommunications reshaped industries, finance, and entertainment, leading to globalization. The knowledge economy emerged, emphasizing information processing and services over traditional manufacturing.

Industry 4.0, also known as the Fourth Industrial Revolution, encompasses a range of technologies that are transforming manufacturing and industrial processes. Robotics and IoT are increasingly used in manufacturing for tasks ranging from assembly and material handling to quality inspection (Wong et al., 2021). Collaborative robots, or cobots, work alongside human workers, enhancing productivity and safety on the factory floor. The current era is characterized by the convergence of digital, physical, and biological technologies, i.e., Industry 5.0. AI, robotics, biotechnology, and the IoT are reshaping industries and societies (Noble et al., 2022). Industry 5.0 represents a paradigm shift that aims to foster collaboration between humans and machines, combining the strengths of both to drive innovation, productivity, and sustainability in manufacturing and industrial processes. Figure 3.2 describes the various applications of AI in the Industry 5.0 revolution. Themes such as automation, big data, and sustainability are driving numerous innovations and challenging existing economic and social structures. AI and machine learning (ML) algorithms are used to automate decision-making processes and optimize manufacturing operations. AI-driven predictive maintenance can anticipate equipment failures, while ML algorithms optimize production schedules and quality control.

3.3 ARTIFICIAL INTELLIGENCE APPLICATIONS IN INDUSTRY 5.0-BASED SUSTAINABLE MANUFACTURING

AI/ML techniques play a critical role in Industry 5.0 through the optimization of manufacturing operations, process and product design, computational experiments, scientific ML, and industrial automation (Fahle et al., 2020). Manufacturing applications such as predictive maintenance, quality assurance, and process optimization can benefit greatly from AI. The concept of predictive maintenance is based on the analysis of sensor data from equipment to predict potential equipment failures and schedule routine maintenance to prevent unneeded downtime. Several manufacturing industries, including aerospace, chemicals, electronics, and consumer goods, use AI and ML (Lorido-Botran, 2023). Ensuring the health and safety of customers requires quality assurance. In addition to lowering expenses and waste, preventing quality failures can increase customer satisfaction (Goh et al., 2021). In the semiconductor manufacturing industry, random defects in electron microscope pictures and wafer maps were identified by computer vision models utilizing Convolutional Neural Network (CNN) that were created with the help of AutoML. These defects are significant indicators of semiconductor performance. Both small-scale manufacturing processes and larger-scale operations like building layouts have benefited from optimization (Kamoshida, 2019). There are more variables and interdependencies due to the growing diversity and complexity of industrial jobs, workflows, and supply chains. The use of AI/ML approaches has replaced or supplemented traditional optimization algorithms in industrial processes and procedures. Digital twins powered by AI and ML may potentially help automate processes to enable intelligent and autonomous production (Zhou, 2020).

3.4 ARTIFICIAL INTELLIGENCE APPLICATIONS IN SUPPLY CHAIN MANAGEMENT

Supply chain (SC) excellence frequently depends on the organization's capacity to integrate and coordinate the full range of end-to-end processes of obtaining materials or components, transforming them into finished goods, and delivering them to customers in an era of increased demand uncertainty, higher supply risk, and escalating competitive intensity (Kwon et al., 2007). Through electronic media like websites and electronic data interchange, SC partners frequently share information on demand forecasts, collaborative production, and distribution planning to facilitate the coordination and integration of SC activities. The abundance of data in cyberspace makes it ideal for using ML methods like text and web mining (Liang & Huang, 2006). Used agent technology to calculate the ideal order amount for each

SC tier and simulate inventory levels across the SC employing prior expert knowledge to improve the forecasting of the demand.

3.5 ARTIFICIAL INTELLIGENCE APPLICATIONS IN DESIGN AND INNOVATION

As algorithms play a major role in providing problem solutions, human design is becoming more and more about making sense of the challenges and deciding which ones should or could be handled. This change in emphasis necessitates the development of new theories. AI helps an organization overcome the drawbacks of human-intensive design processes by making the process more scalable, expanding its use outside conventional bounds, and strengthening its capacity for quick learning and adaptation (Verganti et al., 2020). In actuality, AI makes it possible to create solutions that are far more user-centered than those that rely on human expertise. AI systems, on the one hand, give businesses numerous possibilities for improving the economy and efficiency of their processes. However, choosing the right AI tools and features for a given use case and addressing the issue of novel forms of human–machine interaction provide significant hurdles for businesses. For AI systems to provide customized, goal-oriented solutions, a variety of AI functions and technologies must be combined. The application of AI generates new activity profiles that call for related competencies to comprehend and operate AI systems. The design potential to relieve employees of routine activities and to support knowledge-intensive activities to promote productivity and innovation has to be identified.

3.6 ARTIFICIAL INTELLIGENCE APPLICATIONS IN HEALTHCARE

AI-enabled wearable devices and sensors can continuously monitor patients' vital signs, activity levels, and other health metrics outside of traditional healthcare settings (Badawy et al., 2023). These devices can detect early warning signs of health problems, monitor chronic conditions, and alert healthcare providers to intervene when necessary (van Leeuwen, 2024). AI algorithms can analyze medical images such as X-rays, magnetic resonance images (MRIs), and computed tomography (CT) scans to assist radiologists in detecting abnormalities, tumors, fractures, and other medical conditions (Esmaeilzadeh, 2024). These algorithms can help improve diagnostic accuracy and speed up the interpretation process. By analyzing patient data such as medical history, lab results, and vital signs, AI algorithms can predict patient outcomes and identify individuals who are at risk of developing certain diseases or experiencing complications (Turchi et al., 2024). This enables healthcare providers to intervene early and provide personalized treatment plans. AI is revolutionizing the drug discovery process by accelerating the identification of potential drug candidates, predicting their efficacy, and optimizing drug designs. AI algorithms can analyze vast amounts of biological data to uncover new therapeutic targets and streamline the development of novel treatments (Das et al., 2024). Virtual health assistants powered by AI technology can provide patients with personalized health advice, medication reminders, and assistance in scheduling appointments. Chatbots equipped with natural language processing capabilities can engage in conversations with patients to answer their medical questions and provide information about symptoms, treatments, and preventive care (Niu et al., 2024). AI algorithms can analyze genomic data to identify genetic variations associated with disease risk, drug response, and treatment outcomes. This enables the practice of precision medicine, where treatments are tailored to individual patients based on their genetic makeup, lifestyle factors, and environmental influences (Sai et al., 2024). AI can optimize healthcare operations by analyzing data related to resource allocation, patient flow, scheduling, and inventory management (Sampene & Nyirenda, 2024). By identifying inefficiencies and predicting future demand, AI algorithms can help healthcare organizations improve their operational efficiency and reduce costs (Gani et al., 2024). AI-powered clinical decision support systems can assist healthcare providers in making evidence-based treatment decisions by analyzing patient data, medical literature, and clinical guidelines (Nasir et al., 2024). These systems can provide recommendations for diagnostic tests, treatment options, and medication dosages, helping to improve patient outcomes and reduce medical errors.

3.7 CONCLUSIONS AND FUTURE SCOPE

In conclusion, the integration of AI in the Industry 5.0 revolution heralds a new era of industrial transformation, characterized by collaboration, innovation, and human–machine synergy. By harnessing the power of AI, businesses can unlock unprecedented opportunities for growth, efficiency, and sustainability. However, realizing the full potential of AI requires addressing challenges such as data privacy, ethical considerations, and workforce reskilling. Nevertheless, the promise of Industry 5.0, driven by AI, offers a pathway to a more resilient, adaptive, and prosperous future for industries worldwide.

REFERENCES

Adel, A. (2022). Future of industry 5.0 in society: Human-centric solutions, challenges and prospective research areas. Journal of Cloud Computing, *11*(1). https://doi.org/10.1186/s13677-022-00314-5

Badawy, M., Ramadan, N., & Hefny, H.A. (2023). Healthcare predictive analytics using machine learning and Deep Learning Techniques: A Survey. *Journal of*

Electrical Systems and Information Technology. https://doi.org/10.21203/rs.3.rs-1885746/v1.

Bigan, C. (2022). Trends in teaching artificial intelligence for Industry 5.0. In: Draghici, A., Ivascu, L. (eds.), *Sustainability and Innovation in Manufacturing Enterprises. Advances in Sustainability Science and Technology*. Springer, Singapore. https://doi.org/10.1007/978-981-16-7365-8_10

Chandel, A., & Sharma, B. (2023). Technology aspects of artificial intelligence: Industry 5.0 for organization decision making. In: Garg, L., et al. (eds.), *Information Systems and Management Science*. ISMS 2021. Lecture Notes in Networks and Systems, vol. 521. Springer, Cham. https://doi.org/10.1007/978-3-031-13150-9_7

Chen, L., et al. (2024). Smart mining with autonomous driving in industry 5.0: Architectures, platforms, operating systems, foundation models, and applications. *IEEE Transactions on Intelligent Vehicles*, 1–11. https://doi.org/10.1109/tiv.2024.3365997

Chi, H.R., et al. (2023). A survey of network automation for industrial internet-of-things toward industry 5.0. *IEEE Transactions on Industrial Informatics*, 19(2), 2065–2077. https://doi.org/10.1109/tii.2022.3215231

Das, S.K., et al. (2024). AI in Indian Healthcare: From roadmap to reality. *Intelligent Pharmacy*. https://doi.org/10.1016/j.ipha.2024.02.005

Esmaeilzadeh, P. (2024). Challenges and strategies for wide-scale artificial intelligence (AI) deployment in healthcare practices: A perspective for healthcare organizations. *Artificial Intelligence in Medicine*, 151. https://doi.org/10.1016/j.artmed.2024.102861

Fahle, S., Prinz, C., & Kuhlenkötter, B. (2020). Systematic review on machine learning (ML) methods for manufacturing processes—identifying artificial intelligence (AI) methods for Field Application. *Procedia CIRP*, 93, 413–418. https://doi.org/10.1016/j.procir.2020.04.109

Gani, R., et al. (2024). Impact of AI in healthcare services: Analysis Using Medical Synthetic Data. *2024 International Conference on Advances in Computing, Communication, Electrical, and Smart Systems (iCACCESS) [Preprint]*. https://doi.org/10.1109/icaccess61735.2024.10499613

Goh, G.D., Sing, S.L., & Yeong, W.Y. (2021). A review on machine learning in 3D printing: Applications, potential, and challenges. *Artificial Intelligence Review*, 54, 63–94. https://doi.org/10.1007/s10462-020-09876-9

Johri, P., et al. (2021). Sustainability of coexistence of humans and machines: An evolution of industry 5.0 from industry 4.0. *2021 10th International Conference on System Modeling & Advancement in Research Trends (SMART)*. https://doi.org/10.1109/smart52563.2021.9676275

Kamoshida, R. (2019). Job-shop scheduling incorporating dynamic and flexible facility layout planning. *Journal of Industrial and Intelligent Information*, 7(1), 12–17. https://doi.org/10.18178/jiii.7.1.12-17

Kwon, O., Im, G.P., & Lee, K.C. (2007). MACE-SCM: A multiagent and case-based reasoning collaboration mechanism for supply chain management under supply and demand uncertainties. *Expert Systems with Applications*, 33, 690–705.

Leeuwen, van., et al. (2024) Clinical use of artificial intelligence products for radiology in the Netherlands between 2020 and 2022. *European Radiology* 34 (1), 348–354. https://doi.org/10.1007/s00330-023-09991-5

Liang, W., & Huang, C. (2006). Agent-based demand forecast in multi-echelon supply chain. *Decision Support Systems*, *42*, 390–407.

Lorido-Botran, T. (2023). Scaling the metaverse: An AI perspective. *Companion of the 2023 ACM/SPEC International Conference on Performance Engineering.* https://doi.org/10.1145/3578245.3584920

Mathur, A., Dabas, A., & Sharma, N. (2022). Evolution from Industry 1.0 to Industry 5.0. 2022 4th International Conference on Advances in Computing, Communication Control and Networking (ICAC3N), Greater Noida, India, pp. 1390–1394. https://doi.org/10.1109/ICAC3N56670.2022.10074274

Nasir, S., Khan, R.A., & Bai, S. (2024). Ethical framework for harnessing the power of AI in healthcare and beyond. *IEEE Access*, *12*, 31014–31035. https://doi.org/10.1109/access.2024.3369912

Nguyen, H.D., & Tran, K.P. (2023). Artificial intelligence for smart manufacturing in Industry 5.0: Methods, applications, and challenges. *Springer Series in Reliability Engineering*, 5–33. https://doi.org/10.1007/978-3-031-30510-8_2

Niu, S., et al. (2024). Enhancing healthcare decision support through explainable AI models for risk prediction. *Decision Support Systems*, *181*, 114228. https://doi.org/10.1016/j.dss.2024.114228

Noble, S.M., et al. (2022). The Fifth Industrial Revolution: How harmonious human–machine collaboration is triggering a retail and service [r]evolution. *Journal of Retailing*, *98*(2), pp. 199–208. https://doi.org/10.1016/j.jretai.2022.04.003

Olaizola, I.G. (2022). Artificial intelligence from Industry 5.0 perspective: Is the technology ready to meet the challenge? *Proceedings of the Workshop of I-ESA'22*, March 23–24, 2022, Valencia, Spain.

Sai, S., et al. (2024). Generative AI for transformative healthcare: A comprehensive study of emerging models, applications, case studies, and limitations. *IEEE Access*, *12*, 31078–31106. https://doi.org/10.1109/access.2024.3367715

Sampene, A.K., & Nyirenda, F. (2024). Evaluating the effect of artificial intelligence on pharmaceutical product and drug discovery in China. *Future Journal of Pharmaceutical Sciences*, *10*(1). https://doi.org/10.1186/s43094-024-00632-2

Turchi, T., et al. (2024). Pathways to democratized healthcare: Envisioning human-centered AI-as-a-service for customized diagnosis and Rehabilitation. *Artificial Intelligence in Medicine*, *151*, 102850. https://doi.org/10.1016/j.artmed.2024.102850

van Leeuwen, K.G., et al. (2024). How AI should be used in radiology: Assessing ambiguity and completeness of intended use statements of commercial AI products. *Insights into Imaging*, *15*(1). https://doi.org/10.1186/s13244-024-01616-9

Verganti, R., et al. (2020). Innovation and design in the age of artificial intelligence. *Journal of Product Innovation Management*, *37*(3) https://doi.org/10.1111/jpim.12523

Wong, Y.J., et al. (2021). Toward industrial revolution 4.0: Development, validation, and application of 3D-printed IOT-based water quality monitoring system. *Journal of Cleaner Production*, *324*, 129230. https://doi.org/10.1016/j.jclepro.2021.129230

Zhou, G., et al. (2020). Knowledge-driven digital twin manufacturing cell towards intelligent manufacturing. *International Journal of Production Research*, *58*(4), 1034–1051. https://doi.org/10.1080/00207543.2019.1607978

Chapter 4

An Explorative Study on the Use of Artificial Intelligence in Oman High School Education

Pooja Chhabra and Sameer Babu M

4.1 IMPORTANCE OF ENGLISH FOR ACADEMIC PURPOSES

In a contemporary world saturated with information and global commerce, the need for a universal language is imperative. English has assumed this role, being widely recognized as the language of science, technology, commerce, and communication, as well as the language of computing. It has become a secondary literacy, as evidenced by the phrase 'computer literate' (Mocanu et al., 2012).

In Asia, English proficiency is considered essential for participation in the global economy, leading to its inclusion in the curricula of many Asian nations. For example, the number of English learners in China surpasses that of North America (Robertson et al., 2013). Moreover, the Japanese government recognizes the importance of English for individual and national advancement (Morita, 2013).

English has become the primary medium of instruction in universities worldwide, and scholars emphasize its importance for higher education, technology, and mobility. Baniabdelrahma, Abdallah Ahmad (2006) underscores its necessity for academic progress in many developing countries. Additionally, English predominates in academic literature, with 28% of annually published books being in English (Baniabdelrahman, Abdallah Ahmad 2006). Figure 4.1 illustrates the percentages of countries adopting English as a language, even those without direct historical or colonial ties to English-speaking nations like England or America. This suggests a global trend where countries see the value in embracing English to tap into its international economic opportunities and growing readership.

4.2 ARTIFICIAL INTELLIGENCE IN EDUCATION

AI in Education (AIEd) technology applications and educational benefits - AI technology brings virtually have unlimited possibilities to education. The 40 articles investigated a wide variety of AI applications in education, including the following types of learning technology.

DOI: 10.1201/9781003494027-4

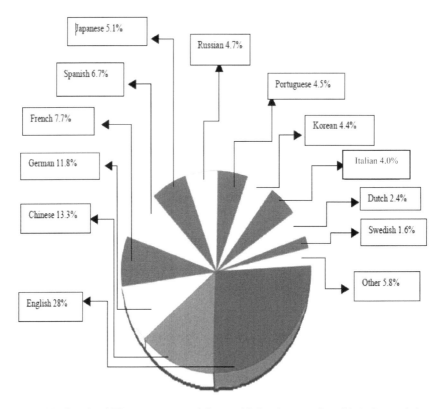

Figure 4.1 (Pie chart) The proportion of the world's books annually published in each language (Graddol, 1998 as cited in Zughoul, 2003).

4.2.1 Chatbot

There is a solitary study specifically addressing the use of chatbots in education, albeit not directly associated with learning outcomes. In a 12-week experiment involving 122 students, researchers examined the impact of chatbot partners versus human partners on students' interest in foreign language courses (Fryer et al., 2017). Results revealed a decline in students' interest after just 1 week with a chatbot, contrasting with the findings under human partner conditions.

4.2.2 Expert system

Research in the field of AIEd suggests that dynamic, comprehensive expert systems have the potential to enhance pedagogical planning and maximize the effectiveness of learning management systems (LMS) in teaching and learning (Dias et al., 2015). For instance, Dias and colleagues conducted a study on interaction quality in a blended learning environment involving 1037 students and 75 professors across multiple courses in an academic year. Their research demonstrated that the structural characteristics of an expert system can accurately model user interactions within an LMS, thereby enhancing teaching and learning experiences.

In a more recent investigation by Hwang et al. (2020), the effects of a sophisticated expert system on elementary students' math learning outcomes in Taiwan were examined. Results indicated that students in the experimental group exhibited superior performance compared to those in the control groups in terms of mathematics learning achievement. Furthermore, the adaptive learning model incorporating affective and cognitive performance analysis was found to effectively alleviate math anxiety among fifth graders in Taiwan.

4.2.3 Intelligent Tutors or Agents

Intelligent tutors or agents play a crucial role in delivering personalized, timely, and suitable materials, guidance, and feedback to learners. Despite their immense potential, research presents mixed findings regarding their impact on learning outcomes. For instance, several studies have investigated the effects of Teachable Agents (TA) (Chin et al., 2013; Matsuda et al., 2020). These studies indicated that TA enhanced learning among elementary students across various grades and facilitated the acquisition of new science content outside of AI software usage.

Research also indicates that TAs with comparable self-efficacy levels to target students may enhance performance in mathematics (Tarning et al., 2019). However, McCarthy et al. (2018) found that metacognitive prompts from intelligent tutoring systems did not enhance student performance, emphasizing the importance of practice and actionable feedback for improving reading comprehension.

4.2.4 Machine Learning

While machine learning has broad applications, only a limited number of research studies met the criteria for comprehensive analysis in this investigation. Notably, this AI technology proved effective in evaluating shifts in learning styles among English as Second Language/English as Foreign Language (ESL/EFL) students across various grade levels (Wei et al., 2018).

4.2.5 Personalized Learning Systems or Environments

Personalized learning systems or environments (PLS/E) have demonstrated effectiveness in fostering interactions (Xu & Wang, 2006) and enhancing e-learning experiences (Cheung et al., 2003; Köse, 2017; Köse & Arslan, 2016; Xu & Wang, 2006). In a study by Turkish researchers Köse and Arslan (2016), the impact of PLS on 110 undergraduate students over two semesters in computer programming courses was examined, revealing that the PLS system facilitated desirable learning outcomes and improved overall learning experiences.

A study involving high school students in the USA (Walkington & Bernacki, 2019) indicated that linking mathematics to students' personal interests outside of school could enhance learning within an intelligent tutoring system. This underscores the potential of highly tailored personalization to promote learning and contribute to student success.

4.2.6 Visualizations and Virtual Learning Environments

Accompanying the rise of virtual reality (VR) technologies, research has begun to explore the potential advantages of incorporating visualizations and Virtual Learning Environments (VLE) with AI in educational settings. Evidence suggests that students find the learning experience enjoyable in VLEs and perceive them as beneficial for learning and collaboration. Similarly, teachers have observed increased student engagement in learning activities within VLEs (Griol, Molina, & Callejas,2014). In Australia, a technology integrating AI and virtual reality (VR) has proven effective in enhancing learning experiences and captivating the interest of young learners (Ijaz et al., 2017). Moreover, undergraduate students utilizing AI and VR demonstrated improved comprehension skills compared to traditional learning methods (Ijaz et al., 2017).

Research focusing on a smart glass system also confirmed that AI technology with visualizations served as a valuable social communication aid for both children and adults with autism (Keshav et al., 2017).

4.3 ENGLISH LANGUAGE TEACHING IN OMAN

In Oman, the rationale behind language instruction has emerged as a pivotal aspect of the nation's trajectory. Since 1970, Oman has forged robust economic ties with various non-English-speaking countries worldwide, including Portugal, Germany, Cyprus, Austria, Russia, Belgium, Greece, Sweden, France, Turkey, and Italy. Consequently, English has become a common language for facilitating communication among these nations and with other non-native English speakers (Al-Jadidi, 2009), thus serving as a primary driver for English language acquisition in Oman.

Moreover, two additional factors have spurred the integration of English into the Omani school curriculum. Firstly, globalization has compelled Oman to prepare its populace for participation in the international labor market shaped by the contemporary global economy. Consequently, the nation recognized the necessity of equipping its citizens with essential abilities and skills, including proficiency in the English language. The Omani government acknowledges the pivotal role of English communication skills in the modern global economy, particularly in fields such as science and technology as well as in academia and business (Al Abrawi, 2017).

Secondly, the drive for 'Omanisation' has played a role, as the government has sought to decrease reliance on expatriate workers. Notably, Oman has never been under British rule and thus lacks a historical basis for English to be an official language or for English-medium schooling. However, since the 1970s, English has seamlessly integrated itself as a foreign language in schools and a formal mode of communication in both government and private sectors, receiving economic, legislative, and political backing from the Omani government and becoming entrenched in various domains such as education, media, and commerce (First, 2019; Al-Busaidi, 1995).

In Oman, the Ministry of Education has recognized the imperative of significant enhancements to English language education (Al-Jadidi, 2009). Policy dictates that English should solely be utilized for official purposes within the country. Consequently, English instruction is provided for both general and specialized purposes across various institutions. Moreover, English serves as the language of science across government organizations.

Furthermore, English serves as the medium of instruction for all academic pursuits at Institutes of Health Sciences and Sultan Qaboos University to equip learners with the requisite skills for completing assignments and research papers (ibid.). Notably, these institutions procure English Language Teaching (ELT) resources primarily from the UK and the USA, supplemented by a few locally produced materials (ibid.). As of the academic year 2019/20, Oman boasts 1166 public-sector schools catering to 634,770 students, complemented by 834 private schools serving 126,003 learners. Private schools typically deliver instruction in English or a combination of Arabic and English, with an increasing number of middle-class parents opting for private schooling due to its perceived benefits (Al-Mahrooqi & Denman, 2018).

Over the past 50 years, Oman has implemented three educational reforms pertaining to English language instruction. Initially, under the auspices of 'General Education' in the 1970s, followed by the institution of 'Basic Education' in 1998, and most recently, the 'Integrated Curriculum' in 2006 (Al Abrawi, 2017). The General Education curriculum, which initially adopted a teacher-centered approach with English instruction commencing in Grade 4, later shifted to the introduction of English instruction from Grade 1 with the implementation of Basic Education (ibid.).

The objective of Basic Education was to instill Islamic values while encouraging interaction with the global community, thereby promoting equal educational opportunities nationwide (Al Abrawi, 2017). The curriculum sought to improve attitudes toward language learning by integrating engaging materials such as games and songs to develop students' English language skills (MoE, 2010). The integrated curriculum was introduced during the 2006/07 academic year, aiming to teach Mathematics and Science in English. Initially implemented in four schools in the Muscat region, this system expanded to 42 schools across Oman by the academic year 2010/11. However, it was subsequently discontinued due to complaints from educators regarding an overloaded curriculum and challenges associated with teaching Mathematics and Science in English (Al Abrawi, 2017). Consequently, the Ministry reverted all Primary Schools to the Basic Curriculum. Figure 4.2

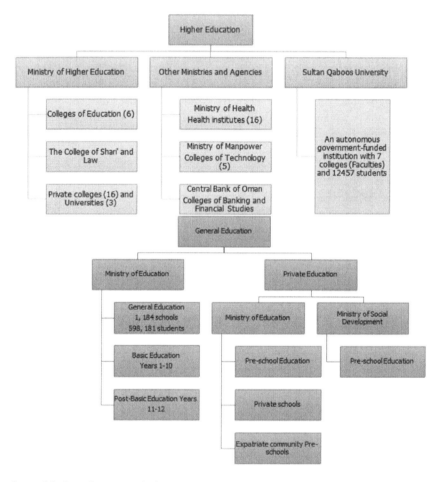

Figure 4.2 Oman's system of education.

*Ministry of Education & UNESCO, 2004. A Summary of the Strategy for Education in the Sultanate of Oman – 2006–2020.

Oman. Ministry of Education, Oman. Ministère de l'éducation, Omán. Ministerio de Educación, International Conference on Education, 48th, … Geneva. (n.d.). Inclusive education in the Sultanate of Oman: national report of the Sultanate of Oman. Retrieved from https://unesdoc.unesco.org/notice?id=p::usmarcdef_0000186279

illustrates Oman's Education System proposed by the Ministry of Education and UNESCO in 2004 that summarizes the educational strategies in the Sultanate of Oman from 2006 to 2020.

4.4 A CASE STUDY OF GRADE 11 WRITING CURRICULUM IN THE SULTANATE OF OMAN UNDER THE INFLUENCE OF AI

I recently conducted a case study of grade 11 writing curriculum in public schools of Muscat, Sultanate of Oman. A qualitative approach was employed which involved conducting interviews with teachers and class observations. The case study methodology enables the researcher to uncover important factors that emerge from the open-minded process (Yin, 2009). Through this method, I was able to explore the types of AI writing tools used by EFL teachers, as well as their perceptions about the impact of these tools on students' writing quality, particularly in terms of content and organization.

4.5 RESEARCH PARTICIPANTS

The research participants involved eight teachers from two different public schools, one being a girls' school and the other for boys. The selection was done through the approval from the Ministry of Education, Sultanate of Oman. Lessons of four teachers were observed: two male teachers and two female teachers in both schools. The selection was primarily based on their practical experience with AI writing tools in the classroom, accessibility, willingness to participate, and teaching expertise in advanced courses such as opinion essays and narrative essay writing.

All the recruited teachers possessed at least 10 years of experience in teaching EFL and a minimum of 2 years of experience in applying AI writing tools in their classrooms. Participants were designated as T1, T2, T3, and T4 to maintain confidentiality. These AI tools served as supplementary teaching resources, aiding in grammar checking, paraphrasing, generating content, and offering suggestions to enhance the clarity and coherence of students' writing. This ensured teachers' and students' experiences transcend mere theoretical knowledge, encapsulating practical applications of AI tools in real-life teaching scenarios. Table 4.1 provides the demographic information of the male and female teachers who were the subjects of the case study.

4.6 INSTRUMENT

I used interviews as my research instrument to collect the required data for this study. This instrument allowed me to gather rich, qualitative data

Table 4.1 Demographic information of participants

Respondent	Age (Years)	Gender	Teaching Experience (Years)	Qualification	Experience in Using AIWT* (Years)
T1	37	Male	8	B.A.	≥1
T2	42	Male	11	M.A.	2
T3	39	Female	10	B.A.	2
T4	37	Female	8	B.A.	≥1

* AI writing tools

related to EFL teachers' use of AL writing tools, and their perspectives on the impact of these tools on students' writing quality, specifically in terms of content and literature. Works by Nazari et al. (2021), Dale and Viethen (2021), Lee and Zhao (2022) served as valuable references.

4.7 DATA ANALYSIS

I employed a thematic case study to analyze the qualitative data. As outlined by Braun and Clarke (2006), the goal of this thematic analysis was to identify patterns and connections among themes extracted from the qualitative data.

4.8 FINDINGS

4.8.1 Types of AI Writing Tools Used by EFL Teachers

Based on class observations and interviews, it was concluded that four teachers integrated various AI writing tools into their instructional methods. These tools were selected and utilized according to their specific functionalities and suitability for the teachers' teaching contexts and goals. Table 4.2 functions as a comprehensive inventory, showcasing the diverse array of AI writing tools employed by these English as a Foreign Language (EFL) instructors. This table effectively outlines the names of the AI tools utilized and their respective purposes, offering an encompassing overview of the present utilization of AI in EFL instruction.

4.8.2 Teachers' Perspectives toward the Impact of AI Writing Tools

This section shows the summary of findings from my interviews regarding the teachers' perspectives on the impact of AI writing tools on students'

Table 4.2 Types of AI writing tools were used by the EFL teachers

Initial of Teachers	Types of AI Writing Tools Used	Purpose of Use
T1	ChatGPT	Assists students in brainstorming and organizing thoughts prior to writing; stimulates conversations with students.
	QuillBot	Used to teach paraphrasing skills and improve sentence flow.
T2	QuillBot	Facilitates teaching of paraphrasing and the avoidance of plagiarism.
	Wordtune	Aids students in improving sentence structure and in expressing their ideas more clearly.
	Paperpal	Performs grammar checks and provides real-time corrections.
T3	Essay Writer	Guides students in structuring their essays logically and suggests areas for improvement.
	QuillBot	Used for practice in paraphrasing.
	Wordtune	Improves sentence structure and clarity.
	ChatGPT and Jenni	Facilitates brainstorming and aids students in organizing their thoughts effectively.
T4	Wordtune	Refines sentence structure and enhances coherence.
	QuillBot	Teaches students to paraphrase their sentences.
	Copy.ai and Essay Writer	Helps students craft compelling pieces of writing in various styles and formats.

writing quality, specifically focusing on the aspects of content and organization.

4.8.3 Idea Generation

The response from the four teachers highlighted a wide range of opinions on the impact of AL writing tools on students' creativity and ability to generate ideas. T1 for example, emphasized that AI writing tools stimulated students, creativity and idea expansion.

4.8.3.1 Vocabulary and Language Use

Teachers acknowledged both benefits and potential downsides in relation to the vocabulary and language use. For instance, T2 suggested that these tools recommend synonyms or more advanced words, thereby expanding their lexical repertoire and enabling them to express their ideas more precisely and engagingly. T1, however, had induced a note of caution. She had expressed worries that AI tools could sometimes result in over-complicated language. T3 presented a more balanced view, acknowledging the usefulness of AI writing tools as a learning resource for expanding vocabulary.

However, he also cautioned against an over-reliance on these tools. T4 had a totally different perspective, focusing on the role of AI tools in reducing repetition in students writing, thereby enhancing the quality of their writing and prompting students to incorporate new vocabulary into their writing and then in real-life situations.

4.8.3.2 Coherence and Logical Flow

Each teacher provided a comprehensive picture of the benefits of AI writing tools in enhancing students' writing skills. T1 said that these tools help in suggesting better phrasing and enhancing coherence which subsequently improves students' confidence in their writing abilities. T2 offered both perspectives. She highlighted the positive assistance but also showed concerns about over-dependence on these tools. T3 provided his insights on the instant feedback provided by the AI writing tools, whereas T4 highlighted the role of AI tools as personalized writing tutors.

4.8.3.3 Use of Transition Words and Phrases

All four teachers stressed the significant improvement in the use of transition words and phrases. T3 also provided a before-and-after comparison and T4 stressed that the AI tools would be beneficial for mitigating common writing issues and enhancing overall writing quality.

So the study proved that the impact of AI writing tools on students' writing was positive, although with a note of caution about potential over-reliance on the tools, hinting at the importance of maintaining a balance in their use.

4.9 EXPECTED CHALLENGES IN AI IMPLEMENTATION

Although the case study showed many positive aspects of using AI writing tools in the classroom, some natural and expected challenges cannot be ignored.

AI has the power to radically change teaching practices, but its effective integration in educational contexts comes with its set of challenges. There are major challenges associated with the adoption of AI in education: data privacy concerns, the risk of algorithmic bias, and infrastructural resource constraints.

4.9.1 Data Privacy

One of the primary concerns associated with incorporating AI into education revolves around data privacy. AI-driven educational tools gather extensive student data, including personal information and detailed learning data. While this data enables AI's personalized features, it raises concerns regarding data access and safeguarding. The risk of potential misuse or breaches of this data is particularly concerning, especially for young learners who may not fully comprehend the consequences of sharing their data.

4.9.2 Algorithmic Bias

Another significant hurdle in the adoption of AI is algorithmic bias. As AI algorithms are created by humans, there is a risk that unintentional biases could be embedded within the system if not consciously addressed. These biases can manifest in various forms, such as biased content recommendations or assessment evaluations. To mitigate such biases, it is essential to use diverse and inclusive datasets when training AI systems.

4.9.3 Infrastructural Resources

The successful implementation of AI in education necessitates significant infrastructure resources, encompassing both physical hardware for AI applications and dependable access to internet and electricity. However, in many regions, particularly rural or underserved areas, these resources may be lacking or unreliable. This presents a risk of widening educational disparities, with students in resource-rich areas benefiting from AI-powered education while those in resource-poor areas are disadvantaged. Additionally, there is the challenge of ensuring that educators receive adequate training to effectively utilize AI tools. Without proper training and ongoing support, the integration of AI into teaching practices may prove ineffective or even harmful.

4.10 BENEFITS OF AI TECHNOLOGIES IN EDUCATION

In general, the benefits of AI technologies in Education are undeniable.

4.10.1 Changing the Traditional Role of a Teacher

As far back as 1991, Garito emphasized the transformative impact of AI on the traditional role of educators (Garito, 1991). More recently, scholars have highlighted AI's potential to enhance teaching and learning practices. Through scalable applications, AI is revolutionizing educational methodologies worldwide, including in regions like the Global South and in emerging educational formats such as MOOCs, blended learning, flipped classrooms, and others.

A recent review by Zawacki-Richter et al. (2019) provides an overview of various AIEd applications for diverse purposes, such as learner profiling, performance prediction, assessment, evaluation, personalization, adaptive learning, and more. AI systems can analyze student input and offer immediate corrective feedback (Mirzaeian et al., 2016; Roschelle et al., 2020), automated scoring and formative assessments (Zhu et al., 2020), and assist students with revisions during the learning process (Lee et al., 2019). Intelligent tutoring systems can identify learners' strengths and weaknesses in their knowledge base (Zawacki-Richter et al., 2019). Moreover, intelligent feedback systems can evaluate not only what is learned but also how people learn (Cutumisu et al., 2019). Machine learning, for instance, can accurately predict at-risk or high-achieving college students, enabling educators to intervene effectively for student success (Chui et al., 2020).

Figure 4.3 Practical examples of AIEd applications.

The advancement of AIEd underscores the need for more empirical studies focusing on AI technologies in real educational settings to address specific educational needs. To illustrate the current utilization of AI technologies for various educational purposes, Figure 4.3 provides practical examples of AIEd applications from reviewed research articles. This serves as a snapshot of AIEd's current practices with educational aspirations, potentially prompting further research in the field. With educational objectives in mind, the examples in Figure 4.3 demonstrate how students and educators can benefit from AI-enhanced learning systems or experiences.

4.10.2 Practical Implications for AIEd

The progress of AI offers promising prospects for education, presenting a diverse range of technologies, features, and functionalities. Yet, realizing its full potential necessitates bridging the gap between AI technological advancements and their practical applications in education. Various AIEd technologies demonstrate or hold potential benefits for education, catering to learners through improved engagement, tailored learning materials, metacognitive support, enriched environments, and enhanced outcomes. Similarly, educators and administrators benefit from predictive analytics, progress monitoring, personalized resources, and data analysis for assessment and administrative purposes. AI-enhanced learning environments further enrich LMS through expert systems, visual feedback, and immersive tools.

Figure 4.4 acts as a pragmatic roadmap for both AI technology developers and educational stakeholders, facilitating effective collaboration to leverage AI innovations in educational settings. It empowers technological innovators to work alongside educators, ensuring the effective utilization of AI technologies, while educators can identify suitable AI solutions for diverse educational needs without delving into technical complexities. Moreover, Figure 4.4 highlights key challenges in AIEd adoption, such as cost, ethics, privacy concerns, lack of educator guidelines, and limited AI expertise among educators, fostering dialogue among stakeholders to address these issues and foster collaborative efforts in AIEd research, development, implementation, and assessment.

4.11 CONCLUSION

In conclusion, the high school writing curriculum in the Sultanate of Oman faces several anticipated challenges with the arrival of AI. These challenges stem from the intersection of educational reforms, technological advancements, and the unique cultural and linguistic context of Oman.

The introduction of AI into the writing curriculum brings forth concerns regarding resistance to change among educators, uncertainties regarding

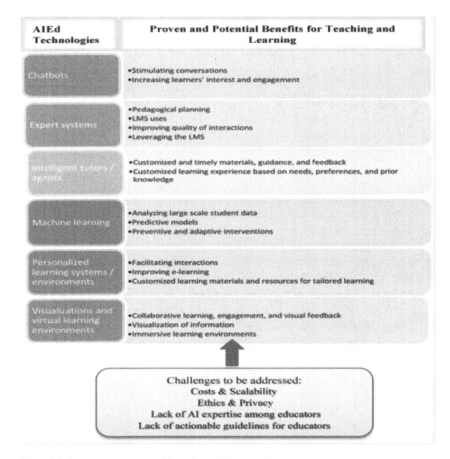

Figure 4.4 Proven and potential benefits of AI technologies.

the reliability and fairness of AI in evaluating writing proficiency, and the potential depersonalization in feedback mechanisms. Furthermore, the cultural and linguistic nuances specific to Oman may impact the effectiveness and acceptance of AI tools in writing instruction and assessment.

Despite these challenges, this case study serves as a crucial step toward understanding the complexities of integrating AI into educational practices in Oman. By critically examining these challenges, educators, policymakers, and curriculum developers can gain insights into adapting writing instruction to leverage AI's benefits while addressing its limitations effectively.

Moving forward, further research and collaboration are essential to develop effective strategies that harness AI's potential to enhance writing education while ensuring that it aligns with the goals and values of Oman's educational system. Through ongoing dialogue and adaptation, Oman can navigate the challenges posed by AI integration, ultimately improving the quality of writing education for high school students across the Sultanate.

REFERENCES

Al Abrawi, N. A. S. (2017). The challenges of designing and delivering an appropriate English as a modern foreign language curriculum for primary school-aged children in Oman [Doctoral dissertation]. Liverpool John Moores University. https://researchonline.ljmu.ac.uk/id/eprint/8085/

Al-Busaidi, K. (1995). English in the labour market in multilingual Oman with special reference to Omani employees [Unpublished PhD thesis]. University of Exeter, UK.

Al-Jadidi, H.S. (2009). Teaching English as a foreign language in Oman: An exploration of English language teaching pedagogy in tertiary education [Unpublished doctoral dissertation]. Victoria University, Melbourne, Australia.

Al-Mahrooqi, R. & Denman, C. (Eds.) (2018). English education in Oman: Current scenarios and future trajectories (Vol. 15). Springer.

Al-Mahrooqi, R. (2012). English communication skills: How are they taught at schools and universities in Oman? English Language Teaching, 5(4), 124–130.

Al Mashaikhi, M., Yusof, S. M., Tahir, L. M., & Atan, N. A. (2020). Blended Learning in English Teaching and Learning: An Overview of Current Practice in Oman. Innovative Teaching and Learning Journal, 4(2).

Baniabdelrahman, A. Ahmad (2006). The Effect of Using Authentic English Language Materials on EFL Students' Achievement in Reading Comprehension. http://dx.doi.org/10.12785/JEPS/070111

Braun, V., & Clarke, V. (2006). Using thematic analysis in psychology. Qualitative Research in Psychology, 3(2), 77–101.

Cheung, B., Hui, L., Zhang, J., & Yiu, S. (2003). SmartTutor: An intelligent tutoring system in web-based adult education. Journal of Systems and Software, 68(1), 11–25. https://doi.org/10.1016/s0164-1212(02)00133-4.

Chin, D. B., Dohmen, I. M., & Schwartz, D. L. (2013). Young children can learn scientific reasoning with teachable agents. IEEE Transactions on Learning Technologies, 6(3), 248–257. https://doi.org/10.1109/tlt.2013.24

Chui, K. T., Fung, D. C. L., Lytras, M. D., & Lam, T. M. (2020). Predicting at-risk university students in a virtual learning environment via a machine learning algorithm. Computers in Human Behavior, 107, 105584. https://doi.org/10.1016/j.chb.2018.06.032.

Cutumisu, M., Chin, D. B., & Schwartz, D. L. (2019). A digital game-based assessment of middle-school and college students' choices to seek critical feedback and to revise. British Journal of Educational Technology, 50(6), 2977–3003. https://doi.org/ 10.1111/bjet.12796.

Dale, R., & Viethen, J. (2021). The automated writing assistance landscape in 2021. Natural Language Engineering, 27(4), 511–518.

Dias, S. B., Hadjileontiadou, S. J., Hadjileontiadis, L. J., & Diniz, J. A. (2015). Fuzzy cognitive mapping of LMS users' Quality of Interaction within higher education blended-learning environment. Expert Systems with Applications, 42(21), 7399–7423. doi:10.1016/j.eswa.2015.05.048

First, E. E. (2019). EF EPI: EF English Proficiency Index: A Ranking of 100 Countries and Regions by English Skills: EF Education First.

Fryer, L. K., Ainley, M., Thompson, A., Gibson, A., & Sherlock, Z. (2017). Stimulating and sustaining interest in a language course: An experimental comparison of Chatbot and Human task partners. Computers in Human Behavior, 75, 461–468. https://doi.org/ 10.1016/j.chb.2017.05.045

Garito, M. A. (1991). Artificial intelligence in education: Evolution of the teaching-learning relationship. British Journal of Educational Technology, 22(1), 41–47. https://doi.org/10.1111/j.1467-8535.1991.tb00050.x.

Graddol, D. 1998. The future of English. London: The British Council.

Griol, D., Molina, J. M., & Callejas, Z. (2014). An approach to develop intelligent learning environments by means of immersive virtual worlds. Journal of Ambient Intelligence and Smart Environments, 6(2), 237–255.

Hwang, G., Sung, H., Chang, S., & Huang, X. (2020). A fuzzy expert system-based adaptive learning approach to improving students' learning performances by considering affective and cognitive factors. Computers and Education: Artificial Intelligence, 1(1), 100003. https://doi.org/10.1016/j.caeai.2020.100003

Ijaz, K., Bogdanovych, A., & Trescak, T. (2017). Virtual worlds vs books and videos in history education. Interactive Learning Environments, 25(7), 904–929. https://doi. org/10.1080/10494820.2016.1225099.

Keshav, N. U., Salisbury, J. P., Vahabzadeh, A., & Sahin, N. T. (2017). Social communication coaching smartglasses: Well tolerated in a diverse sample of children and adults with autism. JMIR MHealth and UHealth, 5(9). https://doi.org/10.2196/ mhealth.8534.

Köse, U. (2017). An augmented-reality-based intelligent mobile application for open computer education. *Mobile Technologies and Augmented Reality in Open Education Advances in Educational Technologies and Instructional Design*, 154–174. https://doi. org/10.4018/978-1-5225-2110-5.ch008.

Köse, U., & Arslan, A. (2016). Intelligent e-learning system for improving students' academic achievements in computer programming courses. International Journal of Engineering Education, 32, 185–198.

Lee, H. S., Pallant, A., Pryputniewicz, S., Lord, T., Mulholland, M., & Liu, O. L. (2019). Automated text scoring and real-time adjustable feedback: Supporting revision of scientific arguments involving uncertainty. Science Education, 103(3), 590–622. https://doi.org/10.1002/sce.21504.

Matsuda, N., Weng, W., & Wall, N. (2020). The effect of metacognitive scaffolding for learning by teaching a teachable agent. International Journal of Artificial Intelligence in Education, 30, 1–37.

Marzuki, U. W., Rusdin, D., Darwin, & Indrawati, I. (2023). The impact of AI writing tools on the content and organization of students' writing: EFL teachers' perspective. Cogent Education, 10(2), https://doi.org/10.1080/2331186X.2023.2236469

McCarthy, K. S., Likens, A. D., Johnson, A. M., Guerrero, T. A., & McNamara, D. S. (2018). Metacognitive overload!: Positive and negative effects of metacognitive prompts in an intelligent tutoring system. International Journal of Artificial Intelligence in Education, 28(3), 420–438.

McKenney, S., & Reeves, T. C. (2018). Conducting educational design research. NYC: Routledge.

Mirzaeian, V. R., Kohzadi, H., & Azizmohammadi, F. (2016). Learning Persian grammar with the aid of an intelligent feedback generator. Engineering Applications of Artificial Intelligence, 49, 167–175. https://doi.org/10.1016/j.engappai.2015.09.012.

Mocanu, M., Vasiliu, E., & Stancu, A. M. R. (2012). Some considerations on globalisation and its effects on education. Globalisation, 4(1), 104.

Education for All Education for All Movement EFA, Oman. Ministry of Education, Oman. Ministère de l'éducation, & Omán. Ministerio de Educación. (n.d.). From access to success: Education for All (EFA) in the Sultanate of Oman, 1970-2005;a contribution to the celebration of UNESCO's 60 year anniversary. Retrieved from https://unesdoc.unesco.org/notice?id=p::usmarcdef_0000157711

MoE (2010). The English language curriculum framework. English language curriculum development section, Grades 1–10. Oman: Sultanate of Oman Ministry of Education.

Morita, L. (2013). Japanese university students' attitudes toward globalisation, intercultural contexts and English. World Journal of English Language, 3(4), 31.

Nazari, N., Shabbir, M. S., & Setiawan, R. (2021). Application of artificial intelligence powered digital writing assistant in higher education: Randomized controlled trial. Heliyon, 7(5), e07014.

Pal, S. (2023). Artificial intelligence: Reshaping the topography of pedagogic practices – a comparative study on curriculum construction, teaching modalities, and evaluation techniques. AGPE: The Royal Gondwana Research Journal of History, Science, Economic, Political and Social Science, 4(7), 13–20. Retrieved from www.agpegondwanajournal.co.in/index.php/agpe/article/view/283

Robertson, P., Nunn, R., & Al-Mahrooqi, R. 2013. The Asian EFL Journal Quarterly. Asian EFL Journal, 15(3).

Roschelle, J., Lester, J., & Fusco, J. (Eds.). (2020). AI and the future of learning: Expert panel report [Report]. Digital Promise. https://circls.org/reports/ai-report.

Tarning, B., Silvervarg, A., Gulz, A., & Haake, M. (2019). Instructing a teachable agent with low or high self-efficacy–does similarity attract? International Journal of Artificial Intelligence in Education, 29(1), 89–121.

Umale, J. (2011). Pragmatic failure in refusal strategies: British versus Omani interlocutors. Arab World English Journal, 2(1), 18–46.

Walkington, C., & Bernacki, M. L. (2019). Personalizing algebra to students' individual interests in an intelligent tutoring system: Moderators of impact. International Journal of Artificial Intelligence in Education, 29(1), 58–88.

Wei, Y., Yang, Q., Chen, J., & Hu, J. (2018). The exploration of a machine learning approach for the assessment of learning styles changes. Mechatronic Systems and Control (Formerly Control and Intelligent Systems), 46(3), 121–126. https://doi.org/ 10.2316/journal.201.2018.3.201-2979.

Xu, D., & Wang, H. (2006). Intelligent agent supported personalization for virtual learning environments. Decision Support Systems, 42(2), 825–843. https://doi.org/ 10.1016/j.dss.2005.05.033.

Yin, R. K. (2009). Case study research: Design and methods (4th ed.). Sage Publications.

Zawacki-Richter, O., Marín, V. I., Bond, M., & Gouverneur, F. (2019). Systematic review of research on artificial intelligence applications in higher education – where are the educators? International Journal of Educational Technology in Higher Education, 16(1), 1–27. https://doi.org/10.1186/s41239-019-0171-0.

Zhang, K., & Aslan, A. B. (2021). AI technologies for education: Recent research & future directions. Computers and Education: Artificial Intelligence, 2, 100025. https://doi.org/10.1016/j.caeai.2021.100025

Zhao, X. (2022). Leveraging artificial intelligence (AI) technology for English writing: Introducing wordtune as a digital writing assistant for EFL writers. RELC Journal, 54(3). https://doi.org/10.1177/00336882221094089

Zhu, M., Liu, O. L., & Lee, H.-S. (2020). The effect of automated feedback on revision behavior and learning gains in formative assessment of scientific argument writing. Computers & Education, 143, 103668. https://doi.org/10.1016/j.compedu.2019.103668.

Zughoul, M. 2003. Globalisation and EFL/ESL pedagogy in the Arab world. Journal of Language and Learning, 1(2), 146.

Chapter 5

Evolving Industries
A Journey from Industry 1.0 to 5.0

Sanjeev Kumar, Geeta Tiwari, and Neeraj Tiwari

5.1 INTRODUCTION

5.1.1 Overview of the Historical Evolution of Industrial Revolutions

The development of industry from the harnessing of steam power in Industry 1.0 to the smart manufacturing of Industry 4.0 has been a journey of technological advancement and societal transformation. Every era, set apart by innovations like large-scale manufacturing and digitalization, has pushed economic development and reshaped how businesses work. Presently, as Industry 5.0 weaving machines, the focus shifts toward a future where innovation upgrades human capacities as opposed to supplanting them, promising mutual benefits for businesses and society. Figure 5.1 depicts the timeline transformation of Industry 1.0 to 5.0.

5.1.2 Importance of Understanding the Implications on Society and the Economy

Understanding the effect of modern transformations on society and the economy is critical for bits of knowledge into cultural development and economic progress. Each industrial revolution has generally changed ways of life, work, and collaborations, reshaping accepted practices and economic systems. Analyzing these impacts helps policymakers, businesses, and individuals anticipate and explore challenges and opportunities presented by technological advancements, promoting inclusive development and sustainable development.

5.1.3 Transition to Industry 5.0 as a Focal Point for Discussion

Industry 5.0 imprints a shift toward human-centric manufacturing, integrating innovation with human skills like creativity and empathy. This

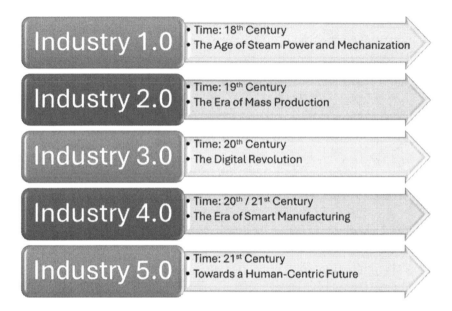

Figure 5.1 Timeline for the transformation of Industry 1.0 to Industry 5.0.

approach improves efficiency and safety through collaborative robots, augmented reality (AR), and advanced interfaces. It tends to address societal challenges like joblessness and working environment safety, underlining the requirement for figuring out its subtleties for a prosperous and reasonable future industrial landscape.

5.2 INDUSTRY 1.0: THE AGE OF STEAM POWER AND MECHANIZATION

5.2.1 Introduction to Industry 1.0

The presentation of Industry 1.0 denoted a turning point in mankind's set of experiences, described by the tackling of steam power and the beginning of motorized creation. This time, which arose in the late eighteenth century, saw a significant change in assembling processes, as conventional craftsmanship gave an approach to motorized frameworks driven by steam motors. The reception of steam power empowered phenomenal degrees of efficiency, upsetting ventures like materials, mining, and transportation. This shift prodded financial development as well as reshaped social designs and metropolitan scenes, establishing the groundwork for the industrialization that would characterize the advanced period. Understanding the beginning and elements of Industry 1.0 is fundamental for fathoming the resulting floods of modern upheaval and their getting through influence on society and the economy.

5.2.2 Characteristics of Steam Power and Mechanized Production

During Industry 1.0, steam power and computerized manufacturing established the foundation for modern upheaval. Steam engines revolutionized industries, improving efficiency and effectiveness. They fueled hardware in plants, mines, and transportation, thereby changing assembling cycles and driving economic development. Automated production presented standardized parts, streamlining production and reducing costs, prompting the ascent of large-scale factories and critical societal changes, including urbanization and altered work relations, making way for additional modern and cultural advancements.

During Industry 1.0, the qualities of steam power and motorized creation changed in different areas, with striking models showing their effect:

1. **Textile Industry:** The material business encountered a critical transformation with the presentation of steam-controlled winding around looms. One unmistakable model is the power loom developed by Edmund Cartwright in 1785. This development empowered the motorized winding of material on a much bigger scope than customary hand looms, prompting expanded effectiveness and efficiency in material creation.
2. **Mining Industry:** Steam power reformed mining tasks by empowering the utilization of steam motors to siphon water out of mines and power hardware for removal. James Watt's superior steam motor, licensed in 1769, played a pivotal part in making profound mining tasks doable. The utilization of steam-fueled siphons and hardware considered further and greater mining tasks, prompting expanded extraction of coal, iron, and different minerals.
3. **Transportation:** The improvement of steam-controlled trains reformed transportation during Industry 1.0. George Stephenson's "Rocket," worked in 1829, is a popular illustration of an early steam train that exhibited the practicality of steam-controlled rail travel. Steam-fueled trains worked with quicker and more proficient transportation of products and individuals, interfacing with far-off locales, and driving financial development and urbanization.
4. **Agriculture:** Steam power likewise reformed agribusiness through the creation of steam-fueled rural apparatus. One outstanding model is the steam-controlled work vehicle, which supplanted animal-fueled furrows and expanded the proficiency of cultivating activities.

Moreover, steam-fueled sifting machines and gatherers changed reaping processes, prompting expanded farming efficiency and the automation of provincial economies.

5.2.3 Challenges and Innovations during This Phase

Industry 1.0 set apart by steam power and motorization brought huge challenges and innovations, establishing the foundation for modern industrialization. Changing from manual to automated production caused broad job displacement and social distress, requiring workforce change and retraining. However, it additionally prodded remarkable advancements like James Watt's steam engine and automated manufacturing, showing innovation's significant effect on efficiency and shaping the future of work and society.

5.3 INDUSTRY 2.0: THE ERA OF MASS PRODUCTION

5.3.1 Introduction to Industry 2.0

The Preamble to Industry 2.0 addresses a pivotal stage in the Industrial Revolution, characterized by the rise of large-scale manufacturing methods and mechanized manufacturing systems. Initiated by figures like Henry Portage, it moved from labor-intensive methods to standardized, automated processes, reforming both assembling and societal accessibility to goods. However, it likewise brought challenges, for example, labor issues, safety concerns, and environmental impacts, requiring regulatory and social reforms.

5.3.2 Shift toward Mass Production Techniques

Industry 2.0 denoted a significant transition to large-scale manufacturing, driven by innovations like mechanical production systems spearheaded by Henry Ford. This revolutionized efficiency across sectors, empowering quick production of goods on a massive scale, from vehicles to gadgets. While supporting economic development and consumer accessibility, it presented difficulties, such as labor conditions, environmental effects, and product homogenization. Despite these challenges, it laid the foundation for current assembling works, shaping subsequent industrial revolutions.

5.3.3 Impact on Economies of Scale and Consumer Goods Availability

Industry 2.0, described by the period of large-scale manufacturing, significantly affected economies of scale and the accessibility of customer merchandise. With the execution of normalized creation procedures, for example, Henry Portage's sequential construction system, makers had the option to accomplish uncommon degrees of proficiency and efficiency. This brought about huge expense decreases in the manufacturing cycle, considering the large-scale manufacturing of merchandise at lower costs. Thus, buyer products turned out to be more available to everybody, prompting expanded utilization and further developed ways of life. The economies of scale accomplished through large-scale manufacturing drove down costs for makers as well as animated financial development by setting out new business sectors and opening doors for extension. In general, Industry 2.0 assumed a crucial part in forming present-day customer culture and reforming the worldwide economy.

During Industry 2.0, the adoption of large-scale manufacturing revolutionized economies of scale and the availability of purchaser products. Henry Ford's Model T production exemplified this, radically decreasing expenses incurred per unit and making vehicles reasonable to the majority. Essentially, the textile industry saw expanded results and lower costs through mechanized production, prompting far and wide accessibility of affordable clothing. These models outline how the enormous scope of fabricating changed consumer product accessibility, shaping consumption patterns and economic development.

5.3.4 Emerging Challenges and Complexities

The change to Industry 2.0 delivered a rush of arising difficulties and intricacies that reshaped the modern scene. While large-scale manufacturing strategies altered assembling cycles and prompted remarkable degrees of efficiency, they additionally acquainted new obstacles with survival. One critical test was the ascent of work issues, including worries over working circumstances, wage inconsistencies, and specialist abuse. The shift toward normalized creation techniques additionally presented natural difficulties, like expanded contamination and asset consumption, as industrial facilities increased creation to fulfill developing needs. Also, the dependence on large-scale manufacturing prompted worries about item quality and security, as organizations focused on volume over fastidious craftsmanship. These arising difficulties and intricacies highlighted the requirement for imaginative arrangements and administrative systems to guarantee that the advantages of Industry 2.0 were offset with social and natural contemplations.

78 Artificial Intelligence and Communication Techniques

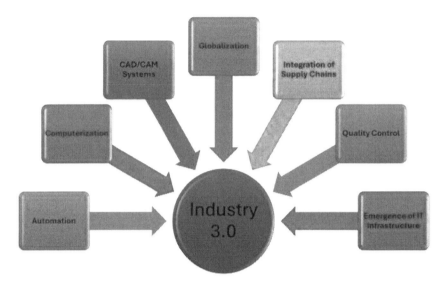

Figure 5.2 Key features of Industry 3.0.

5.4 INDUSTRY 3.0: THE DIGITAL REVOLUTION

5.4.1 Introduction to Industry 3.0

The presentation of Industry 3.0 denoted an extraordinary stage in the development of modern unrest, described by the beginning of the Computerized Upset. This time saw the fast headway and boundless reception of advanced innovations, essentially adjusting the way organizations had worked and collaborated with their surroundings. Key developments like figuring innovation, broadcast communications foundation, and information examination prepared for exceptional degrees of availability, mechanization, and data handling. Industry 3.0 introduced a period of digitized creation processes, supply chains, and correspondence organizations, empowering associations to accomplish more elevated levels of effectiveness, nimbleness, and development. As organizations embraced digitalization, they acquired the capacity to advance tasks, smooth out direction, and open new doors for development and seriousness in an undeniably globalized and interconnected commercial center. Figure 5.2 shows the key features of Industry 3.0.

5.4.2 Key Technological Advancements Driving the Digital Revolution

The Computerized Upset of Industry 3.0 was pushed by a few key mechanical progressions that reshaped business tasks and networks on a worldwide scale. At the bleeding edge of this unrest was a leap forward in processing innovation, especially the advancement of chip and coordinated circuits, which empowered the scaling down of personal computers (PCs) and the multiplication of computerized gadgets. Close by this, headways in broadcast communications, including the extension of broadband organizations and the development of versatile advances, worked with constant correspondence and information trade across immense distances. Also, the appearance of programming applications and programming dialects engaged organizations to robotize processes, examine huge measures of information, and foster imaginative answers for complex issues. These mechanical headways established the groundwork for the advanced change of enterprises, driving proficiency gains, cultivating development, and catalyzing financial development in the computerized age.

5.4.3 Transformation of Business Operations and Challenges Faced

The coming of Industry 3.0 flagged a significant change in business operations, driven by the digital revolution. Companies embraced computing, telecommunications, and data analysis to smooth out processes and improve efficiency. Automation extended, yielding significant cost savings and productivity gains. However, challenges arose, including the requirement for workforce upskilling, data security concerns, and enlarging advanced digital disparities. Addressing these demands required comprehensive investment in technological infrastructure and human capital improvement.

During Industry 3.0, exemplified by the Digital Revolution, the retail sector went through huge changes through the reception of e-commerce platforms. Organizations like Amazon changed the retail landscape by utilizing digital technologies to smooth out purchasing processes and offer personalized experiences. In any case, this shift additionally brought challenges, for example, serious contest competition leading to store closures and job losses, the requirement for online protection, that is cybersecurity measures, and augmenting digital splits between retailers, highlighting the dual nature of benefits and challenges in e-commerce adoption.

5.4.4 Importance of Continuous Learning and Adaptation

In Industry 3.0, described by the Computerized Upheaval, the significance of nonstop learning and transformation could not possibly be more significant.

The quick speed of innovative headways, including registering innovation, media communications, and information examination, essentially changed the way organizations worked. To stay cutthroat in this powerful scene, associations, and people, the same needed to embrace a culture of nonstop learning and transformation. New abilities and capabilities were expected to use arising innovations really, while customary jobs advanced or became out of date. Also, the computerized partition broadened, featuring the criticalness of outfitting people with the important advanced proficiency and abilities to partake in the computerized economy. Constant learning and transformation were not just fundamental for exploring the intricacies of Industry 3.0 yet in addition for guaranteeing long-haul pertinence and flexibility in an undeniably digitized world.

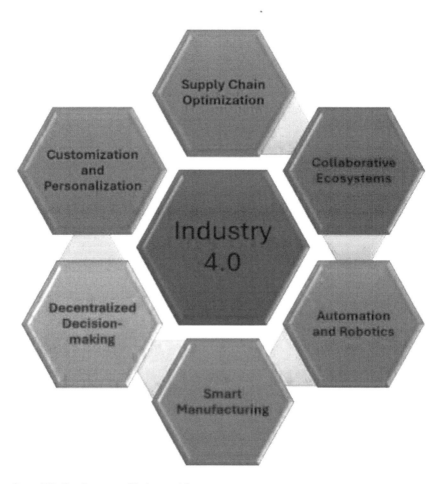

Figure 5.3 Key features of Industry 4.0.

5.5 INDUSTRY 4.0: THE ERA OF SMART MANUFACTURING

5.5.1 Introduction to Industry 4.0

The emergence of Industry 4.0 addresses an essential crossroads in industrial history, denoting the combination of physical and digital systems in manufacturing. Characterized by the integration of cutting-edge innovations like computer-based intelligence, Internet of Things (IoT), and cyber–physical systems, it empowers machines to communicate, analyze data, and settle on independent choices, improving effectiveness, adaptability, and customization in production processes [1]. In any case, challenges, for example, high implementation costs, cybersecurity concerns, and the requirement for workforce upskilling go with its adoption, require cautious navigation of obstacles to unlock its transformative potential. Figure 5.3 shows the key features of Industry 4.0.

5.5.2 Convergence of Physical and Digital Systems

Industry 4.0 imprints a transformative era characterized by the integration of physical and digital systems, proclaiming another wave of development and effectiveness in manufacturing. This convergence facilitates real-time communication between physical machinery and digital stages, empowering data collection and analysis for predictive maintenance, smoothing out production planning, and enhanced decision-making [2]. With advancements in IoT, machine learning, and artificial intelligence (AI), manufacturers can make smart, autonomous plants, reforming and revolutionizing traditional practices and offering valuable opportunities for customization, effectiveness, and supportability across industries.

5.5.3 Advancements in AI, IoT, and Cyber–Physical Systems

Industry 4.0 imprints a transformative era driven by advancements in AI, the IoT, and cyber–physical systems, revolutionizing manufacturing. AI algorithms empower predictive analysis and machine learning for proactive maintenance and optimization. IoT gadgets accumulate real-time data from implanted sensors, working with consistent communication and coordination. Cyber–physical systems integrate physical components with digital advancements for autonomous control, introducing unprecedented efficiency, flexibility, and competitiveness. This intermingling guarantees heightened productivity, innovation, and global market competitiveness.

Instances of headways in man-made brainpower (computer-based intelligence), Internet of Things (IoT), and digital actual frameworks in Industry 4.0 include:

1. **Predictive Maintenance:** Producing organizations use man-made intelligence calculations to break down sensor information from apparatus progressively. By recognizing designs demonstrative of expected disappointments or glitches, prescient support frameworks can plan upkeep exercises before hardware breakdowns happen, limiting margin time and improving efficiency.
2. **Smart Inventory Management:** IoT sensors implanted away offices and stockrooms persistently screen stock levels and track the development of merchandise. Through interconnected frameworks, organizations can computerize stock recharging, expect request changes, and smooth out inventory network activities for improved effectiveness and cost investment funds.
3. **Autonomous Robots:** Cyber–physical systems integrate AI-powered robots into production lines, thereby enabling them to perform repetitive tasks with precision and autonomy. These robots can adapt to dynamic environments, collaborate with human workers, and optimize workflows to maximize productivity while ensuring workplace safety.
4. **Connected Manufacturing Ecosystems:** IoT gadgets and AI-empowered examination stages work with consistent correspondence and information trade among different parts of the assembling biological system, including providers, merchants, and clients. This interconnectedness empowers nimble reactions to advertise requests, without a moment to spare creation booking and upgraded permeability across the worth chain.
5. **Personalized Product Customization:** Simulated intelligence calculations examine client inclinations and verifiable buying information to create customized item suggestions. Through IoT-empowered shrewd assembling processes, organizations can tailor item particulars and designs continuously, offering tweaked answers to meet individual client needs and inclinations.

5.5.4 Barriers to Adoption and Uneven Progress

Regardless of the capability of Industry 4.0 to revolutionize manufacturing through the combination of physical and digital systems, several barriers hinder its far and wide adoption. High upfront costs, particularly challenging for small- and medium-sized enterprises (SMEs), alongside concerns about data security and privacy, make remarkable differences among large and small firms. Compatibility issues, regulatory barriers, and uncertainties surrounding liability and intellectual property rights further impede adoption, necessitating collaborative efforts among policymakers, industry stakeholders, and technology providers to overcome these obstacles and cultivate impartial advancement in Industry 4.0.

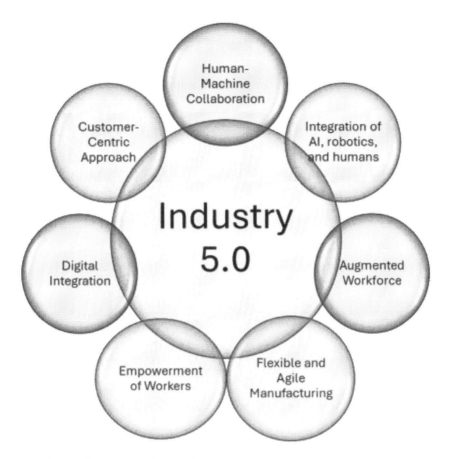

Figure 5.4 Key features of Industry 5.0.

5.6 INDUSTRY 5.0: TOWARD A HUMAN-CENTRIC FUTURE

5.6.1 Introduction to Industry 5.0

Industry 5.0 imprints a takeoff from previous industrial revolutions by prioritizing human-centered approaches to deal with production, stressing collaboration between humans and machines [3]. It perceives the one-of-a-kind characteristics of both, esteeming human inventiveness and critical thinking close by mechanical progressions [4]. By incorporating trend-setting innovations like collaborative robots and AR, it plans to upgrade task execution while advancing persistent learning. This shift toward inclusivity and sustainability holds the possibility to produce value for organizations and society, introducing another time of human-driven manufacturing. Figure 5.4 shows the key features of Industry 5.0.

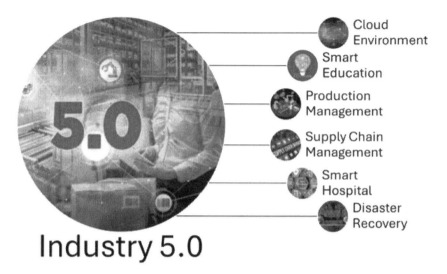

Figure 5.5 Applications of Industry 5.0.

5.6.2 Concept of Human-Centric Manufacturing

Industry 5.0 movements concentrated on human-driven assembly, departing from previous industrial revolutions focused on automation [5]. It stresses the collaboration between humans and innovation, utilizing human inventiveness and flexibility close to machine precision. Using cutting-edge innovations like collaborative robots and AR, it plans to upgrade efficiency and worker fulfillment through comprehensive work environments advancing participation, development, and consistent learning for a sustainable future [6].

5.6.3 Technologies Empowering Workers and Fostering Innovation

Industry 5.0 focuses on human-centered manufacturing, encouraging joint effort among workers and machines through advancements like cobots. These robots work close to people, improving efficiency and safety. Advanced human–machine interfaces, for example, AR and wearables give continuous guidance, encouraging inventiveness and critical thinking among workers. By incorporating human creativity with machine capabilities, Industry 5.0 promotes collaboration, engaging workers to drive organizational achievement.

5.6.4 Value Creation for Businesses and Society

In Industry 5.0, human-centered manufacturing carries huge worth to businesses and society through cutting-edge innovations like collaborative robots, AR, and human–machine interfaces. This approach upgrades efficiency, effectiveness, and development while focusing on specialist prosperity and encouraging a culture of innovativeness. Besides, it underlines inclusivity, variety, manageability, and social obligation, driving monetary development and advancing social union for an additional impartial and prosperous future. Figure 5.3 shows the applications of Industry 5.0.

5.7 LIMITATIONS OF INDUSTRY 5.0

Industry 5.0, or the fifth modern upheaval, is an imagined industry that utilizes a blend of human and machine abilities to add to society. The guiding principles of Industry 5.0 are human-centricity, maintainability, and versatility. Yet, here are a few limitations of Industry 5.0:

1. Unemployment
2. Technical skills/human–machine collaboration
3. Information security
4. Privacy
5. Lack of skilled workers
6. A large budget is required.

5.8 FUTURE RESEARCH OF INDUSTRY 5.0

Innovative work in Industry 5.0, enveloping human–machine joint effort, customization advances, simulated intelligence morals, and manageable assembling rehearses, holds critical commitment for driving development and practical development. Key areas of the center incorporate streamlining the increased labor force, propelling simulated intelligence morals and administration, creating manageable assembling works, guaranteeing digital actual security, cultivating cooperative biological systems, and planning client-driven encounters. These endeavors plan to address difficulties like work removal, moral simulated intelligence advancement, natural effects, network protection dangers, and administrative intricacies. By propelling exploration in these basic regions, Industry 5.0 can understand its capability to upset assembling and drive positive financial effects while moderating dangers and guaranteeing dependable development.

5.9 CONCLUSION

Overall, the changing venture from Industry 1.0 to Industry 5.0 imprints a critical development in the manner businesses work and cooperate with innovation. From the appearance of automation and steam power in Industry 1.0 to the human-driven approach of Industry 5.0, each stage has achieved significant changes underway techniques, labor force elements, and cultural effects. Industry 5.0 addresses a zenith of these headways, where people and machines team up synergistically to drive development, customization, and supportability. It highlights the significance of embracing mechanical advancement while focusing on the prosperity and strengthening of laborers. Additionally, Industry 5.0 stresses adaptability, spryness, and client-centricity, empowering organizations to flourish in a period of quick change and disturbance. As we keep on exploring the intricacies of the cutting-edge modern scene, the standards of Industry 5.0 act as a directing system for encouraging comprehensive development, dependable advancement, and practical turn of events.

REFERENCES

[1] X. Xu, Y. Lu, B. Vogel-Heuser, and L. Wang, "Industry 4.0 and Industry 5.0—inception, conception and perception," *Journal of Manufacturing Systems*, vol. 61, pp. 530–535, Oct. 2021, doi: 10.1016/J.JMSY.2021.10.006.

[2] C. Zhang and Y. Chen, "A review of research relevant to the emerging industry trends: Industry 4.0, IoT, blockchain, and business analytics," *Journal of Industrial Integration and Management*, vol. 5, no. 1, pp. 165–180, Mar. 2020, doi: 10.1142/S2424862219500192.

[3] F. Aslam, W. Aimin, M. Li, and K. U. Rehman, "Innovation in the era of IoT and Industry 5.0: Absolute innovation management (AIM) framework," *Information (Switzerland)*, vol. 11, no. 2, Feb. 2020, doi: 10.3390/INFO11020124.

[4] A. Angelopoulos, E. T. Michailidis, N. Nomikos, P. Trakadas, A. Hatziefremidis, S. Voliotis, and T. Zahariadis, "Tackling faults in the industry 4.0 era—a survey of machine-learning solutions and key aspects," *Sensors (Switzerland)*, vol. 20, no. 1, Jan. 2020, doi: 10.3390/S20010109.

[5] F. Longo, A. Padovano, and S. Umbrello, "Value-oriented and ethical technology engineering in industry 5.0: A human-centric perspective for the design of the factory of the future," *Applied Sciences (Switzerland)*, vol. 10, no. 12, pp. 1–25, Jun. 2020, doi: 10.3390/APP10124182.

[6] S. Nahavandi, "Industry 5.0—a human-centric solution," *Sustainability (Switzerland)*, vol. 11, no. 16, Aug. 2019, doi: 10.3390/SU11164371.

Chapter 6

The Industrial Revolution

From Mechanisation (1.0) to Smart Automation (5.0)

Babita Jha, Sarthak Garg, and Sarthak Dhingra

6.1 INTRODUCTION

Beginning in latter half of the 18th century in Britain, period of time known as the First Industrial Revolution was a time period of major development expanding throughout the broader world. Through the utilisation of water and steam power, it signified the move from manual manufacturing methods to machine production methods. During this time period, the factory system was established, which had a tremendous influence on the production of textiles, iron and steel, coal mining, and shipbuilding. It was the introduction of the spinning jenny, the water frame, and the power loom that brought about significant advances in the textile industry (Zakoldaev, Shukalov & Zharinov, 2019). The growth of industry 4.0 digital production businesses built on paperless and human less technologies is an intriguing path for the industrial economy sector. These technical advancements not only led to a revolution in industrial processes, but they also brought about enormous social and economic developments. These shifts included urbanisation, changes in job patterns, and the beginning of what we now refer to as the modern economy.

Electricity, the internal combustion engine, the assembly line were invented during the Second Industrial Revolution, which passed over from the late 19th century to the start of 20th century. Because of this revolution, significant technical improvements were introduced (Adeyeri, 2018). Through the broad use of electrical power, industries were able to undergo a transformation that enabled longer working hours and machinery that was more flexible. Henry Ford's invention of assembly line for the Model T vehicle exemplified emphasis that was placed during this time period on increased productivity and mass manufacturing. As a result, prices were greatly reduced, and commodities were more accessible to the general populace. In addition, during this time period, the steel, chemical, and electrical sectors witnessed fast development, which laid the foundation for the present industrial landscape. The telegraph, the telephone, and the railway were just some of the innovations in communication and transportation that helped further connect distant markets and made it easier for people throughout the world to trade products and thoughts with one another.

Industry 3.0 was a manufacturing paradigm that emerged in the middle of the 20th century and was distinguished by the use of computers and automation into the production processes (Dolga, 2021). Based on this concept and utilising the same communication channels, Industry 4.0 represents the latest industrial revolution. The growth and implementation of digital logic circuits, microprocessors, and the Internet brought about a transformation in several sectors by making it possible to achieve greater levels in precision, automation, and efficiency. During the production process, robotics started to play an important role by taking over jobs that were both repetitive and dangerous. This led to an increase in both safety and productivity. Supply chain management and inventory control are only two examples of the processes that have been revolutionised by information technology systems, which have also revolutionised how firms operate and compete.

Building on the progress that has been made in digital technology, Industry 4.0 arose at the beginning of the 21st century. Its primary focus is on the integration of sophisticated digital technologies, machine learning, and large amounts of data with physical production and operations (Bai et al., 2020). Additive manufacturing, artificial intelligence (AI), big data and analytics, blockchain, cloud, Industrial Internet of Things (IIoT), and simulation are some examples of Industry 4.0 technology. This revolution is characterised by the Internet of Things (IoT), which is a networking idea in which machines and other pieces of equipment are linked to one another and interact with one another to make choices that are decentralised. The emphasis switched from technology alone to human cooperation with robots, with a particular emphasis on factors such as sustainability, personalisation, and social well-being (Demir, Döven & Sezen, 2019). To reduce its negative influence on environment and to advance the cause of sustainability, Industry 5.0 places a premium on circular economy, renewable energy, and efficient resource utilisation. This necessitates a manufacturing system that is both flexible and adaptive. Industry 5.0 places a significant emphasis on tackling societal concerns, with the goal of ensuring that technology contributes to improvement of society's quality of life, reduction in inequality, and promotion of a more sustainable future (Figure 6.1).

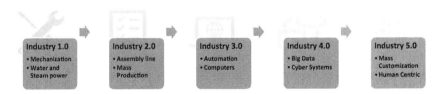

Figure 6.1 Different Phases of Industrialisation.

The progression from Industry 1.0 to Industry 5.0 is illustrative of humanity's unrelenting quest of innovation, efficiency, and, more recently, sustainability and social well-being (Özdemir & Hekim, 2018).

6.2 LITERATURE REVIEW

Industrial development is defined by the advent of disruptive technologies that result in revolutions with significant social and economic impact, as examined (Coelho et al., 2023). Taking into consideration the backdrop of Industry 2.0–4.0, Yin, Stecke and Li (2018) discussed about the production systems with a particular emphasis on the connections that exist between product supply and consumer demand. In Adeyeri's (2018) study, the author discusses how technological progress has resulted in notable improvements and changes in the design and use of equipment, spanning from ancient times to the present day.

The evolution of the production system's organisational model, driven by advancements in science and technology and market demand, has led to changes in production scheduling systems. The progress of production scheduling in relation to the economic and technological aspects of the latest industrial revolution is known as Industry 4.0 (Jiang et al., 2022). The research conducted by Zakoldaev, Shukalov, and Zharinov (2019) offers detailed explanations of the mechanical and assembly processes employed by well-established Industry 3.0 companies and potential Industry 4.0 corporations. The fundamental elements of a smart factory and how they are connected to streamline production processes employing automated and digital technology are outlined. The study done by Jeevitha and Ramya (2018) determines the core concept of Industry 4.0, known as the 'Internet of Things', which involves the widespread interconnection of people, objects, and devices. Industry 4.0 represents the Fourth Industrial Revolution.

Nikolić, Stefanovic and Djapan (2023) in their paper have discussed about Industry 4.0, or the Fourth Industrial Revolution that involves advanced technologies like the IoT, cybernetics, machine learning, and automation in production and commercial operations. Fourth Industrial Revolution, known as Industry 4.0, has presented problems to businesses worldwide (Sairam & Andrews, 2021). Barata and Kayser (2023) described Industry 5.0 as a human-centred approach to technological advancements in the industrial sector. Business are being encouraged to take into consideration the three pillars of human-centricity, sustainability, and resilience as a result of the advent of the idea of Industry 5.0 (Chivilò & Meneghetti, 2023). The Fourth Industrial Revolution, 4.0 (Skobelev & Borovik, 2017), focuses on the technical vision of the economy's future, whereas the Fifth Industrial Revolution, 5.0, emphasises personification and introduces a new paradigm centred around coopetition (Ungureanu, 2020). The intersection of these two revolutions will bring about changes that can acknowledge the value of human capital (Klabusayová & Černek, 2022).

Industry 5.0 is shifting the paradigm by reducing the focus on technology and emphasising that development relies on collaboration between humans and robots (Adel, 2022). However, Murugan (2023) emphasised on the obstacles that are impeding the mainstream implementation of Industry 5.0. The study done by Irpan and Shaddiq (2024) explores the notion of Industry 5.0, focusing on incorporating human-centred values, innovation, and adaptability into manufacturing processes. Demir, Döven and Sezen (2019) focused on the increasing trend in the collaboration between humans and robots in the workplace.

Society 5.0 is a concept that describes the transformation in individuals' lives due to the advancement of the Fourth Industrial Revolution (Al Faruqi, 2019). Industry 5.0 has brought new technologies including predictive maintenance, hyper-customisation, cyber–physical cognitive systems, and collaborative robots (Khan, Haleem & Javaid, 2023). To prioritise human centricity, it is crucial to enhance the skills and competencies of human beings and industrial operators through partnership with digital technology (Alves, Lima & Gaspar, 2023; Golovianko et al., 2023; Joglekar, Kadam & Dharmadhikari, 2023). Industry 5.0 encompasses organisation, management, technology, and performance assessment (Ivanov, 2023). An in-depth analysis is conducted by Alvarez-Aros and Bernal-Torres (2021) on the technical competitiveness and emerging technologies of Industry 4.0 and Industry 5.0 to pinpoint the essential components of both established and emerging economies.

The rapid advancement of the Fourth Industrial Revolution centred on adopting Sector 4.0 technology raised concerns among government and society about the potential degradation of the sector in future. The study done by Saniuk, Grabowska and Straka (2022) propagated the importance of sustainable development (Ghobakhloo, 2020) and the significant role of humans in the expectations of the industry's future growth. The concerns around using of technology from the Fourth Industrial Revolution led to development of principles of Industry 5.0. Rane (2023) talks about this changing environment, vision transformers that have emerged as a vital answer in the fields of AI and computer vision. Industry 5.0, which involves collaboration between humans and robots (Nahavandi, 2019), uses vision transformers to enhance human–machine interaction and improves safety in shared workspaces. Industry 5.0 aims to combine human expertise with advanced machines to create resource-efficient and user-centric manufacturing solutions, beyond the capabilities of Industry 4.0 (Tiwari, Bahuguna and Walker, 2022).

6.3 STAGES OF TRANSFORMATION (INDUSTRY 1.0–1.4)

6.3.1 Industry 1.0

Industry 1.0 refers to the initial phase of Industrial Revolution, which originated in England during the 18th century, specifically spanning from around 1760 to 1840. The industrial revolution had already disseminated to the United States by the latter part of 18th century. Industry 1.0 is characterised by the automation of manufacturing processes and the widespread adoption of steam power. Additionally, it signified the initial significant shift from an economy reliant on manual craftsmanship to one that incorporated the usage of machinery in manufacturing operations. The industries affected by First Industrial Revolution, also known as Industry 1.0, include glass, mining, agricultural, and textile sectors. The defining technologies of Industry 1.0 were the machinery driven by water and steam. An exemplary instance of such machinery is the weaving loom, which was initially devised in 1784.

6.3.2 Industry 2.0

Industry 2.0, often known as the Second Industrial Revolution, commenced in the 19th century, namely in the 1870s. The occurrence was mostly concentrated in Germany, America, and Britain. The primary focus was on industrial operations utilising machinery driven by electrical power. Electric machines exhibited superior efficiency, as well as enhanced ease of operation and maintenance, in comparison to water- and steam-powered machinery. Furthermore, they were highly economical, using fewer resources and human labour compared to the machinery employed in the initial industrial revolution. Industry 2.0 also showcased an enhanced and efficient mass manufacturing method. This occurred subsequent to the establishment of the initial assembly line, which facilitated the production of things on a larger scale and with enhanced quality. Numerous technical systems emerged during the Second Industrial Revolution. The utilisation of electricity facilitated the integration of contemporary production lines in several industries, enabling the implementation of mass production techniques for commodities. Furthermore, the company's technical systems also served as a prototype for Henry Ford, who pioneered the concept of mass manufacturing. He developed a strong fascination with the process of suspending the pigs on conveyor belts at an abattoir in Chicago. Multiple butchers were present, with each assigned to a specific task in the process of slaughtering the pigs.

6.3.3 Industry 3.0

The Third Industrial Revolution is generally known as the 'Digital Revolution' or the 'First Computer Era'. The onset of the Third Industrial Revolution was initiated by the implementation of partial automation, a technical procedure accomplished via the use of basic computers and Programmable Logic Controllers (PLCs). In the latter part of the 20th century, significant progress was achieved in the electronics sector. These electronic gadgets facilitated a limited level of automation in the equipment utilised in production operations. Consequently, this resulted in enhanced precision in manufacturing, heightened velocities, improved proficiency, and, in some manufacturing procedures, even the substitution of a human workforce.

The PLC was developed in the 1960s, marking a significant advancement in the application of electronics for automating tasks. Moreover, the incorporation of electronic gear in industrial processes necessitated the implementation of software systems to control and oversee this electronic hardware. Consequently, this fostered the growth of the software development industry during that time. The software solutions not only enhanced the functionality of electronic devices but also streamlined the implementation of diverse management tasks. During the Third Industrial Revolution, several electronic devices were created. These include integrated circuit chips, digital logic systems, Metal-Oxide-Semiconductor (MOS) transistors, and related technologies such as the Internet, computers, digital cellular phones, and microprocessors. Essentially, the digital revolution converted the current analogue universe into a modern and digital realm.

6.3.4 Industry 4.0

Industrialisation 4.0, the most recent Industrial Revolution, is not predicated on the identification of a novel energy source, but rather on the realm of digital technology or digitalisation. This is a result of the ongoing advancement of computers, sensors, and the widespread adoption of the Internet (Dolga, 2021). The current era is distinguished by the incorporation of communication and advanced information technology into many sectors, which is a consequence of the Third Industrial Revolution. Study done by Xu et al. (2021) talks about Industry 4.0, an initiative from Germany, has become a globally adopted term in the past decade. The manufacturing industry witnessed extensive digitalisation, facilitating efficient and timely dissemination of information to the appropriate individuals. The IIoT facilitates the efficient distribution of information. The (IIoT) has four fundamental components: cloud computing and big data, cyber–physical systems, machine learning and AI, and IoT.

Another crucial facet of the Fourth Industrial Revolution is the heightened focus on sustainability and environmental concerns. Sustainable development, as delineated by the three pillars of sustainability (environmental, economic, and social sustainability), is perceived not only as an imperative to adopt environmentally friendly practices and preserve natural resources for future generations but also as a prospect to enhance manufacturing efficiency and augment business profitability.

6.4 LATEST PHASE OF INDUSTRIALISATION

6.4.1 Industry 5.0

Industry 5.0 denotes the collaboration between humans, automated robots, and intelligent machines. The concept revolves around the utilisation of sophisticated technologies such as the IoT and big data to enhance human productivity and efficiency through the assistance of robots. It introduces a personalised and human element to the core principles of automation and efficiency in Industry 4.0. Historically, robots have been utilised in manufacturing settings to carry out hazardous, repetitive, or physically strenuous tasks, such as welding and painting in automobile plants, as well as the handling of large products in warehouses. Industry 5.0 seeks to integrate the cognitive computing capabilities of increasingly intelligent and interconnected robots with human intellect and resourcefulness in collaborative activities in the workplace. Universal Robots, a Danish firm, established itself as the pioneer in offering industrial robots that operate safely and efficiently in collaboration with people.

6.4.2 Advancements in Industry 5.0

Major advancements mentioned briefly in Figure 6.2 are as follows:

- **Collaborative robots (Cobots):** Cobots are equipped with force sensors, vision systems, and machine learning algorithms that allow them to recognise and react to human presence and motions immediately, guaranteeing safety in operations that involve close contact.
- **Artificial intelligence and machine learning:** Through the analysis of data collected from sensors on machinery, AI has the capability to anticipate when a machine is prone to failure or in need of repair. Utilising machine learning algorithms based on visual or sensor input, flaws or abnormalities in goods may be promptly detected, leading to a substantial enhancement in the precision and efficiency of quality control procedures when compared to manual inspections.
- **Internet of Things:** IoT devices, when used in industrial processes, gather extensive data on many aspects, such as machine performance and environmental conditions. By utilising this data, producers can actively monitor processes in real time and promptly make necessary modifications to enhance productivity and minimise waste.
- **Advanced materials and 3D printing:** The advancement of novel materials can result in goods that possess reduced weight, increased strength, or enhanced sustainability. Biodegradable materials have the ability to lessen the negative effects on the environment, while

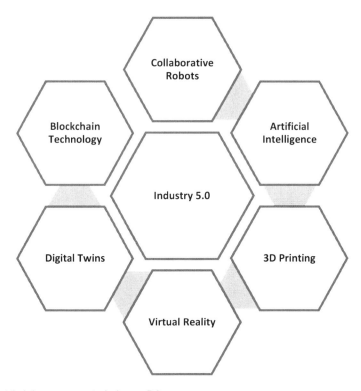

Figure 6.2 Advancements in Industry 5.0.

improved composites have the potential to provide better performance for certain uses.

Digital twins: Digital twins allow firms to model and optimise production processes, product designs, and supply chain activities in a virtual environment.

Augmented reality (AR) and virtual reality (VR): AR has the capability to superimpose digital data, such as assembly instructions or maintenance procedures, onto the real world, aiding workers in carrying out tasks more effectively and with reduced mistakes. VR technology has the capability to replicate intricate settings for the purpose of training, enabling workers to acquire proficiency in handling equipment and executing procedures without being exposed to the hazards typically associated with physical training.

Blockchain technology: The decentralised ledger of blockchain can securely document transactions and the movement of products, creating an unchangeable record of the whole supply chain. This promotes transparency, enabling firms and customers to authenticate and validate the origin and ethical procurement of items.

Integration of energy efficiency and renewable energy: Intelligent energy management systems have the ability to dynamically control energy usage based on current information. These systems may also integrate renewable energy sources such as solar and wind power to reduce reliance on fossil fuels and decrease carbon emissions.

6.4.3 Advantages of Industry 5.0

Enhanced productivity through autonomous manufacturing: There are several advantages to autonomous production. Automation 5.0 enhances productivity by allowing people to concentrate on key activities while robots take care of monotonous and repetitive jobs.

Enhanced client satisfaction: The evolution of the customer experience is progressing swiftly, and it is imperative for every organisation to stay abreast of these changes. Businesses must distinguish themselves by providing a superior experience. Irrespective of whether you provide customised services or goods, it is imperative that your clients are able to enjoy an optimal experience.

Extreme personalisation: Consumers are increasingly seeking personalised and distinctive products and services. Companies must possess the capability to integrate a blend of personal interaction and diverse technology to fulfil the distinct requirements of clients.

Customised merchandise: The emergence of the Fifth Industrial Revolution entails a transition in manufacturing from large-scale production to customisation. This shift will introduce products and services that are specifically customised to an individual's preferences, aspirations, and requirements.

Agile supply chain: A responsive supply chain is an adaptable business strategy specifically developed to fulfil the requirements of diverse enterprises. An adaptable supply chain can promptly accommodate fluctuations in demand, volume, or weather conditions.

Engaging and user-friendly product: The integration of humans and robots in the expanding human-centred manufacturing paradigm offers several advantages. Through the integration of these systems, firms have the ability to enhance competitiveness by increasing productivity and reducing costs.

Centred around humans: Adopting a human-centric approach offers several advantages. Primarily, this design style prioritises the user's experience and progression through a product or service, rather than placing emphasis on the product or service itself. The primary objective of human-centric manufacturing in Industry 5.0 is to enhance the well-being of workers and society. It aims to tackle the evolving skillsets and training requirements of employees while prioritising the overall welfare of the organisation to effectively satisfy the unique demands of customers.

Enhanced resilience: Resiliency mitigates the consequences of individual failures and strengthens a company's capacity to recover from adversity. Considering the ongoing climate catastrophe as well as the uncertainties surrounding wars and pandemics, firms are being challenged to demonstrate their resilience.

Environmentally friendly: A significant advantage of Industry 5.0 is its ability to mitigate its environmental footprint. Sustainability may enhance staff productivity and morale, minimise waste, and enhance product quality. The issue at hand is not about incurring additional expenses but rather about prioritising the sustained prosperity of a corporation in the long run.

Value based: Industry 5.0 is a novel business strategy that prioritises human needs as the central focus of the production process. This industrial method aims to reduce pollution and waste while promoting circular manufacturing processes. The consideration of societal value in Industry 5.0 is crucial for the development and implementation of initiatives.

Cooperative automatons: The advantages of collaborative robots in Industry 5.0 are manifold, ranging from diminishing operator effort to augmenting worker happiness. Due to their ability to carry out a range of repetitive activities, they have the capacity to relieve human workers from laborious and tedious employment.

6.4.4 Challenges

The transition to Industry 5.0 encompasses complex difficulties across several disciplines, such as economics, social dynamics, regulations, and environmental concerns. Below are the challenges the most notable issues associated with the implementation of Industry 5.0.

6.4.4.1 Polarisation of the Workforce and the Skill Gap: As sectors progress, there is an increasing polarisation between high-skilled roles that are well rewarded and low-skilled occupations that are poorly compensated. The existence of this dichotomy poses a risk of widening the gap between various social and economic groups, which in turn results in difficulties in terms of workforce development and social fairness. As a result of the fast advancement of technology, it is essential to have a workforce that is not only technically competent but also flexible enough to accommodate new methods of working in conjunction with intelligent systems.

6.4.4.2 Human–Robot Collaboration and Safety: The incorporation of collaborative robots into the workplace has the intention of enhancing human skills; nevertheless, it also raises new considerations regarding safety and ethics. There is a need for comprehensive standards and policies that are still in the process of being developed to ensure the physical safety of people, protect against malfunctions, and handle the psychological implications of working beside robots.

6.4.4.3 Data Integrity and Management: In Industry 5.0, where decision-making is becoming increasingly dependent on data analytics and AI, the integrity of data and the management of data are of the utmost importance. The challenges include assuring the quality of the data, protecting the data from being breached, and handling the enormous volumes of data that are created by devices that are connected to one another.

6.4.4.4 Cybersecurity Dangers: The linked structure of Industry 5.0 makes the dangers of cybersecurity threats far more severe. To protect critical infrastructure and sensitive data from cyberattacks that are becoming increasingly sophisticated, it is necessary to implement advanced security solutions, maintain continual awareness, and take a proactive mindset towards cybersecurity.

6.4.4.5 Barriers to Investment and Adoption: The deployment of technologies for Industry 5.0 demands a significant investment, which can be particularly prohibitive for small and medium-sized enterprises (SMEs) and startups with limited resources. Barriers to adoption are created as a result of the high expenses associated with updating infrastructure, acquiring new technology, and training people. This has the potential to expand the gap between major organisations that are financially stable and smaller businesses.

6.4.4.6 Considerations Regarding Regulations and Ethical Issues: It is necessary for regulatory frameworks to develop to handle new ethical considerations, liability difficulties, and compliance standards as the roles that automation and AI play in decision-making become increasingly prominent. The regulatory problems that are presented by Industry 5.0 include, but are not limited to, determining accountability in automated systems, preserving individuals' privacy, and ensuring that everyone has equal access to technology.

6.4.4.7 Corporate Model Transformation: Industry 5.0 requires a change in corporate strategy to allow mass personalisation and a customer-centric approach. This shift is necessary to support Business Model Transformation. This transition necessitates the development of dynamic company models that are capable of responding quickly to developing technical breakthroughs and shifting tastes among consumers. The creation and implementation of these models call for innovation not only in the products and services that are offered, but also in the organisational structures and operational procedures that are utilised.

6.4.4.8 Concerns Regarding the Environment and Long-Term Sustainability: Even though Industry 5.0 places a strong emphasis on environmental responsibility and sustainability, making progress towards these objectives in the face of fast technical improvement and growing resource consumption offers a substantial challenge. The development of creative solutions that decrease waste, improve energy efficiency, and minimise the carbon footprint of manufacturing processes is necessary to strike a balance between the expansion of industry and the protection of the environment.

6.5 CONCLUSION

It is abundantly obvious that each phase of the Industrial Revolution has been a cornerstone in redefining the outlines of society, economics, and technology while we traverse the transformational journey from Industry 1.0 to the coming horizon of Industry 5.0. The IoT, AI, and big data analytic technologies have all made significant strides in recent years. This period not only facilitated the development of intelligent manufacturing, but also laid the groundwork for predictive maintenance and customisation on a scale that had never been seen before. We are currently experiencing a paradigm change in the industrial process, which is moving towards a more human-centric approach. This transition is occurring as we stand on the brink of Industry 5.0. Personalisation, sustainability, and social responsibility are the focal points of this new era, which is characterised by the symbiotic partnership between individuals and smart technology. As we traverse the complexity of Industry 5.0, the lessons that we have learnt from previous industrial revolutions will be extremely helpful in utilising technical developments to create a society that is more inclusive, sustainable, and affluent for future generations.

REFERENCES

Adel, A. (2022). Future of industry 5.0 in society: Human-centric solutions, challenges and prospective research areas. Journal of Cloud Computing, 11(1), 1–15.
Adeyeri, M. K. (2018). From industry 3.0 to industry 4.0: smart predictive maintenance system as platform for leveraging. Arctic Jj, 71(11), 64–81.
Al Faruqi, U. (2019). Future service in industry 5.0. Jurnal Sistem Cerdas, 2(1), 67–79.
Alvarez-Aros, E. L., & Bernal-Torres, C. A. (2021). Technological competitiveness and emerging technologies in industry 4.0 and industry 5.0. Anais da Academia Brasileira de Ciências, 93, 7.

Alves, J., Lima, T. M., & Gaspar, P. D. (2023). Is industry 5.0 a human-centred approach? A systematic review. Processes, 11(1), 193.

Bai, C., Dallasega, P., Orzes, G., & Sarkis, J. (2020). Industry 4.0 technologies assessment: A sustainability perspective. International journal of production economics, 229, 107776.

Barata, J., & Kayser, I. (2023). Industry 5.0–Past, present, and near future. Procedia Computer Science, 219, 778–788.

Chivilò, M., & Meneghetti, A. (2023). An industry 5.0 perspective on feeding production lines. Sustainability, 15(22), 16088.

Coelho, P., Bessa, C., Landeck, J., & Silva, C. (2023). Industry 5.0: The arising of a concept. Procedia Computer Science, 217, 1137–1144.

Demir, K. A., Döven, G., & Sezen, B. (2019). Industry 5.0 and human-robot co-working. Procedia Computer Science, 158, 688–695.

Dolga, V. I. (2021). From Industry 3.0 to Industry 4.0. https://scholar.google.com/scholar?hl=en&as_sdt=0%2C5&q=Dolga%2C+V.+I.+%282021%29.+From+Industry+3.0+to+Industry+4.0.&btnG=

Ghobakhloo, M. (2020). Industry 4.0, digitization, and opportunities for sustainability. Journal of Cleaner Production, 252, 119869.

Golovianko, M., Terziyan, V., Branytskyi, V., & Malyk, D. (2023). Industry 4.0 vs. Industry 5.0: co-existence, transition, or a hybrid. Procedia Computer Science, 217, 102–113.

Irpan, M., & Shaddiq, S. (2024). Industry 4.0 and Industry 5.0—Inception, conception, perception, and rethinking loyalty employment. International Journal of Economics, Management, Business, and Social Science (Ijembis), 4(1), 95–114.

Ivanov, D. (2023). The industry 5.0 framework: Viability-based integration of the resilience, sustainability, and human-centricity perspectives. International Journal of Production Research, 61(5), 1683–1695.

Jeevitha, T., & Ramya, L. (2018). Industry 1.0 to 4.0: The Evolution of Smart Factories. APICS Mag.

Jiang, Z., Yuan, S., Ma, J., & Wang, Q. (2022). The evolution of production scheduling from Industry 3.0 through Industry 4.0. International Journal of Production Research, 60(11), 3534–3554.

Joglekar, S., Kadam, S., & Dharmadhikari, S. (2023). Industry 5.0: Analysis, applications and prognosis. The Online Journal of Distance Education and e-Learning, 11(1), 1.

Khan, M., Haleem, A., & Javaid, M. (2023). Changes and improvements in Industry 5.0: A strategic approach to overcome the challenges of Industry 4.0. Green Technologies and Sustainability, 1(2), 100020.

Klabusayová, N., & Černek, M. (2022, September). From Industry 4.0 to Society 5.0: Starting points, relationship, problems. In Proceedings of the International Scientific Conference (p. 159). Vysoká Škola Prigo.

Murugan, P. (2023). Industry 5.0: Revolutionizing the future of manufacturing. MDIM Journal of Management Review and Practice, 6, 7.

Nahavandi, S. (2019). Industry 5.0—A human-centric solution. Sustainability, 11(16), 4371.

Nikolić, J., Stefanovic, M., & Djapan, M. (2023). Industry 4.0 and Industry 5.0–Opportunities and Threats. International Journal of Recent Technology and Engineering, 8(5), 3305–3308.

Özdemir, V., & Hekim, N. (2018). Birth of industry 5.0: Making sense of big data with artificial intelligence, 'the internet of things' and next-generation technology policy. OMICS: A Journal of Integrative Biology, 22(1), 65–76.

Rane, N. (2023). Transformers in Industry 4.0, Industry 5.0, and Society 5.0: Roles and Challenges. SSRN Electronic Journal. https://ssrn.com/abstract=4609915 or http://dx.doi.org/10.2139/ssrn.4609915

Sairam, S. S., & Andrews, F. (2021) An Overview on the Future of Industry 5.0. https://scholar.google.com/scholar?hl=en&as_sdt=0%2C5&q=An+overview+of+the+future+of+Industry+5.0+Sairam&btnG=

Saniuk, S., Grabowska, S., & Straka, M. (2022). Identification of social and economic expectations: Contextual reasons for the transformation process of Industry 4.0 into the Industry 5.0 concept. Sustainability, 14(3), 1391.

Skobelev, P. O., & Borovik, S. Y. (2017). On the way from Industry 4.0 to Industry 5.0: From digital manufacturing to digital society. Industry 4.0, 2(6), 307–311.

Tiwari, S., Bahuguna, P. C., & Walker, J. (2022). Industry 5.0: A macroperspective approach. In Handbook of Research on Innovative Management Using AI in Industry 5.0 (pp. 59–73). IGI Global.

Ungureanu, A. V. (2020, August). The transition from industry 4.0 to industry 5.0. The 4Cs of the global economic change. In 16th Economic International Conference NCOE 4.0 2020 (pp. 70–81). Editura Lumen, Asociatia Lumen.

Xu, X., Lu, Y., Vogel-Heuser, B., & Wang, L. (2021). Industry 4.0 and Industry 5.0—Inception, conception and perception. Journal of Manufacturing Systems, 61, 530–535.

Yin, Y., Stecke, K. E., & Li, D. (2018). The evolution of production systems from Industry 2.0 through Industry 4.0. International Journal of Production Research, 56(1–2), 848–861.

Zakoldaev, D. A., Shukalov, A. V., & Zharinov, I. O. (2019, June). From Industry 3.0 to Industry 4.0: Production modernization and creation of innovative digital companies. In IOP Conference Series: Materials Science and Engineering (Vol. 560, No. 1, p. 012206). IOP Publishing.

Chapter 7

Enhancing the Digital Economy in the Context of the Fourth Industrial Revolution

The Case of Vietnam

Nguyet Thao Huynh and Chien-Van Nguyen

7.1 INTRODUCTION

With the rapid development of the economy along with strong changes in information and communication technology, the application of modern technology aims to improve efficiency in state management as well as bring benefits for the people. One of those special needs is to develop a digital economy based on non-cash payment (NCP) activities. In fact, NCP is a process of overall and comprehensive change of organizations and individuals in transaction methods based on new technology.

To create a legal foundation for NCP activities, the Government issued Decree No. 101/2012/ND-CP dated November 22, 2012 on NCPs; Decree No. 80/2016/ND-CP dated July 1, 2016 amending and supplementing a number of articles of Decree No. 101/2012/ND-CP. In addition, the Prime Minister also issued Decision No. 2545/QD-TTg dated December 30, 2016 approving the Project to develop NCPs in Vietnam for the period 2016–2020; Decision No. 241/QD-TTg dated February 23, 2018 approving the Project to promote payments via banks for public services: taxes, electricity, water, tuition, hospital fees, and payment of safety programs. social life; Directive No. 22/CT-TTg dated May 26, 2020 on promoting the implementation of NCP development solutions in Vietnam. With the above guidelines and policies and timely direction from all levels and sectors, NCP activities have developed rapidly with very remarkable results. During the 2020–2021 period, the Covid pandemic has turned the lives of most people upside down, from health, work, education, communication to daily living habits, including the need for online transactions, online shopping.

To create a positive change in NCPs in the economy with high growth rates, strongly applying the achievements of the fourth industrial revolution, meeting the needs of NCPs conveniently, on October 28, 2021, The Prime Minister issued Decision No. 1813/QD-TTg approving the Project to develop NCPs in Vietnam for the period 2021–2025; Decision No. 316/QD-TTg dated March 9, 2021 on approving the pilot implementation of using telecommunications accounts to pay for goods and services of small value. Although many outstanding results have been achieved, contributing to changing people's habits, non-commercial marketing activities still have some limitations such as: fear when using new technology; worried about safety and security issues when using non-commercial payment methods; the payment network and infrastructure of service providers are mostly concentrated in urban areas, while the payment has not met people's expectations in rural areas; some NCP services have not been designed to target customers, especially customers in rural areas and customers with limited technological knowledge; the problem of high-tech crime is increasingly numerous and sophisticated.

7.2 LITERATURE REVIEW

An outstanding research in Vietnam was conducted by Bui et al. (2022) on factors affecting the acceptance of using NCP among Ho Chi Minh City's consumers with research content on effectiveness and risk factors that impact users' decisions to use NCPs. Risk factor, specifically "risk is considered unluckiness and loss. When customers who use NCP will encounter many risks in information, systems, services and products, thereby losing reputation and consumer trust."

In addition, Ho et al. (2022) also analyzed the service element in using NCPs, including accompanying services when using payment types via electronic banking, cards, or wallet. The article also indicates most people use NCPs in purchase transactions such as online shopping, hotel payments, or watching movies because there are more promotions in using NCPs.

According to Nguyen (2021), payment infrastructure serving NCPs is also increasingly improved. Banks have researched and applied many new and modern technologies and solutions for NCP. Further, technical infrastructure and technology serving NCP, especially electronic payments, continue to be invested. The interbank electronic payment system, financial switching, and electronic clearing system are operated stably, smoothly, and safely. Based on a study in the US, Roubini ThoughtLab (2018) confirmed that realizing the benefits of technical payments, and technological factors that influence the decision to use NCPs, specifically: digital payment technology is a key enabler of smart cities and can contribute significant benefits to consumers, businesses, governments, and economies.

Widjaja (2016) pointed out that when developing a NCP system, service providers must prioritize focusing on investing in financial infrastructure. In addition, service providers should improve new payment methods toward the public. Cash may be completely replaced by electronic payment systems in the future, payment tools suitable for modern lifestyles (Harasim, 2016) as well as the fourth industrial revolution. Therefore, preparing financial institutions and infrastructure for this form of payment is very necessary and needs to start slowly because cash payment has almost become a cultural feature many people around the world (Busse et al., 2020).

7.3 DATA SELECTION AND METHODOLOGY

The data source used in this study was collected in Vietnam in 2023. The study conducted a questionnaire survey using the convenience sampling method in Vietnam, but also considered diversity in age, income, level, gender, etc., thereby analyzing factors affecting the use of NCP by local people to make appropriate recommendations.

With the convenience sampling method, Hair et al. (2006) indicated that choosing an appropriate sample size is good for data processing using exploratory factor analysis (EFA). Specifically: a ratio of 5:1 (5 survey subjects/1 observed variable) is considered minimally acceptable, a ratio of 10:1 is considered good, and a ratio of 20:1 is acceptable. In addition, Tabachnick and Fidell (2007) showed that the empirical formula often used to calculate sample size for multiple regression is $n > 50 + 8p$ in which n is the sample size and p is the number of independent variables in the model. Accordingly, the number of research samples required is at least $n = 50 + 8 \times 5 = 90$ observations. From the above two methods, to ensure a good sample size for the study, the author decided to choose a sample size of $n = 220$.

From the previous studies based on Bui et al. (2022), Roubini ThoughtLab (2018), the authors built a hypothetical model for the study as follows (Figure 7.1):

The authors put forward the following constructive hypotheses:

Dependent variable: Y *(SD)*—Decision on using NCP

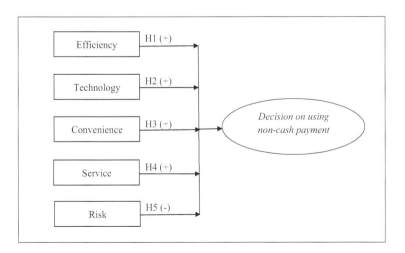

Figure 7.1 Proposed model.

Source: Authors' analysis (2023).

Independent variables:

(i) *H1 (+) – Efficiency (HQ):* has a positive relationship with the decision to use NCP.
(ii) *H2 (+) – Technology (CN):* has a positive relationship with the decision to use NCP.
(iii) *H3 (+) – Convenience (TT):* has a positive relationship with the decision to use NCP.
(iv) *H4 (+) – Service (DV):* has a positive relationship with the decision to use NCP.
(v) *H5 (–) – Risk (RR):* has a negative relationship with the decision to use NCP.

7.4 RESULTS AND DISCUSSION

7.4.1 Frequency Analysis

According to survey statistics in Table 7.1, the proportion of men and women participating in the survey is relatively equal with men being 50.91% and women being 49.09%. The age group participating in the survey is also quite diverse: the majority is the age group 31–45 years old, reaching 55%

Table 7.1 Frequency analysis

Items		Number	Percentage
Gender	Male	112	50.91
	Female	108	49.09
Age	18–30 years old	79	35.91
	31–45 years old	121	55.00
	46–60 years old	8	3.64
	> 61 years old	12	5.45
Qualification	Graduate	112	50.91
	Postgraduate	75	34.09
	Vocational/College	13	5.91
	high school/secondary school	20	9.09
Occupation	State agencies sector	35	15.91
	Firm/Business	84	38.18
	Technical sector	21	9.55
	Finance – banking	24	10.91
	Students	17	7.73
	Others	39	17.73
Income	Over 20 million	48	21.82
	From 10 to 15 million	48	21.82
	From 15 to 20 million	28	12.73
	From 5 to 10 million	80	36.36
	5 million and less	16	7.27

Source: Authors' analysis (2023).

of the total votes, followed by the age group 18–30 years old, accounting for 35.91% of the total votes. Both these high-proportion groups are in the prime working age. The remaining two groups are 46 years old or older, accounting for about 9%. The survey results also showed that people with a university degree or higher accounted for 85% of the survey sample and they were more interested in non-commercial payment methods. The educational status in the province has gradually improved, so it is not difficult to explain why the number of survey subjects with college degrees or less only accounts for 15% of the total survey sample. In addition, the majority of occupations in the survey sample with the highest proportion of 84% belong to the industry group of the business sector. This result shows the suitability of the data, because due to the nature of work, having to proactively make flexible payments through many forms, most businesses no longer use cash for direct transactions like before.

The income of the surveyed people is also an important factor in analyzing the topic, through data collected from the group with income within the range of 5–10 million Vietnamese Dong (VND), accounting for 36.36%. The two groups with an equal proportion of 21.82% constitute the group with income 10–15 million VND and the group over 20 million VND. Next is the income group from 15 to under 20 million VND, accounting for 12.73% of the survey number. The lowest 7.27% belong to the group under 5 million VND.

7.4.2 Descriptive Statistics

With values of variables ranging from 1 to 5, Table 7.2 shows the average value of each observed variable, in which the lowest average value is the observed variable DV5 (3.31) and the highest value is the variable passing close to TT2 (4.23). The other values also take values between 1 and 5 and have an average value around 4 and a low standard deviation (SD).

Table 7.2 Descriptive statistics

Scale	Minimum	Maximum	Mean	Standard deviation
HQ1	1	5	4.14	1.051
HQ2	1	5	4.03	1.011
HQ3	1	5	4.00	1.066
HQ4	1	5	4.02	1.040
HQ5	1	5	3.96	1.051
HQ6	1	5	3.98	1.025
CN1	1	5	4.09	0.942
CN2	1	5	4.02	0.791

(continued)

Table 7.2 (Cont.)

Scale	Minimum	Maximum	Mean	Standard deviation
CN3	1	5	4.01	1.014
CN4	1	5	4.13	0.939
CN5	1	5	4.15	0.953
CN6	1	5	3.93	0.967
TT1	1	5	3.77	1.149
TT2	1	5	4.23	1.048
TT3	1	5	4.09	0.982
TT4	1	5	4.14	1.031
TT5	1	5	4.17	0.976
DV1	1	5	3.60	1.217
DV2	1	5	3.73	0.978
DV3	1	5	3.66	1.023
DV4	1	5	3.90	1.046
DV5	1	5	3.31	0.954
RR1	1	5	3.58	1.010
RR2	1	5	3.78	1.060
RR3	1	5	3.52	1.022
RR4	1	5	3.85	1.104
RR5	1	5	3.49	1.027
SD1	1	5	3.96	0.745
SD2	1	5	3.94	0.825
SD3	1	5	4.03	0.764
SD4	1	5	3.93	0.805
SD5	1	5	3.95	0.772

Source: Authors' analysis (2023).

Therefore, it can be predicted that surveyors highly appreciate factors such as efficiency, technology, service, risk, and convenience to the effectiveness of not using cash and developing the current digital economy.

Based on the results detailed in Table 7.2, it can be concluded that the rating level of each observed variable is acceptable, the service (DV) and risk (RR) variables are rated lower than the observed variables. The remaining factors include efficiency factor (HQ), convenience factor (TT), and technology factor (CN). Among them, the most stable are the observed variables that are the deciding factors in using non-trading payment methods (SD).

7.4.3 The Test of Cronbach's Alpha and EFA

Testing the reliability of the scale is very necessary in analysis. To test the reliability of the scale of factors in the study, the authors calculated the Cronbach's alpha coefficient of the scale and considered the item–total correlations. Variables with item–total correlations coefficients less than 0.3 will be eliminated and scales with Cronbach's alpha reliability of 0.6 or higher will be accepted for analysis in the next steps. In addition, the

Table 7.3 The Test of KMO and Bartlett

Kaiser–Meyer–Olkin Measure of Sampling Adequacy		0.885
Bartlett's Test of Sphericity	Approx. Chi-Square	2514.417
	df	300
	Sig.	0.000

Source: Authors' analysis (2023).

authors conducted EFA with oblique rotation using Promax, extracted variance Principal Axis Factoring, applied Kaiser–Meyer–Olkin (KMO) and Bartlett's test methods to measure the compatibility of survey variables.

With the results in Table 7.3, we have: KMO value = 0.885 > 0.5 is appropriate. Further, Sig. (Bartlett's Test) = 0.000 < 0.05 proves that overall the observations are correlated with each other. This reflects that the selection of variables is relatively appropriate and should be implemented in future studies, and this may suggest that an increase in non-cash use may come from convenience, risk, efficiency, technology, and service.

From the test results in Table 7.4, there are five factors extracted based on the eigenvalue criterion of 1.363 > 1, so the analysis of EFA is well - discussed. Further, the total variance extracted by these factors is 62.499% > 50%; thus, the 5 extracted factors explain 62.499% of the data variation of the 25 observed variables participating in EFA. This explanation rate is relatively high and can produce more reliable results, so that the results of this study can be more representative not only for Vietnam but also for countries similar to Vietnam. According to Widjaja (2016), service providers must prioritize focusing on investing in financial infrastructure, or digital payment technology can contribute significant benefits to consumers, businesses, governments, and economies. Therefore, reliable results can help develop the digital economy in developing countries.

Table 7.5 shows that the factor loading coefficients of the observed variables in the rotated matrix are all greater than 0.5, so these observed variables all meaningfully contribute to the model according to Hair et al. (2006). Therefore, the scales of the independent variables are qualified and will be kept the same to continue for subsequent analysis.

7.4.4 Correlation Matrix

Pearson correlation analysis aims to analyze whether the relationship between independent variables and dependent variables is linearly correlated or not, and at the same time identify multicollinearity between observed variables.

Table 7.6 shows that the correlation coefficients are all less than 0.8; therefore, there is possibility of multicollinearity occurring. In particular, the correlation results also predict the positive impact of HQ (Efficiency), Technology

Table 7.4 Total variance explained

Component	Initial Eigenvalues			Total Variance Explained					
				Extraction Sums of Squared Loadings			Rotation Sums of Squared Loadings		
	Total	% of Variance	Cumulative %	Total	% of Variance	Cumulative %	Total	% of Variance	Cumulative %
1	7.742	30.969	30.969	7.742	30.969	30.969	3.695	14.780	14.780
2	3.123	12.491	43.460	3.123	12.491	43.460	3.129	12.516	27.296
3	1.919	7.675	51.135	1.919	7.675	51.135	3.126	12.503	39.800
4	1.478	5.910	57.045	1.478	5.910	57.045	2.979	11.918	51.717
5	1.363	5.454	62.499	1.363	5.454	62.499	2.695	10.781	62.499
6	0.842	3.370	65.868						
7	0.750	3.001	68.870						
8	0.677	2.708	71.578						
9	0.646	2.586	74.164						
10	0.611	2.443	76.607						
11	0.559	2.235	78.841						
12	0.534	2.136	80.978						
13	0.508	2.033	83.011						
14	0.483	1.934	84.944						
15	0.468	1.871	86.816						
16	0.452	1.807	88.623						
17	0.423	1.692	90.315						
18	0.404	1.616	91.931						
19	0.368	1.472	93.402						
20	0.332	1.328	94.730						
21	0.310	1.241	95.970						
22	0.283	1.132	97.102						
23	0.258	1.033	98.136						
24	0.251	1.002	99.138						
25	0.216	0.862	100.000						

Source: Authors' analysis (2023).

Table 7.5 Rotated component matrix

Items	Rotated Component Matrix[a]				
	1	2	3	4	5
HQ4	0.766				
HQ1	0.733				
HQ3	0.712				
HQ2	0.698				
HQ5	0.684				
HQ6	0.679				
CN3		0.777			
CN6		0.743			
CN1		0.735			
CN4		0.665			
CN5		0.653			
RR4			0.829		
RR1			0.786		
RR5			0.782		
RR2			0.768		
RR3			0.723		
TT2				0.806	
TT5				0.738	
TT3				0.723	
TT4				0.704	
TT1				0.614	
DV3					0.783
DV2					0.746
DV5					0.694
DV4					0.687

Source: Authors' analysis (2023).

Table 7.6 Correlation matrix

Variable	SD	HQ	CN	TT	DV	RR
SD	1	0.605[a]	0.522[a]	0.686[a]	0.562[a]	−0.332[a]
HQ	0.605[a]	1	0.567[a]	0.480[a]	0.502[a]	−0.077
CN	0.522[a]	0.567[a]	1	0.436[a]	0.470[a]	0.007
TT	0.686[a]	0.480[a]	0.436[a]	1	0.529[a]	−0.067
DV	0.562[a]	0.502[a]	0.470[a]	0.529[a]	1	−0.037
RR	−0.332[a]	−0.077	0.007	−0.067	−0.037	1

a Correlation is significant at the 0.01 level (two-tailed).

Source: Authors' analysis (2023).

(Technology), TT (Convenience), QC (Service) on digital economic development while RR (Risk) has a negative impact on digital economic growth in Vietnam. Indeed, digital economic development can bring risks to consumers, especially risks coming from the security of private information, from the process of manipulating online services, or the ability to receive user's information.

7.4.5 Regression Results

Based on the analyses presented in this study, the dependent variable is the decision to use NCPs. The authors used SPSS 20.0 software to support the construction of multivariate linear regression equations. The regression method used here is the ordinary least squares method.

- HQ = mean(HQ1,HQ2,HQ3,HQ4,HQ5,HQ6)
- CN = mean(CN1,CN3,CN4,CN5,CN6)
- TT = mean(TT1,TT2,TT3,TT4,TT5)
- DV = mean(DV2,DV3,DV4,DV5)
- RR = mean(RR1,RR2,RR3,RR4,RR5)

The proposed model is: **SD = f(HQ, CN, TT, DV, RR)**

The results of Table 7.7 show that all variables (HQ, CN, TT, DV, RR) have an impact on the dependent variable with the t-test Sig. of each independent variable being less than 0.05. In addition, the variance inflation factor (VIF) coefficients of the independent variables are all less than 5, concluding that no multicollinearity occurs. Further, the standardized Beta coefficients of the independent variables are all greater than 0 and reach the Sig. value and is also less than 5%. This proves that the factors in the research model all have a positive influence on the decision to use NCP with a significance level of 95%.

Table 7.7 Regression results

Model	Unstandardized Coefficients		Standardized Coefficients			Collinearity Statistics	
	B	Standard error	Beta	t	Sig.	Tolerance	VIF
1 (Constant)	1.804	0.196		9.200	0.000		
HQ	0.177	0.040	0.227	4.429	0.000	0.577	1.734
CN	0.120	0.042	0.143	2.889	0.004	0.617	1.621
TT	0.334	0.039	0.417	8.589	0.000	0.643	1.554
DV	0.117	0.039	0.150	3.029	0.003	0.616	1.623
RR	-0.212	0.029	-0.282	-7.201	0.000	0.988	1.012

a. Dependent variable: SD.

Source: Authors' analysis (2023).

The results are as follows: the factor that has the strongest impact on the user's decision to use NCP is convenience, followed by the risk factor with a negative value, showing that users are concerned about risk. The risk of using non-trading methods is quite high, affecting their decisions. According to Bauer (1960), the "risk perception factor for user behavior" was first proposed. He believes that "beliefs about perceived risk are also a determining factor in consumer behavior and it is also an important factor affecting the transformation of human behavior." The remaining factors include efficiency, technology, and service in descending order, but all have a certain impact on users' usage decisions.

Table 7.8 and Table 7.9 indicate that factors have an impact on the decision to use NCP in the case of Vietnam, in which factors that have an impact in descending order such as service, technology, efficiency, risk, and

Table 7.8 The impact levels of factors

Number	Factor	Abbreviation	Order of impact level (descending)
1	Convenience	TT	5
2	Risk	RR	4
3	Efficiency	HQ	3
4	Technology	CN	2
5	Service	QC	1

Source: Authors' analysis (2023).

Table 7.9 Analyzing for the hypotheses

		Results	
No.	Hypotheses	Sig.	Conclusion
H1	The Efficiency positively impacts the decision to use non-cash payment.	0.000 (<0.05)	Accept the hypothesis at the 5% significance level.
H2	The Technology positively impacts the decision to use non-cash payment.	0.004 (<0.05)	Accept the hypothesis at the 5% significance level.
H3	The Convenience positively impacts the decision to use non-cash payment.	0.000 (<0.05)	Accept the hypothesis at the 5% significance level.
H4	The Service positively impacts the decision to use non-cash payment.	0.003 (<0.05)	Accept the hypothesis at the 5% significance level.
H5	The Risk negatively impacts the decision to use non-cash payment.	0.000 (<0.05)	Accept the hypothesis at the 5% significance level.

Source: Authors' analysis (2023).

convenience. This reflects the decision not to use cash for reasons ranging from service quality, technology quality, efficiency, risk, to convenience. This explains the need for Vietnam in particular and other countries in general to improve service quality, technology, efficiency, risk, and convenience to increase digital economic development. In fact, Duc et al. (2023) indicate that the scale of Vietnam's digitalized economy grew significantly in the period of 2007–2019, and accounted for 4.90% gross domestic product (GDP) to 11.56% GDP. Therefore, the digital spillover effect in Vietnam has significantly changed so far.

7.5 CONCLUSIONS AND FUTURE SCOPE

With this study, we present factors that affect NCP in Vietnam and these factors, through surveys and data, have proven to be meaningful to users' decisions and usage behavior.

As for the convenience factor, most users know and favor NCP because of the convenience, and it has the strongest impact on people's decision to use it. Therefore, to increase the number of users of NCPs, it is necessary to disseminate more widely about this feature and highlight the benefits of using NCPs such as not having to move to the place where people need to pay. In addition, users can make transactions at home, at work, at a coffee shop, even abroad, with just a smartphone, space—location is no longer an issue in online payments. It does not take much time when customers have to work too hard in moving to the bank, waiting for their turn to make transactions, presenting requests to bank staff, filling out information on the request form, waiting for the teller to confirm the payment, submitting to the controller to approve the transaction, and returning documents to the customer to complete a transaction. Further, people can actively make transactions at any time of the day; this is suitable for everyone, especially those who have to go to work during office hours, at the same time as the bank's business hours. Banking transactions mean interruption of work at work or personal matters; you can manage personal and business money without manual management. The money in the account is always transparent without fear of errors or losses in counting. In addition, the added convenience is that ATM users do not need to carry money with them, they do not worry about losing cash, and have peace of mind when their accounts have a layer of password security.

Regarding risk factors, in addition to communicating to the people, there needs to be specific instructions on the level of safety when using noncommercial payment methods. Most people are concerned and the risk has a significant impact on their decision to use NCP methods. In addition, fear of losing money when not proactively making transactions; or the wrong transaction, not the right object or amount of money to be transacted. In fact, this is a subjective factor, requiring users of NCPs to operate correctly as a prerequisite, and other objective factors will be determined by service providers such as banks, financial institutions, and intermediary channels. Complete payment of their risk management process. These organizations must be responsible when there are risks arising from their own service providers, not only responsible to customers (users of NCPs), but also responsible before the law and the State bank—the direct management agency.

Efficiency is the third factor out of five factors affecting users' decision to use NCP methods. By participating in NCP centers, survey subjects have seen the outstanding efficiency between online and cash payments, especially large value transactions with remote partners, and the use of cash payments becomes bulky, heavy, and lacks safety in transportation. In addition, NCP helps users manage income and expenditure more effectively down to the odd number, easily querying statements without needing to keep records. Buying and selling directly or indirectly through e-commerce platforms or businesses doing business with many partners becomes easier and faster when diversifying payment methods, no longer 100% dependent on cash.

As regards technology factor, today, young people's access to technology has become easy because of the remarkable progress of science and technology. Along with intellectual development, children are able to know the information technology, smartphones, and other electronic devices. Not only that, for older people who have become accustomed to advanced technology, it is no longer rare to see images of elderly people using phones, tablets, laptops, etc. Depending on the level of acceptance and passionate about technology, Vietnamese people have made certain progress and information technology has become a factor influencing local users' decision to use NCP methods.

The service factor is the factor that has the lowest impact on the decision to use NCPs of Vietnamese people, which proves that the services accompanying NCPs are not very important from the user's point of view. The criteria of convenience, efficiency, and technology have almost satisfied users' expectations, so they do not place much importance on accompanying services, although this factor still holds a certain significance in the decision of non-custodial payment center users.

Theoretically, the research contributes academically to the province in developing and promoting larger-scale research projects on NCP in Vietnam. Through data analysis and regression results, a partial picture of payment activities has been shown about the influence of the independent variable on the dependent variable.

In terms of practice, the study identifies five factors that influence users' decisions to use NCP methods. These typical factors have given certain significance in influencing local people's decision to use or not use non-commercial payment methods. In addition, the study has partly communicated and promoted non-commercial payment activities in the community, when giving questionnaire information to survey respondents with diverse occupations, living areas, ages, qualifications, and income. With the proposed recommendations, the authors hope to quickly become an effective, convenient, and popular payment channel in Vietnam.

The authors propose a number of recommendations to develop NCP centers in Vietnam, specifically:

i. Encourage businesses to pay salaries to employees via card accounts. Every business makes salary payments via banks, electronic salary payments, or at least manual salary reports from card account to card account for small and medium-sized businesses. This synchronizes digitization on a large scale to help workers and laborers in the area all have cards and be able to use payment cards (ATM cards).
ii. Promote NCP in two key and most important sectors of every country, which are health and education. Recommending that patients with NCP receive medical examination and treatment at all medical facilities in the provinces. On the other hand, school tuition and fees are also completely replaced by non-commercial payment methods.
iii. For businesses, business units such as supermarkets, restaurants, eateries, small businesses, business households inside and outside the market, units that declare and pay taxes should thoroughly implement this through channels as online, electronic payment. It is evident that the customers have the conditions to perform non-commercial payment when using services or purchasing goods at these places.
iv. Expenditures and revenues from the state budget for people in the provinces are encouraged to be spent through NCP such as social insurance payments, unemployment benefits, maternity support, pensions, and benefits. Supporting poor households, spending on policy family support, spending on scholarships, collecting taxes, collecting fees, etc. all use NCP so that the cash flow of transactions goes through the account; this will stimulate people to use it directly or indirectly with NCP.
v. Increase the ability to manage risks in the digital economy, especially attacks on network infrastructure that negatively affect consumers. Therefore, countries need to improve legal institutions in the new context, while improving risk management capacity to improve safety in transactions and in the economy.

REFERENCES

Bauer, R.A. (1960). Consumer Behavior as Risk Taking. In: Hancock, R.S. (ed.), Dynamic Marketing for a Changing World, Proceedings of the 43rd Conference of the American Marketing Association, pp. 389–398.

Bui, T.K.H., Pham, T.A.T., Tran, T.N.T, Nguyen, A.T., Ngo, T.T.V. (2022). Factors affecting the acceptance of non-cash payments by consumers in Ho Chi Minh City. Journal of Corporate Finance, 9, 22–25.

Busse. (2020). Cash, Cards or Cryptocurrencies? A Study of Payment Culture in Four Countries. 2020 IEEE European Symposium on Security and Privacy Workshops (EuroS&PW), pp. 200–209.

Duc, D.T.V., Dat, T.T., Linh, D.H., & Phong, B.X. (2023). Measuring the digital economy in Vietnam. Telecommunications Policy, 102683.

Hair, J.F., Black, W.C., & Babin, B.J. (2006). Multivariate Data Analysis. 6th edition. Upper Saddle River, NJ: Prentice Hall.

Harasim, J. (2016). Europe: The Shift from Cash to Non-Cash Transactions. In: Górka, J. (eds) Transforming Payment Systems in Europe. London: Palgrave Macmillan Studies in Banking and Financial Institutions, Palgrave Macmillan. https://doi.org/10.1057/9781137541215_2

Ho, H.P.C., Nguyen, K.T., & Quach, D.T. (2022). Factors affecting the choice of cashless payment in Vietnam. Journal of Information System and Technology Management, 15–23, 128–1666.

Nguyen, T.T.H. (2021). Non-cash payment in Vietnam: Current situation and solutions. Journal of Economics and Forecasting, 22, 21–24.

Roubini ThoughtLab. (2018). Cashless Cities: Realizing the Benefits of Digital Payments. https://thoughtlabgroup.com/visa-cashless-cities-realizing-benefits-digital-payments/, accessed on June 14, 2024.

Tabachnick, B.G., & Fidell, L.S. (2007). Using multivariate statistics (5th ed.). Allyn & Bacon/Pearson Education.

Widjaja, E.P.O. (2016). Non-cash payment options in Malaysia. Journal of Southeast Asian Economies, 33(3), 398–412.

Chapter 8

Machinery to Mind
Navigating the Transformation from Industry 1.0 to Industry 5.0

M. C. Shanmukha, K. C. Shilpa, and A. Usha

8.1 INTRODUCTION

Industry has been an extended and exciting path of development, with new manufacturing inventions and advancements appearing at every turn. Every industrial revolution, starting with Industry 1.0 and extending with the approaching Industry 5.0, has been distinguished by a distinct set of advancements as shown in Figure 8.1 (Yasemin, 2022).

Over a period of technical advancement, the industrial landscape has changed dramatically over time. Industries constantly changed, from the beginning of Industry 1.0, which was defined by mechanization and steam power, to the present Industry 4.0, which is centered on automation, data interchange, and artificial intelligence. This chapter examines the path from Industry 1.0 to the emerging frontier of Industry 5.0, emphasizing the potential and difficulties that come with such a rapid shift. Industry 5.0 aims to create a comprehensive manufacturing strategy by showcasing a future where technology and humans work together harmoniously. Innovation, sustainability, and resilience are combining in this exciting time.

8.2 INDUSTRY 1.0: THE ERA OF MECHANIZATION

The 18th century saw the start of Industry 1.0, which was the shift from a handicraft economy to a manufacturing sector that made use of new technology and machinery. The invention of steam power, which raised output and volume, was the main driving force behind this revolution. Examples include the invention of the flying shuttle in 1733 to facilitate the process of weaving cloth, the construction of the first textile mill in the United States in 1790, and the patenting of the cotton gin in 1794.

The pivotal components of Industry 1.0 marking the beginning of mechanization and the first industrial revolution are steam engines and water wheels. The idea behind steam engines is the transformation of thermal energy into mechanical work. Steam engines and water wheels played a

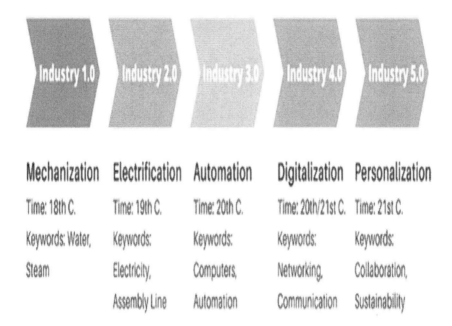

Figure 8.1 Evolution of industrial revolution.

crucial role in the development of Industry 1.0. Because they supplied the energy required to ignite mechanized manufacturing and industrialization. The modern workplace was ultimately shaped by these advances in technology that served as the foundation for further industrial revolutions.

8.2.1 Limitations of Industry 1.0

- **Human-centric workforce:** Human labor was used to do the majority of jobs manually. Workers and craftsmen engaged directly in the production processes, often doing laborious and repetitive jobs.
- **Simple mechanization:** Even while mechanization increased significantly with the advent of steam engines and water wheels, the machinery employed remained relatively simple.

- **Lack of connectivity:** There was a lack of real-time coordination or communication between factories or between various stages of production, and factories worked as largely distinct units.
- **Technological constraints:** The level of accuracy of early machines was restricted and it impacted the consistency and reliability of mass production of goods. The care and maintenance of early machines caused significant problems. Production was frequently interrupted by failures and ineffectiveness.

8.3 INDUSTRY 2.0: THE AGE OF MASS PRODUCTION

The establishment of assembly lines and electricity production in the 19th century represented the beginning of the Second Industrial Revolution. The latter, period spanning from 1870 to 1914, introduced existing structures like railroads and telegraphs into new industries. Perhaps the most distinctive feature of the era was the development of mass manufacturing as the primary avenue of total production. Factory electrification significantly improved the manufacturing level (Sharma and Singh, 2020). Railroads became more economically integrated as a result of heavy steel production, which paved the way for mass production (Linderman et al., 2006). Workforce organization innovation was successful in the identical manner as mechanical innovation. Division of labor was also implemented during this time, and assembly line mass production gained popularity. Frederick Taylor, an American mechanical engineer, pioneered the use of labor analysis techniques to maximize workplace and employee productivity. Automation and technological breakthroughs were greatly accelerated during Industry 2.0, which was defined by the extensive use of electricity. However, it possesses certain drawbacks and difficulties.

8.3.1 Limitations of Industry 2.0

- **Limited flexibility in production:** Even though Industry 2.0 used electricity to enable mass production, the procedures were often inflexible and required a lot of time and work. Factories were designed to produce a single product on a vast scale; therefore it became difficult to adjust for variations in the product or changes in demand. This made it more difficult to effectively produce smaller batches or react quickly to changes in the market.
- **Centralized control and decision-making:** Because decisions were made at the highest levels, Industry 2.0's centralized command structures resulted in delayed reactions to local issues and shifts in the market. At the operational level, this hierarchical approach limited innovation and problem-solving by restricting worker empowerment.

- **Standardization challenges:** Even with Industry 2.0's effort for standardization, it was difficult to guarantee total consistency amonggoods. The supply chains dependability was impacted by product variations and reliance on standardized items, which made it difficult to manage variations in raw materials and supplier capabilities.
- **Environmental impact:** Industry 2.0 significantly advanced the use of more electricity, but sustainability was not given importance, and there was a general lack of awareness of the effects on the environment. Large amounts of trash were produced by mass industrial methods, which increased resource depletion and pollution.

8.4 INDUSTRY 3.0: THE RISE OF COMPUTERS

The digital revolution, also known as Industry 3.0, began in the 1970s with the development of computers and memory-programmable devices. An age marked by the broad adoption of information technology and the worldwide interconnection of economies was brought in by the Third Industrial Revolution, or Industry 3.0. Mass manufacturing and the application of integrated circuit chips, or digital logic, are the key points of this specific phase; related technologies include computers, digital cell phones, and the internet (Pathak et al., 2021; He et al., 2017). Technology advancements are changing not only company processes but also traditional products. Technology is being transformed into digital format by the digital revolution.

Through improved economic interconnectedness and smooth cross-border flows made possible by developments in trade agreements, communication, and transportation, the Information Age hastened globalization. The expansion of information and communication technologies, such as the internet, made it possible for people to communicate globally in real time, share data, and conduct cooperative economic activities. With the transition from analog to digital technologies during the Digital Revolution, processing power and data storage capacity increased dramatically, stimulating innovation in industries like entertainment and telecommunications.

8.4.1 Limitations of Industry 3.0

- **Interoperability challenges:** Compatibility problems frequently result from the integration of diverse technologies, such as robots, Programmable Logic Controllers (PLCs), and computer systems, making it difficult for different systems to communicate with each other smoothly.
- **High implementation costs:** Some organizations, particularly smaller ones, faced financial difficulties when they had to make large upfront investments in new hardware, software, and employee training to implement Industry 3.0 technology.

- **Cyber security concerns:** Industries have become increasingly vulnerable to cyber threats such as illicit activity, subversion, and data breaches due to the extensive use of networked systems and the internet.
- **Dependency on technology:** As industries grew more dependent on networked systems, major downtime, and output losses might result from technological disruptions such as system breakdowns or cyber-attacks.

8.5 INDUSTRY 4.0: THE ERA OF SMART MANUFACTURING

The industrial processes that make use of more advanced computers, robots, and communication technologies are referred to as Industry 4.0. An Internet of Things (IoT) network, which is a collection of devices that share and process data, is now connected to automated and robot-assisted processes. Industry 4.0 requires transforming ordinary machines into self-aware, self-learning machines to enhance their overall functionality and maintenance management in conjunction with external interactions (Lee et al., 2014). Industry 4.0 improves supply and demand processes, adaptability, and resource efficiency. As a result, cities, industries, and potentially intelligent machinery and items become smarter.

Organizations, workers, and society at large will all see significant changes as a result of the Fourth Industrial Revolution. Employees perceptions of work and its meaning are evolving, and organizations now need to strike a balance between the benefits of technology innovation, new job categories, and employment models, and local and global power dynamics.

8.5.1 Limitations of Industry 4.0

- **Adverse effects of exchanging data in a competitive setting:** As Industry 4.0 is adopted, internet connectivity will facilitate data interchange through end-to-end, vertical, and horizontal integration.

This suggests two problems.

  - The high level of transparency could lead to hazards like industrial spying and cyber-attacks, as well as other issues like access and data rights.
  - The companies that build platform standards would impede other companies that have formed Unique Selling Propositions (USPs) and might even force them out of business (Zhou et al., 2017).
- **Requirement for highly qualified workers:** The adoption of Industry 4.0 will necessitate the requirement for workers with several skill

sets. The availability of employees with these skills will be a challenge (Bonekamp and Sure, 2015).
- **Initial cost is high:** When Industry 4.0 is implemented in a company, end-to-end, vertical, and horizontal integration will occur (Wang et al., 2016). A significant initial financial and temporal investment is necessary to design and build the architecture in accordance with business needs.

8.6 THE EMERGENCE OF INDUSTRY 5.0

Industry 5.0 is bringing about a paradigm shift and resolution by placing less emphasis on technology and presuming that human–machine interaction is the cornerstone for success. Employing specific goods is one way that the industrial revolution increases consumer satisfaction. Due to recent advances in technology, Industry 5.0 is required in todays marketplace to give manufacturing both competitive advantages and economic growth (Adel, 2022). According to the industrial revolution 5.0, human–machine interaction is increasing the efficiency of industrial output. The manufacturing sectors increasing productivity can be attributed to both human labor and universal robotics (Majumdar et al., 2021).

Industrial 5.0 is taking the place of previous innovations, and because of its extreme precision and the reduction of human labor time and effort through machine work, it is an efficient process (Adadi and Berrada, 2018). Because it necessitates accurate monitoring and repair of the malfunctions in the smart equipment, predictive maintenance is necessary for smart sensors, the IoT, and customized software. The machines will probably break down subsequently, but maintaining it will stop it from arising (Ali et al., 2022; Sherburne, 2020). Local production, in conjunction with fresh initiatives, is what sustains economies (Javaid and Haleem, 2020; Saraswat, et al., 2021). Corporate technologies are redefining the trend with Industry 5.0. It causes the development of sustainable policies, such as waste management and minimal waste generation, which can increase the effectiveness of businesses.

Collaborative robots perform hazardous and repetitive activities, while people focus on creativity and efficient business solutions (Yu et al., 2018). These skills boost business productivity and motivate staff to put in long hours and generate high-caliber work. A human-centered approach prioritizes the needs of people over the needs of the industrial process. Producers need to acknowledge the benefits that technology can provide to people and concentrate on how it can fulfill the needs of workers, not the other way around. Technology needs to solve privacy and autonomy concerns. Innovations in technology are not allowing for the kind of personalization that customers want. Workers are part of Industry 5.0, which can benefit from the promise of technology (Babamiri et al., 2019). It looks for methods to deliver fresh concepts that may result in the creation of individualized products.

8.6.1 Innovations and Modernizations

Industry 5.0 is being embraced by a number of industries, including healthcare, manufacturing, textiles, education, and food. The products are discussed at Bundesgartenschau (Llorens Duran et al., 2019), a woolen pavilion that showcases the robot hand made by a joint venture of the enterprises. The KR 500 FORTEC robot is the item being used. It can perform a wide range of architectural tasks, including securing down pieces, assembling them, and letting the robots cooperate.

The majority of industries are moving toward the smart social factory by implementing Industry 5.0. As the project of the month, Repsol's intelligent management initiative was chosen to enhance comprehension of Industry 5.0 ideas. The company uses robotic process automation and blockchain technologies to increase business security and productivity. Repsol's first cobot, the automated guided vehicle, is utilized for logistics jobs like lab visualization, raw material transfer from the warehouse, and waste deposition. Repsol works on the Block Lab project, where the business uses a blockchain feature to transfer sensitive data (Li, 2020). The goal of the project is to manage 10,000 samples of safety-related material annually in an efficient manner.

8.6.2 Applications of Industry 5.0

Smart hospital: Building a real-time, smart hospital is the aim of Industry 5.0. Technology can provide options for remote monitoring in the healthcare sector. It is crucial to raise the standard of living for doctors. This intelligent healthcare device allows doctors to focus on treating infected patients during the COVID-19 epidemic while also providing effective data for better treatment. Additionally, it helps medical students get the training they need in case of a COVID-19 epidemic. Machine learning (ML) is used in medical imaging, natural language processing, and genetic data (Nahavandi, 2019). The primary areas of focus are illness diagnosis, detection, and prediction.

Manufacturing industry: Industry 5.0 is recognized as a new paradigm for production that places a strong emphasis on human–machine collaboration. Industry 5.0 is all about using the collective intelligence of human inventiveness with ever more accurate machinery. Reusing and recycling resources is how manufacturing develops sustainable processes. It is also essential to lessen the harmful environmental effects of the manufacturing industry. Through additive manufacturing, greater customization is required to optimize waste and resource efficiency. Industry 5.0 is removing repetitive tasks from human labor, revolutionizing production processes globally.

Supply chain management: The value of human and more intelligent machine cooperation such as COBOTS is emphasized by supply chain 5.0 (Zambon et al., 2019). Industry 5.0 aims to meet customer needs for hyper-personalization and hyper-customization, which necessitate a fusion of machine intelligence and human creativity. For supply chain management with large production volumes and standardized processes, robots are essential. This presents a difficulty as each product needs the right guidance in order for the robots to operate as intended (Wang et al., 2020).

8.6.3 Technologies of Industry 5.0

The technologies used in Industry 5.0 include cloud computing, blockchain, big data analytics, IoT, and sixth-generation (6G) networks.

Cloud computing: Cloud computing is the supply of computing services, including networks, databases, software, and intelligence analytics, among others. This technique provides economies of scale and innovative efficiency. This technology stores and manages data on remote servers over the internet, from which users can access the data. It provides apps, storage, and processing power on demand for computer needs. The industrial cloud is the name of the virtual environment that provides support for industry applications. Cloud providers create apps like IoT monitoring tools that are utilized online and on mobile devices. The cloud also makes it easier to employ Application Programming Interfaces (APIs), which automate the process of normalizing data from many sources that are used in data production.

Collaborative robots: The need for human involvement to expand on previous iterations is one of the main effects of Industry 5.0 and collaborative robotics. Industry 5.0 and collaborative robots signal a new era in robotics plus production. Industry 5.0 seeks to restore human interaction in development and manufacturing by granting human operators access to robots sophisticated functionality and capacity for heavy lifting. Crucial activities can be completed by humans with great skill. Human–robot and human–machine collaboration are described in Industry 5.0. By making use of cutting-edge technology like the IoT, robots are increasing human productivity. It enhanced business productivity and added a human touch to automation in Industry 4.0.

Analysis of big data: The utilization of three-dimensional symmetry in innovation ecosystem architecture is made possible by the ground-breaking technology referred to as Industry 5.0. Big data analytics is the process of examining vast amounts of data to identify patterns, market trends, and other information (Matheus et al., 2019). It employs a potent analytical method with both organized and semi-structured data, among other sorts of data. Using traditional technologies to handle and store massive data amounts is necessary. It is used as real-time data to enhance the competitive advantages of the business sector, with a focus on potentially providing recommendations for predictive discoveries. The organization employs discrepancies to find big data analytics using a list of the issues primary causes. Most businesses employ big data analytics to inform their strategic decisions. Even with big data, Industry 5.0 faces difficulties if comprehensive manufacturing cycle data is not collected.

Blockchain: The digital ledger uses records referred to as blocks to record the transaction data; this is a distributed, decentralized system. It is a shared ledger that makes keeping track of assets in the company network and documenting transactions easier. The information powers the business. As a result, blockchain technology provides data by storing finished and shared information in an immutable ledger that is accessible to network participants. Blockchain technology helps customers by enabling them to track orders, payments, production, and other related details. The network participants have dispersed ledger records of transactions to avoid redundant work and records in the database system. To speed up transactions, a smart contract is deployed on a blockchain and designed to execute automatically. These are referred to as corporate terms and conditions, and they include things like paid travel insurance.

Beyond 6G: It is a 6G standard for developing wireless technologies that communicate with mobile data networks. According to Raj et al. (2020), 6G organizations are anticipated to be significantly more variable than their predecessors. It is highly probable that they will provide applications that surpass the current boundaries of portable use, such as always-present instant messaging, Virtual Reality and Augmented Reality (VR/AR), and pervasive IoT information. Flexible organization managers will typically accept flexible decentralized strategies for 6G, with innovations in short-range communication, artificial intelligence, flexible edge processing, and blockchain enabling close authorization, spectrum sharing, infrastructure sharing, and intelligent automated management. 6G networks are expected to satisfy the standards of the intelligent information society and offer ultrahigh reliability for industry 5.0.

8.6.4 Limitations of Industry 5.0

1. **Need for skilled workers:** Industry 5.0 requires new knowledge and skills from its personnel. Human operators must change as technology advances to manage more difficult jobs and work well with automated systems.
2. **Data and system security:** Industry 5.0 is largely dependent on networked systems and data. It is essential to guarantee the integrity and security of this networked infrastructure. There are serious hazards associated with cyber security attacks, data breaches, and system weaknesses.
3. **Ethical implications of artificial intelligence:** Ethical questions are raised by Industry 5.0's use of AI. These include possible discrimination, prejudices, and privacy concerns related to AI algorithms. Therefore, it becomes crucial to strike a balance between automation and human supervision.
4. **Transition challenges:** It can be difficult to transition from current production systems to Industry 5.0 because of things like the lack of autonomy in existing systems and the difficulty in obtaining high-quality data. Supporting heterogeneous data sources is still a challenge.

8.6.5 Future Direction

Human–machine interaction is the term used to describe communication between humans and machines via a user interface. As they enable people to operate machines using instinctive and natural behaviors, natural user interfaces, such as gestures, are utilized to attract attention. It is the way that Industry 5.0 will go in the future as it keeps people at the center of the system and allows for the integration of new technologies. People can even learn more about people's motives and behaviors through the user interface. Quantum computing is a type of computing that performs computations by utilizing the collective properties of quantum states, such as interference coherence. Quantum computing is carried out using those devices referred to as quantum computers. It is performing calculations based on the probability of the objects state before measurement.

8.7 CONCLUSION

Our current state, known as Industry 5.0, will see machines becoming sophisticated enough to carry out complicated tasks on their own. In tandem with people, these will use state-of-the-art technology and computer power to deliver precision and speed. Industry 5.0 is predicated on finding the optimal balance between human ingenuity and robotization, fusing human inventiveness with the precision, intelligence, and exactness of technology. The most recent phase has a strong emphasis on human-centricity, resilience, and sustainability. By creating cutting-edge technologies with the needs of people in mind, Industry 5.0 empowers employees rather than replaces them. It also supports the sustainability and resilience of the industry.

In conclusion, the transition from Industry 1.0 to Industry 5.0 presents an infinite number of possibilities and challenges for various businesses. A precise compromise between employing cutting-edge technologies and ensuring that an approach is human-centric is required to navigate this transition. Businesses can thrive in the rapidly changing setting of Industry 5.0 and contribute to a prosperous and equitable future by embracing forward-thinking initiatives and learning from the past.

REFERENCES

1. Sharma, A., & Singh, B. J. (2020). Evolution of industrial revolutions: A review. *International Journal of Innovative Technology and Exploring Engineering*, 9(11), 66–73.
2. Linderman, K., Schroeder, R. G., & Choo, A. S. (2006). Six Sigma: The role of goals in improvement teams. *Journal of Operations Management*, 24(6), 779–790.
3. Pathak, A., Kothari, R., Vinoba, M., Habibi, N., & Tyagi, V. V. (2021). Fungal bioleaching of metals from refinery spent catalysts: A critical review of current research, challenges, and future directions. *Journal of Environmental Management*, 280, 111789.
4. He, D., Ma, M., Zeadally, S., Kumar, N., & Liang, K. (2017). Certificateless public key authenticated encryption with keyword search for industrial internet of things. *IEEE Transactions on Industrial Informatics*, 14(8), 3618–3627.
5. Lee, J., Kao, H. A., & Yang, S. (2014). Service innovation and smart analytics for Industry 4.0 and big data environment. *Procedia CIRP*, 16, 3–8.
6. Zhou, W., Piramuthu, S., Chu, F., & Chu, C. (2017). RFID-enabled flexible warehousing. *Decision Support Systems*, 98, 99–112.
7. Bonekamp, L., & Sure, M. (2015). Consequences of Industry 4.0 on human labour and work organisation. *Journal of Business and Media Psychology*, 6(1), 33–40.
8. Wang, S., Wan, J., Li, D., & Zhang, C. (2016). Implementing smart factory of Industrie 4.0: an outlook. *International Journal of Distributed Sensor Networks*, 12(1), 3159805.
9. Adel, A. (2022). Future of Industry 5.0 in society: human-centric solutions, challenges and prospective research areas. *Journal of Cloud Computing*, 11(1), 40.
10. Majumdar, A., Garg, H., & Jain, R. (2021). Managing the barriers of Industry 4.0 adoption and implementation in textile and clothing industry: Interpretive structural model and triple helix framework. *Computers in Industry*, 125, 103372.

11. Adadi, A., & Berrada, M. (2018). Peeking inside the black-box: A survey on explainable artificial intelligence (XAI). *IEEE Access*, 6, 52138–52160.
12. Ali, M. H., Issayev, G., Shehab, E., & Sarfraz, S. (2022). A critical review of 3D printing and digital manufacturing in construction engineering. *Rapid Prototyping Journal*, 28(7), 1312–1324.
13. Sherburne, C. (2020). Textile industry 5.0? Fiber computing coming soon to a fabric near you. *AATCC Review*, 20(6), 25–30. https://doi-org.login.ezproxy.library.ualberta.ca/10.14504/ar.20.6.2
14. Javaid, M., & Haleem, A. (2020). Critical components of Industry 5.0 towards a successful adoption in the field of manufacturing. *Journal of Industrial Integration and Management*, 5(03), 327–348.
15. Saraswat, V., Jacobberger, R. M., & Arnold, M. S. (2021). Materials science challenges to graphene nanoribbon electronics. *ACS Nano*, 15(3), 3674–3708.
16. Yu, M., Lou, S., & Gonzalez-Bobes, F. (2018). Ring-closing metathesis in pharmaceutical development: fundamentals, applications, and future directions. *Organic Process Research & Development*, 22(8), 918–946.
17. Babamiri, B., Bahari, D., & Salimi, A. (2019). Highly sensitive bioaffinity electrochemiluminescence sensors: Recent advances and future directions. *Biosensors and Bioelectronics*, 142, 111530.
18. Llorens Duran, J. I. D. (2019). TensiNet Symposium 2019: Softening the habitants: Report. *TensiNews*, 37, 16–23.
19. Li, L. (2020). Education supply chain in the era of Industry 4.0. *Systems Research and Behavioral Science*, 37(4), 579–592.
20. Nahavandi, S. (2019). Industry 5.0—A human-centric solution. *Sustainability*, 11(16), 4371.
21. Zambon, I., Cecchini, M., Egidi, G., Saporito, M. G., & Colantoni, A. (2019). Revolution 4.0: Industry vs. agriculture in a future development for SMEs. *Processes*, 7(1), 36.
22. Wang, S., Wang, H., Li, J., Wang, H., Chaudhry, J., Alazab, M., & Song, H. (2020). A fast CP-ABE system for cyber–physical security and privacy in mobile healthcare network. *IEEE Transactions on Industry Applications*, 56(4), 4467–4477.
23. Matheus, L.E., Vieira, A.B., Vieira, L.F., Vieira, M.A., & Gnawali, O. (2019). Visible light communication: Concepts, applications and challenges. *IEEE Commun Surveys Tutor*, 21(4), 3204–3237.
24. Raj, V., Shim, J. J., & Lee, J. (2020). Grafting modification of okra mucilage: Recent findings, applications, and future directions. *Carbohydrate Polymers*, 246, 116653.
25. Yasemin, www.sente.vc/post/industry-5-0-how-did-we-get-here-a-look-at-industry-1-0-5-0

Chapter 9

The Role of Cutting-Edge Technologies in Revolutionary Industry 5.0

Abhay Bhatia

9.1 INTRODUCTION

Individuals need the right advanced abilities to secure positions and catch open doors. The suggestion is the premise of the European Association's Computerized Single Market Technique, which is essential for an expansive setting pointed toward advancing the unification of European business sectors, with a specific spotlight on advanced innovations. The key design is to advance and permit the most extreme double-dealing of the computerized innovations, seen as an instrument for development. Getting ready for the eventual fate of work is one of the characterizing industry difficulties within recent memory. In seeking this goal, two bearings of mediation can be recognized. One is connected to the business world, with interests in development, and the other is connected to the specialists, with mediation connected with expanding the degree of advanced skill and the development of an undeniably comprehensive society. This progress to progressively shrewd modern frameworks went through the execution of frameworks for gathering and deciphering large information, distributed computing, man-made consciousness, and advanced mechanics. Along these lines, more organizations moved to Industry 4.0, which is based on another age of interconnected machines and brings more elevated levels of robots, independent cycles and hardware, and information trade in assembling. In fact, industry 4.0 is fueled by propels in man-made consciousness (computer-based intelligence), mechanical technology, added production as well as manufacturing, and the web of Internet of Things (IoT).

The landscape of technologies and production departments has undergone significant transformations since the early days of Industry 4.0, leading to the emergence of what we now refer to as Smart Factory. The evolution and integration of technologies have positioned them as key players, yet it has become evident that the human element is indispensable for effective management and optimal outcomes. The development of intelligent machines leveraging the collaborative dynamics between humans and machines offers an opportunity to augment human labor with new robotic and artificial intelligence (AI) tools rather than replacing it outright (Directorate-General for the Internal Market, 2017).

This ethos also propels the forthcoming fifth industrial revolution, known as Industry 5.0, which emphasizes the symbiotic relationship between humans and machines (Directorate-General for Research and Innovation & Vanderborght, 2020). Industry 5.0 expands upon the principles of Industry 4.0, highlighting a human-centric collaborative economy within a world facing population growth, resource constraints, and increasing interconnectedness. Future factories will need to adapt their organizational structures and manufacturing systems to provide workers with more meaningful, valuable, and healthy roles.

A recent vision report from the Industry 2030 high-level industrial roundtable positions the European industry as a global leader committed to delivering value responsibly for society, the environment, and the economy, embracing the concept of Society 5.0 (Directorate-General for Internal Market, Industry, Entrepreneurship and SMEs, 2019).

Society 5.0, unveiled by the Japanese government in 2016 as part of the 5th Science and Technology Basic Plan, envisions a perfect future society powered by scientific and technical breakthroughs and focused on the welfare of its people. To develop the "Super Smart Society," as promoted by Keidanren, the Japanese Federation of Enterprises, operations managers in the Society 5.0 period must find a balance between humans and computers, emphasizing holistic practices and social responsibility. Numerous issues pertaining to technology, socioeconomics, regulation, and governance are raised by this transition: Which abilities require improvement? Which rules need to be defined? What possible effects might AI have? What disputes might develop between AI and humans? (Paschek et al., s.d., 2020).

9.2 INDUSTRY 5.0 AS INDUSTRIAL REVOLUTION

9.2.1 About Industry 5.0

The merging of the real and digital worlds via cyber–physical systems and the IoT uniting people, machines, and gadgets is what defines the Fourth Industrial Revolution. Through the creation of innovative value networks and ecosystems, this horizontal and vertical integration crosses whole value chains, from suppliers to customers, across the product lifetime, and across many functional departments. In the era of Industry 4.0, technological developments allow for value creation that is more robust, flexible, traceable, personalized, high quality, and service-oriented. As a result, these developments are in a position to transition into a new generation characterized by a rise in the cooperation between humans and machines as well as the creation of industrial systems that place a high priority on human-machine interfaces.

During two virtual workshops hosted by the Directorate "Prosperity" of DG Research and Innovation on July 2 and 9, 2020, participants from research and technology companies as well as funding agencies around Europe discussed ideas that eventually led to the notion of Industry 5.0. By ensuring that production respects planetary boundaries and placing the welfare of industry workers at the center of the production process, Industry 5.0 can achieve societal goals beyond job creation and economic growth, as defined by the European Commission, which formally advocated for the Fifth Industrial Revolution. The talks emphasized the need to move the emphasis from individual technologies to a systemic approach and include European social and environmental concerns in technical innovation (Directorate-General for Research and Innovation & Müller, 2020)

Industry 4.0 advancements have facilitated a significant technological transition for organizations, benefiting both those already technologically advanced and those lagging behind. This transformation offers substantial potential for newcomers as well. With its focus on digitalization and transformation, Industry 4.0 can profoundly impact organizations across sectors (Figure 9.1).

In manufacturing, technologies like Additive Manufacturing and Autonomous Robots have enhanced efficiency, allowing machines in Smart Factories equipped with Industry 4.0 capabilities to perform various tasks without constant reprogramming. These digitally advanced factories not only demonstrate resilience but also contribute to sustainability. Similarly, in the service sector, AI presents opportunities to elevate customer service while reducing costs and response times.

Overall, Industry 4.0 serves as a critical facilitator, driving organizations toward novel business models and innovative value-creation methods. This digital transformation has significantly boosted productivity and competitiveness within industries. For instance, there is a shift from contemplating whether to adopt digital technologies to strategizing how to best implement them, with IoT being a prime example of enhancing business outcomes. Additionally, organizations are progressively transitioning from being cautious about cybersecurity technologies to taking proactive measures. However, the adoption of Industry 4.0 technologies necessitates adjustments in terms of personnel, processes, and organizational structures to fully harness their benefits. Despite these challenges, organizations are rapidly adapting and leveraging various Industry 4.0 technologies, resulting in notable service and product innovations (Figure 9.2).

9.2.2 Industry 5.0: The Cutting-Edge Technology

New cutting-edge technologies (CET) were predicted to have the biggest impact on international enterprises in 2020 based on data from Statista, as shown in Figure 9.1. The IoT, cloud computing, AI, and big data/

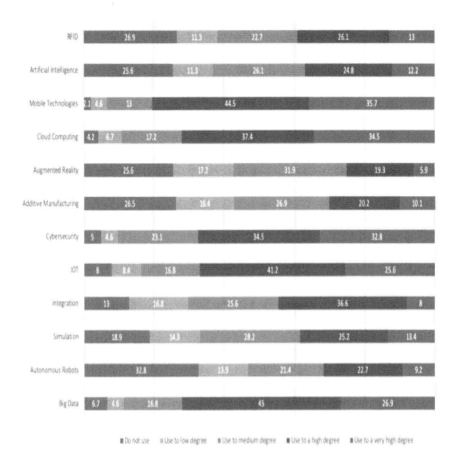

Figure 9.1 Industry 4.0 and its impact on organization before 2021 (Mansoor Ahmed et al. 2022).

analytics—collectively referred to as the "big four" technologies—are critical to Industry 4.0. The most influential Industry 4.0 technology in respondents' firms, according to approximately 72% of them, was the IoT. AI robots, which are widely used in a variety of industries to improve efficiency and supplement human capabilities, were found to be the second most influential technology, closely behind human capabilities (Statista, 2020).

Industry 5.0 shares many core technologies with Industry 4.0 but emphasizes human-centered technologies more strongly. While Industry 4.0 prioritizes smart technology in manufacturing, Industry 5.0 focuses on enhanced collaboration between humans and smart systems. It is important to note that Industry 5.0 is not a replacement or alternative to

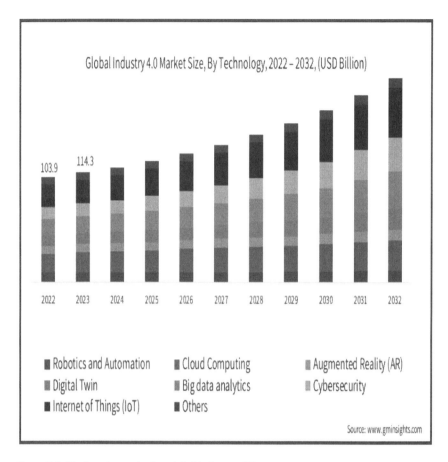

Figure 9.2 Market size to be for global industry 4.0.
(Source: www.gminsights.com)

Industry 4.0 but rather an evolution and natural progression. This paradigm shift underscores the notion that technologies should be oriented toward supporting values, with technological transformations designed in accordance with societal needs, rather than the other way around. This shift is particularly crucial as ongoing societal developments in the Fourth Industrial Revolution that reshaped how value is not for only creation but also used as exchangeable as well as distributed.

Instead of technologies driving societal changes, Industry 5.0 views technologies as components of systems intentionally created to empower ecological and societal ideals. For example, the main goal of technology should be to improve worker capabilities and provide safer, more satisfying work conditions rather than to replace humans on the shop floor. Even though new abilities are needed, everyone involved should see long-term benefits from the collaborative workplace. Atwell (2017) asserts that it can be beneficial to combine the rapid accuracy of industrial automation with the critical and cognitive thinking abilities of people.

Studies, such as one conducted by MIT in 2016, suggest that teams comprising humans and robots working collaboratively can be up to 85% more productive than teams composed solely of humans or robots. Additionally, Østergaard (2017), Chief Technology Officer at Universal Robots, highlighted that the next industrial revolution is essential to meet consumers' demand for personalized products. Human problem-solving skills, value-adding creativity, and uniquely human abilities are pivotal in deeply understanding customers' needs. AI, IoT, and collaborative robots (cobots) saw a sharp increase in investment as businesses realized the advantages of Industry 5.0. The General Directorate for Research and Innovation released a paper in February 2020 that discussed the potential benefits of collaborating robotics for society and the economy while maintaining European ideals. Accenture (NYSE: ACN) predicts that AI stands to significantly boost the Indian economy, potentially adding US$957 billion by reshaping work dynamics to generate favorable outcomes for both businesses and society. According to their report titled 'Rewire for Growth,' AI has the capacity to elevate India's annual growth rate of gross value added (GVA) by 1.3 percentage points, consequently enhancing the country's income by 15% by the year 2035. Accenture India highlighted AI's potential for transformation and compared its effects to the development of computing technology. Menon emphasized the significance of acting now and pointed out that AI has already shown promise in creating substantial socioeconomic benefits for India. She called for prompt investment, pointing out that AI may start a positive feedback loop with the correct resources, enabling people to contribute to long-term social and economic advancement. Specifically, an Accenture study evaluated AI's effects across many businesses. They took into account the GVA growth rate data, which is a good estimate of the GDP of a nation (Figure 9.3).

Despite the concept of sensors being in use for over 15 years, the IoT has recently gained significant traction due to a combination of factors such as decreased sensor prices, improved computational power, advancements in data connectivity, and enhanced machine-to-machine communication. Coined by Kevin Ashton in 1999, IoT represents a paradigm shift in computational technology, enabling devices to communicate with each other via the internet without the need for individual programming for each use case. This technology has particularly taken off in the manufacturing sector,

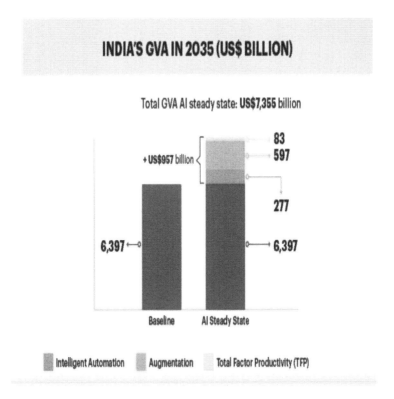

Figure 9.3 Potential impact of AI on real GVA by 2035 in India.
(Source: Accenture 2017)

paving the way for the Industrial Internet of Things (IIoT) and driving the next industrial revolution.

The latest advancements in connected devices, facilitated by smart sensors, have enabled the real-time exchange of data, which is crucial for the digitalization of manufacturing. By connecting various devices across the value chain, these sensors have led to the convergence of the previous. A 2020 survey by Plataine, a provider of IIoT and AI optimization solutions, revealed that IoT adoption in the manufacturing industry tripled compared to 2018. Additionally, 66% of respondents identified IoT as a key technology for their company's future success and profitability. According to the World Economic Forum (WEF), investment in IoT for production doubled from US$35 billion in 2016 to US$71 billion by the end of 2020, with asset tracking, condition-based maintenance, and robotics processing driving these investments.

In the current industrial environment, IoT systems have three primary uses:

1. **Smart enterprise control:** IoT makes it possible to link manufacturing components and smart equipment to a central computing system, which boosts productivity and lowers costs.
2. **Asset performance management:** The IoT enables real-time monitoring of connected machinery, enabling predictive maintenance and precise forecasts of machine breakdowns through the use of wireless sensors, cloud connectivity, and data analytics.
3. **Enhanced operator:** Future workers are anticipated to use IoT technologies to take on specialized jobs, making manufacturing facilities more user-centric and less machine-centric, despite worries that robots may replace people in smart factories.

9.2.3 Industry 5.0: The Way to Enable Technologies

Automation takes care of monotonous, error-prone work; humans are vital for strategy definition, oversight, and creative contribution. This division of labor emphasizes the core idea of Industry 5.0, which is that people lead and technology enables. One of the main features of Industry 5.0 is its enabling technologies, which comprise a sophisticated range of systems that combine integrated, bio-inspired sensors with Industry 4.0 technologies such as smart materials. These technologies fall into six categories, as described by the European Commission's Directorate-General for Research and Innovation and others in 2021, each of which reveals its potential when integrated with others inside technological frameworks:

1. **Individualized human–machine interaction:** This area of study focuses on connecting people and machines and utilizing their respective capabilities.
2. **Bio-inspired technologies and smart materials:** Using materials that prioritize recycling while including sensors and improved functions.
3. **Digital twins and simulation:** Digitally simulate entire systems through the use of sophisticated modeling techniques.
4. **Technologies for efficient data transmission, storage, and analysis:** These are the technologies that deal with these aspects of data management.
5. **AI:** Systems that improve decision-making by imitating cognitive processes in humans.
6. **Technologies for energy storage, renewable energy, efficiency, and autonomy:** inventions meant to increase energy storage capacity, encourage autonomy, and make use of renewable energy sources.

Investigating the main technologies within these six clusters provides deeper insights into their unique characteristics and applications, shedding

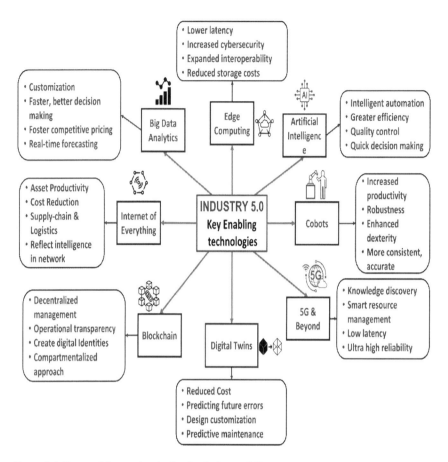

Figure 9.4 Key enabling technologies for Industry 5.0.
(Source: Dimitris Mourtzis et al 2022)

light on their potential to transform industries and drive progress within the context of Industry 5.0 (Figure 9.4). A graphical representation is made on how Industry 5.0 will be impactful to the market.

9.2.3.1 Individualized Human–Machine Interaction

The following technologies are instrumental in supporting humans in both physical and cognitive tasks, facilitating the combination of human innovation with machine capabilities:

1. **Gesture based human interaction:** Systems that can recognize and interpret human speech and gestures in a variety of languages, as well

as anticipate human intentions, are known as multilingual speech and gesture recognition and human intention prediction systems.
2. **Tracking technologies for employee strain and stress:** These are instruments that keep an eye on and assess employees' physical and mental health, spotting symptoms of stress and strain.
3. **Robots:** Cooperative robots made to assist people with a range of jobs and operate side by side with them.
4. **Mixed reality, virtual reality, and augmented reality technologies:** Immersion technologies used for inclusivity promotion and training.
5. **Improving human physical capabilities:** Technologies that improve human physical capabilities include bio-inspired, exoskeletons work gear, and equipment for safety.
6. **Enhancing human capabilities:** Systems aimed at combining the strengths of AI with human cognitive abilities, including decision support systems and technologies for fostering creativity combined with analytical skills.

One recent example of the application of these technologies is the collaboration between Esselunga, Comau (a company under the Italian-French group Stellantis), and Iuvo (a company affiliated with the prestigious Scuola Superiore Sant'Anna University) to introduce a robotic exoskeleton aimed at improving the well-being of workers by alleviating fatigue in the lumbar area. Comau and Iuvo have previously launched the Matext exoskeleton for upper limb support, designed to assist operators during manipulation activities involving raised arms. Data from various companies utilizing this technology indicate a reduction in operator effort by approximately 30% and an increase in productivity by around 10%.

Giacomo Del Panta, Comau's chief customer officer, emphasized the company's commitment to developing new technologies that prioritize better ergonomics and worker well-being in heavy-duty activities. He underscored the importance of prioritizing worker health and well-being, highlighting Comau's dedication to sustainable and human-centered production processes (Carrà, 2022).

9.2.3.2 Bio-Inspired Technologies and Smart Materials

Bio-inspired technologies and processes derived from the concept of biological transformation can be integrated with various properties and functionalities, including:

1. **Self-healing or self-repairing:** Materials capable of repairing damage autonomously, mimicking biological systems.
2. **Lightweight:** Utilization of lightweight materials inspired by natural structures, such as bones or leaves.

3. **Recyclable:** Development of materials that can be easily recycled or repurposed, following nature's efficiency in resource utilization.
4. **Raw material generation from waste:** Technologies that convert waste materials into usable raw materials, mirroring processes found in nature.
5. **Integration of living materials:** Incorporation of living organisms or components into manufacturing processes to enhance functionality or sustainability.
6. **Embedded sensor technologies and biosensors:** Incorporation of sensors into materials or products to monitor performance, detect changes, or interact with the environment.
7. **Adaptive/responsive ergonomics and surface properties:** Materials capable of adjusting their properties in response to environmental conditions or user interactions.
8. **Materials with intrinsic traceability:** Materials that inherently carry information about their origin, composition, or processing history.

The deployment of sensors offers several operational efficiencies across various industries, including reduced labor, logistics, and quality control costs. Sensors enhance inventory counting, material sorting, and automation, leading to increased productivity. Moreover, sensors aid in identifying root errors in manufacturing processes and drive improvements in product design. For instance, smart-sensor technology enables assembly lines for wearables to transmit real-time images to design engineers, facilitating immediate identification and resolution of manufacturing issues, thereby saving time.

Leading players like Mitsubishi are investing in automation systems to capitalize on the benefits of embedded sensors throughout the manufacturing process. These benefits include:

- **Improved operational efficiency:** Sensor-enabled labor monitoring optimizes workforce assignments and enables real-time quality inspection to address manufacturing issues promptly.
- **Enhanced asset management:** Critical equipment is monitored using sensors to anticipate and prevent potential interruptions.
- **Real-time tracking:** Radio frequency identification (RFID)-based sensors facilitate touchless tracking and item identification, reducing the possibility of loss or shrinking of inventory.
- **Product design insights:** Connected products equipped with sensors provide valuable data on customer behaviors and preferences, enabling more responsive product development.

By leveraging bio-inspired technologies and embedded sensor systems, companies can enhance efficiency, sustainability, and innovation across their operations (Boukhalfa, 2021).

9.2.3.3 Digital Twins as well as Simulation

Digital twins and simulation technologies combine various concepts such as AI, the IoT, metaverse, and virtual reality and augmented reality (VR/AR) to create digital representations of real-world objects, systems, or processes. While simulations help in understanding potential real-world scenarios, digital twins allow for comparison and assessment of what may happen alongside what is currently happening. Both technologies optimize production processes, test products, and detect potential harmful effects. Key applications include:

1. **Digital twins of goods and procedures:** Building digital copies of real goods and production procedures to assess and enhance efficiency.
2. **Virtual simulation and testing:** To find possible problems and boost productivity, products and processes are simulated and tested in virtual environments.
3. **Multi-scale dynamic modeling and simulation:** To comprehend the behavior and performance of complex systems, modeling and simulation are used at various scales.
4. **Environmental and social impact simulation and measurement:** Using simulation and measurement to evaluate the effects of products and activities on the environment and society.
5. **Cyber–physical systems and digital twins of complete systems:** Digital twins and physical systems are integrated to track and improve system performance as a whole.
6. **Scheduled maintenance:** Predicting maintenance requirements and optimizing schedules with digital twins.

Boeing's integration of digital twins into design and production processes is a notable example. By assessing how materials would perform throughout an aircraft's lifecycle, Boeing improved the quality of certain parts by up to 40%.

Businesses are investigating the use of digital twins for applications ranging from digital agriculture and precision medicine to engineering design across a variety of industries. However, the applications that are already in use are frequently quite tailored and restricted to high-value use cases, like jet engine operations, industrial facilities, and power plants.

In the manufacturing industry, car giants like Tesla utilize digital simulation for their cars by collecting data from sensors installed in and out of the vehicles and uploading it to the cloud. This data enables the company's AI algorithms to predict potential faults and breakdowns, minimizing the need for owners to visit servicing stations for repairs and maintenance. This approach reduces servicing costs for Tesla, enhances user experience, and increases customer satisfaction, thereby fostering repeat business (Marr, 2022).

9.2.3.4 Transmission of Data, Analysis and Storage

Energy-efficient and secure data transmission, storage, and analysis technologies play a critical role in modern enterprises, characterized by properties such as:

1. **Networked sensors:** Utilizing sensors interconnected through networks to collect and transmit data efficiently.
2. **Data and system interoperability:** Ensuring compatibility and seamless communication between different data sources and systems.
3. **Scalable, multi-level cybersecurity:** Implementing robust security measures at multiple levels to protect data from cyber threats.
4. **Cybersecurity/safe cloud IT infrastructure:** Establishing secure cloud-based infrastructure to store and process data while safeguarding against unauthorized access or breaches.
5. **Big data management:** Implementing strategies and technologies to manage and analyze large volumes of data efficiently.
6. **Traceability (data origin and fulfillment of specifications):** Ensuring transparency and accountability in data handling, including tracking data origin and compliance with specifications.
7. **Data processing for learning processes:** Leveraging data processing techniques to extract insights and support learning and decision-making processes.
8. **Edge computing:** Distributing data processing and storage closer to the source of data generation, enabling faster response times and reducing latency.

Many multinational organizations leverage large amounts of data to uncover patterns and trends through thorough analysis, which enables them to make informed decisions quickly and meet customer needs promptly.

For example, Amazon, a prominent e-commerce platform, extensively collects customer data to understand purchasing behaviors and preferences. This data is utilized in algorithms for social media advertising, customer relationship management, and personalized recommendations. By leveraging big data analytics, Amazon enhances the overall shopping experience and increases customer satisfaction, ultimately driving sales and profitability.

Similarly, Apple utilizes advanced technology and data analytics to understand consumer behavior and preferences. This data-driven approach informs the design and development of new products and services, ensuring alignment with customer needs and preferences. For instance, the Apple Watch not only serves as a wearable device but also collects data to inform future product enhancements and features (Kumari, 2021).

9.2.3.5 Artificial Intelligence

AI, which often encompasses advanced correlation analysis technologies, requires further development in several key areas:

1. **Causality-based AI:** AI systems should move beyond correlation-based analysis to understand causal relationships and effects.
2. **Revealing relations and network effects:** AI should be capable of uncovering and illustrating relationships and network effects beyond simple correlations.
3. **Adaptability to new or unexpected conditions:** AI systems should be able to respond to novel or unforeseen situations autonomously, without human intervention.
4. **Swarm intelligence:** Drawing inspiration from collective behavior in natural systems, AI should leverage swarm intelligence for decentralized decision-making and problem-solving.
5. **Brain–machine interfaces:** Advancing AI technologies to interface directly with human brains for enhanced communication and interaction.
6. **Individual, person-centric AI:** Developing AI systems that cater to individual preferences, needs, and behaviors.
7. **Informed deep learning:** Integrating expert knowledge with AI algorithms to enhance learning and decision-making capabilities.
8. **Skill matching of humans and tasks:** AI systems should effectively match human skills with appropriate tasks to optimize productivity and performance.
9. **Secure and energy-efficient AI:** Ensuring the security and energy efficiency of AI systems to mitigate risks and environmental impacts.
10. **Handling complex, interrelated data:** AI should be equipped to analyze and identify correlations among diverse and dynamic data sources within complex systems.

As it can take time to navigate between programs and context, productivity depends on making pertinent information easily accessible. Many of these tasks are automated by AI-powered knowledge management systems like Microsoft Viva and Guru, giving employees fast access to the data they require. The key to AI's efficacy in knowledge management, according to Deep Analysis founder and principal analyst Alan Pelz-Sharpe, is its scalability. Teams can remain cohesive and productive by adding collaborative knowledge bases to real-time communication. For example, Guru's app for Microsoft Teams integrates verified knowledge directly into the Teams workflow, streamlining both asynchronous and real-time collaboration. Teams can spend more time providing value and less time looking for information when they have access to shared knowledge and AI-driven updates (Microsoft Teams, 2021).

9.2.3.6 Efficiency of Technology, Storage, Renewables, and Autonomy

To achieve emission neutrality, technologies and properties focused on energy efficiency and renewable energy utilization are crucial. These include:

1. **Integration of renewable energy sources:** Incorporating renewable energy sources such as solar, wind, and hydroelectric power into energy systems to reduce reliance on fossil fuels.
2. **Support of hydrogen and power-to-X technologies:** Investing in technologies that enable the production and utilization of hydrogen and other synthetic fuels derived from renewable energy sources.
3. **Smart dust and energy-autonomous sensors:** Deploying energy-efficient sensors and smart devices that can operate autonomously and efficiently, minimizing energy consumption.
4. **Low energy data transmission and data analysis:** Utilizing energy-efficient technologies for data transmission and analysis to minimize energy usage in digital processes.

For instance, Greenled Industry, an Italian startup specializing in smart lighting solutions, focuses on minimizing power consumption while maximizing light intensity. Their smart lights incorporate motion sensors and lighting controls to adjust brightness based on room occupancy, activity levels, and time of day. By integrating LED technology with intelligent lighting systems, the Greenled Industry not only saves energy but also reduces maintenance costs and contributes to a more sustainable environment.

These technologies not only contribute to emission neutrality but also promote the well-being of people, foster innovation, and encourage environmentally conscious business practices. By leveraging state-of-the-art infrastructures and embracing new technologies, companies can improve their environmental footprint while enhancing overall quality of life.

9.3 THE ROLE OF WORKERS IN THE INDUSTRY 5.0

The transition from Industry 4.0 to Industry 5.0 marks a significant shift from mass automation towards enhancing the capabilities of human workers. This transformation, driven by digital, data-driven, and interconnected technologies, has profound implications for both industry and society. While these advancements bring numerous benefits, such as increased productivity and more ergonomic working environments, they also pose challenges, including stress and work-related diseases.

Human–machine interfaces and robot collaboration are examples of Industry 5.0 technologies that use interactive processes to support workers and increase efficiency. The focus is on developing more secure, fulfilling, and ergonomic work settings where people can use their imagination to solve problems and develop their abilities. Prioritizing technologies that are in line with human values and needs over those that are just technical or commercial is the fundamental tenet of Industry 5.0.

Nonetheless, research indicates that a sizable portion of automation deployments has fallen short in terms of dependability, quality, and flexibility. This emphasizes how crucial it is to take human-related concerns and variables into account while adopting new technologies. Technology may make workers feel irritated, abandoned, or overwhelmed, which could result in resistance or poor implementation. The way technology is used, corporate norms, and employee attitudes all influence how it affects mental health at work.

This article aims to explore how the role of workers evolves in environments utilizing these technologies. Specifically, it seeks to understand whether workers are empowered in their industrial work and attracted to new high-tech environments. By examining these dynamics, the chapter aims to shed light on the implications of technological advancements on the workforce and inform strategies for successful technology adoption and workforce adaptation.

9.3.1 Work-Related Stress and Disease

Although the idea of stress is frequently connected to negative, some tension is really important to focus on tasks at work and accomplish goals. According to the World Health Organization, mental health refers to a state of well-being that enables people to reach their full potential, manage stress, work effectively, and give back to their communities.

In engineering, the term "stress" refers to the maximum amount of strain that a human organism can withstand before malfunctioning. Hans Selye first distinguished between two types of stress in the 1950s: negative stress, or distress, and pleasant stress, or eustress. While distress is linked to unfavorable circumstances that cause worry and discomfort, eustress is the result of difficult circumstances and is viewed as helpful.

The European Agency for Safety and Health at Work emphasizes how digital technologies are changing the nature of work and the need to focus on developing digital futures that are safe and healthy.

Numerous studies that looked at how automation technologies affected the workplace found that employees were less motivated, less informal learners, and less cooperative across disciplines. Rapid technological development has also resulted in a lack of situational awareness, a rise in uncertainty, and mistrust of automation.

One significant risk associated with technology use is internet-related issues, including technological dependence, lack of work-life balance, and inappropriate behavior in the workplace. Technostress phenomena can arise from constant connectivity, leading to feelings of detachment from reality and cognitive overload.

As jobs become more complex, workers may experience distress about job insecurity and insufficient training, fearing automation may replace their roles. This psychological stress can manifest as anxiety, mental fatigue, frustration, or isolation, affecting cognitive workload and potentially leading to accidents in industrial settings.

Age can also influence reactions to technology adoption, with older workers initially experiencing more frustration but ultimately achieving satisfactory performance with adequate training and experience.

In conclusion, the adoption of new technologies in the workplace can have both positive and negative impacts, depending on deployment strategies, monitoring processes, and supportive policies. It is essential to assess the effects on health and safety and provide appropriate training to ensure sustainable employment.

9.3.2 Emerging Risks

Undoubtedly, innovation is being considerably influenced by new technology, which is also changing the modern workplace. It is imperative to comprehend the current and future effects of these technologies on the health and safety of workers.

Under the Scientific Research activities in association with BRIC2019, Inail provided funding for the ID 50 project, "Risk analysis and mitigation tools for the protection of the health and safety of workers in work contexts subject to digital transformation." This project's main goals were to perform a thorough investigation of the potential effects of digitization on workers' health and safety, to identify areas that needed immediate attention, and to create targeted risk reduction solutions. The technologies that are most commonly used in the industrial sector were highlighted, especially those that support Industry 5.0.

This chapter is especially interested in the assessment of psychological and organizational hazards. Organizational solutions, policies, practices, and standards that are unrelated to employees' activities give rise to organizational dangers. Conversely, workers' subjective views of their jobs and interactions with digital technologies are the source of psychological risks.

The project's outcomes are divided into six technological areas, each of which has its own set of conclusions. These results clarify the numerous ways that workers' health and safety are impacted by digitalization and offer insightful information for addressing and reducing related hazards.

9.3.2.1 Wearables the Smart Ones

Smart wearables play a crucial role in monitoring working conditions and ensuring worker safety in production environments. These intelligent electronic devices, which can be worn on the body or incorporated into clothing, provide real-time data and can send alarm signals when risky situations arise.

GPS monitoring of workers' positions and the monitoring of biological and physical parameters such as heart rate, number of steps, and activity levels are among the most common solutions implemented with smart wearables. Additionally, smart helmets and smart belts are increasingly being used to enhance safety in workplaces. Smart helmets can detect environmental parameters like brightness, temperature, and humidity, while also featuring LED lights to signal dark areas and sound amplifiers for audible alarm signals. Smart belts, equipped with RFID technology, help control workers' access to different areas and report hazards like falls and improper machine contact.

However, the adoption of smart wearables also introduces organizational hazards related to prolonged device use. Issues such as muscle fatigue, postural damage, muscle destabilization, and technostress can arise from extended wear. Moreover, reliance on machine monitoring for safety conditions may decrease overall surveillance levels and infringe on workers' privacy due to continuous monitoring and data collection. Without proper training, workers may not fully understand device behaviors, increasing their vulnerability to hazards. Additionally, continuous use of various technologies may lead to addiction and separation anxiety from the devices.

9.3.2.2 Cobots and Robots

The implementation of automation and robotics in production systems offers numerous benefits, including minimizing the need for workers to operate in hazardous environments, improving precision and efficiency in routine tasks, and increasing accessibility for individuals with physical impairments. However, these technologies also introduce potential hazards for users, particularly in scenarios involving human–machine collaboration or close proximity.

Organizational hazards associated with automation and robotics include repetitive tasks performed at the pace of machines, leading to fatigue, musculoskeletal stress, psychological stress, and physical overload. The reduction of human activities in such environments can result in cognitive underload and decreased concentration levels among operators. Additionally, workers may experience mental stress when operating in close proximity to machines, thus increasing the risk of collisions or accidents.

Collisions between operators and robots, as well as unpredictable machine behavior, can pose significant risks. Inadequate robot movement fluency may cause discomfort, cognitive stress, and collisions. Employees may perceive the implementation of robots and cobots as a threat, fearing job redundancies, unpredictable machine behavior, and dependency on external repair workers. Clear communication and training are essential to mitigate these risks and ensure worker safety.

Psychological hazards arise from reduced human interaction and increased reliance on machines, leading to social isolation, feelings of inferiority, and subordination to machines. Workers may experience increased psychophysical stress, fear, and insecurity due to the variability and unpredictability of robots and cobots. Although standards exist for implementing these technologies, there is a lack of comprehensive guidelines addressing all hazards associated with their use in industrial settings. Addressing these issues requires careful consideration of training, communication, and safety protocols to ensure the well-being of workers in automated environments.

9.3.2.3 Virtual Reality/Augmented Reality

The use of virtual reality (VR) and augmented reality (AR) technologies provides a number of advantages for helping employees with diverse jobs. To safeguard the health and safety of employees, it also brings up new risks that must be taken into consideration.

The usage of AR and VR technologies in the workplace might present organizational dangers such as uneven illumination, which can lead to eye strain, glare damage, and discomfort as employees adjust to shifting light levels. Prolonged usage of AR and VR devices can cause eye strain, nausea, vertigo, disorientation, motion sickness, headaches, social distancing, elevated heart and respiration rates, stomach damage, distraction-related injuries, and unpredictability in the long run when it comes to musculoskeletal effects. Furthermore, a high amount of information might result in cognitive overload, screen latency can give rise to migraines, and the overlap between virtual images and real objects may cause eye fatigue.

These technologies are frequently employed to provide real-time training to employees, which can lower operator capabilities and pose significant hazards. As AR and VR gadgets can take pictures and record movies, inadequate training could result in inappropriate device handling and privacy issues.

Psychological risks can also occur from workers abusing these gadgets excessively, leading to addiction and separation anxiety. Regular usage of technology can lead to social isolation and techno-stress.

Research on how AR/VR gadgets affect the musculoskeletal system has shown that using these devices while engaging in specific activities or adopting bad postures might lead to musculoskeletal diseases over time.

In conclusion, even if AR and VR technologies have many advantages, it is critical to fully address the risks involved to protect the health and safety of employees. To effectively limit hazards, this calls for the implementation of appropriate training, ergonomic considerations, and privacy safeguards.

9.3.2.4 ExoSkeleton

The implementation of exoskeletons in the workplace presents both benefits and potential hazards for workers' health and safety. While exoskeletons can assist workers in reducing muscle tension and preventing work-related musculoskeletal disorders (WRMSDs), they also introduce new risks that need to be addressed.

One concern is the limited mobility of operators wearing exoskeletons, which may make them unable to avoid collisions with falling objects or cause improper movements leading to muscle damage or overexertion. Additionally, the weight and dimensions of exoskeletons can lead to complications such as musculoskeletal issues, muscular fatigue, pressure injuries, nerve compression, respiratory fatigue, discomfort, cardiovascular issues, and spine overload or damage. Increased directional load may also result in dynamic event-related injuries.

From an organizational perspective, concerns arise regarding privacy violations due to the devices' ability to monitor personal data like localization. The increased physical capabilities of exoskeletons may lead to cognitive overload, while inadequate training may cause fears and insecurities among operators.

Psychological hazards include excessive reliance on exoskeletons leading to decreased attention to security measures and muscle density loss. Operators may also fear stigmatization in the workplace or being perceived as technology-dependent.

In conclusion, while exoskeletons offer potential benefits for worker health and safety, they also introduce significant hazards, especially in the long term. Further studies and practical experiences are needed to better understand and address these risks to ensure the safety of workers using exoskeleton technologies.

9.3.2.5 Digital Twin

The main issues related to digital twin technological solutions, particularly those associated with Internet of Simulation (IoS) solutions, primarily revolve around implementation efficiency rather than health and safety hazards for workers. Some of the key challenges include:

1. **Choice of simulation objective:** Determining the specific objective of the simulation can be challenging and may require careful consideration of various factors such as the desired outcomes, complexity of the system being simulated, and available resources.
2. **Trade-off between quality and execution speed:** There is often a trade-off between the desired quality of the simulation and the speed at which it can be executed. Balancing these factors is essential to ensure that the simulation provides accurate results within a reasonable timeframe.
3. **Cost assessment:** Conducting economic feasibility analyses is crucial to understanding the cost-effectiveness of implementing simulation solutions. This involves assessing factors such as the initial investment

required, ongoing maintenance costs, and potential savings or benefits derived from the simulation.

While these challenges are significant in terms of implementing digital twin and IoS solutions effectively, the analysis of existing studies has not revealed significant health and safety hazards for workers associated with these technologies. Therefore, while there may be technical and operational challenges to address, workers' health and safety do not appear to be a primary concern in this context.

9.3.2.6 Wireless Communication Technology

The use of wireless technologies in the workplace introduces new potential hazards for workers, which can be categorized into organizational and psychological levels:

1. **Organizational hazards:**
   - **Decrease in supervision:** With health and safety conditions monitored by wireless devices, there may be a general decrease in direct supervision of workers. This could lead to situations where workers are less closely monitored for potential hazards or unsafe behavior, thereby increasing the risk of accidents or injuries.
2. **Psychological hazards:**
   - **Psychological pressure:** Constant monitoring by technology, such as wireless communication devices, can create significant psychological pressure on workers. The awareness of being under constant surveillance may lead to feelings of stress, anxiety, or paranoia among workers, impacting their mental well-being and overall job satisfaction.

Additionally, specific wireless technologies like RFID and Bluetooth Low Energy (BLE) can introduce their own set of hazards. These may include issues related to data privacy and security, electromagnetic radiation exposure, and potential interference with medical devices or sensitive equipment.

Overall, while wireless technologies offer many benefits in terms of connectivity and efficiency, it is essential for employers to carefully consider and mitigate the potential hazards they may introduce to ensure the health, safety, and well-being of workers in the workplace.

9.4 PATH TOWARD SOCIETY 5.0

Society 5.0 represents a vision of a society that embraces digital transformation to address challenges and create value in a manner that prioritizes diversity, decentralization, resilience, and sustainability (Figure 9.5). Here are some key points regarding Society 5.0:

1. **Problem-solving and value creation:** Society 5.0 emphasizes solving problems and creating value, with a focus on satisfying individual needs and leveraging digital technologies to meet diverse demands.
2. **Liberation from disparity:** Unlike Society 4.0, where wealth and information were concentrated in limited hands, Society 5.0 aims for the distribution and decentralization of wealth and information throughout society, providing opportunities for participation to anyone, anytime, anywhere.
3. **Liberation from resource and environmental constraints:** In Society 5.0, there is a shift away from models with high environmental impact and mass consumption of resources. Data utilization, energy efficiency, and decentralization offer alternatives to traditional energy networks, reducing dependency on scarce resources.

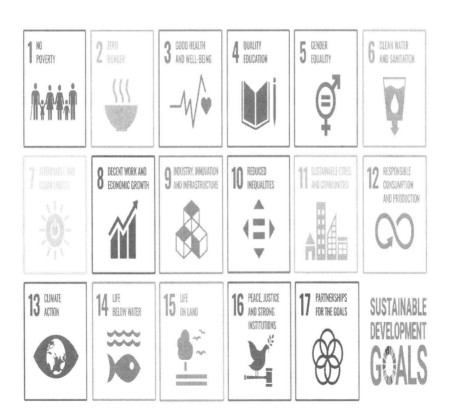

Figure 9.5 Goals towards sustainable development.

4. **Digital transformation and distribution of abilities:** Digital transformation democratizes access to advanced skills and abilities previously limited to experts. AI modules and services enable more people to participate in manufacturing and service provision, contributing to value creation and problem-solving.
5. **Contribution to sustainable development goals (SDGs):** Society 5.0 aligns with the 17 SDGs adopted by the United Nations, particularly goals related to gender equality (Goal 5), decent work and economic growth (Goal 8), and innovation in industry and infrastructure (Goal 9). By leveraging digital transformation and fostering creativity and diversity, Society 5.0 aims to contribute to a sustainable future on a global scale.

Overall, Society 5.0 represents a paradigm shift toward a more inclusive, sustainable, and technologically empowered society, where individuals and communities are empowered to address challenges and create value collaboratively.

9.5 CONCLUSION

The chapter provides a comprehensive analysis of the transition from Industry 4.0 to Society 5.0, focusing on the impact of new technologies on manufacturing firms and workers. Here's a summary of the key findings and conclusions:

1. **Technology adoption trends:** The analysis reveals that while the adoption of enabling technologies of Industry 4.0 is ongoing, many firms prioritize infrastructure investments over application technologies like IoT and robotics. This suggests a gradual approach to digital transformation, with firms focusing on building foundational capabilities before implementing advanced technologies.
2. **Effects on productivity and growth:** The empirical evidence demonstrates a positive association between technology adoption and firm productivity and growth. Firms that invest in digital technologies experience increased added value and revenues per employee. Furthermore, the analysis highlights strong complementarities between skills and technology adoption, emphasizing the importance of human capital in driving digital transformation.
3. **Impact on workers:** The chapter discusses the challenges and stressors that workers may experience due to the changing roles and increased reliance on technology. Psychological stress, cognitive workload, and negative attitudes toward technology are identified as potential barriers to successful digital transformation. However, the study

suggests that recognizing the value of the workforce and investing in skills development can mitigate these challenges.
4. **Limitations and future directions:** The current chapter acknowledges limitations in the available data, particularly regarding Industry 5.0, as surveys and projects on this topic are still in the early stages. Future research could benefit from more recent and comprehensive data sets to further explore technology adoption behaviors and their impact. Additionally, expanding the analysis to other industries and incorporating workers' perspectives would provide a more holistic understanding of the digital transformation process.

Overall, the chapter highlights the importance of a strategic approach to digital transformation, emphasizing the need for firms to invest in both technology and human capital to achieve inclusive and sustainable growth in the digital era.

REFERENCES

Anitec-Assinform. (2021). Il Digitale in Italia 2021. https://ildigitaleinitalia.it/il-digitale-initalia-2019/il-digitale-in-italia-2021.kl

Atwell, C. (2017). Yes, Industry 5.0 is Already on the Horizon. Machine Design. www.machinedesign.com/automation-iiot/article/21835933/yes-industry-50-isalready-on-the-horizon

Bhatia, A., Kumar, A., Verma, P., Kumar, M., & Kumar, J. (2023). "Cyber Threat: A Review on Dark Sides of Dark Web." *2023 3rd International Conference on Advancement in Electronics & Communication Engineering (AECE)*. IEEE.

Bhatia, A., Kumar, A., & Bhatia, P. (2024). "Data Privacy and E-Consent in the Public Sector." *The Ethical Frontier of AI and Data Analysis*. IGI Global. 118–137.

Boukhalfa, S. (2021). *The Applications of Embedded Sensors in the Manufacturing Value Chain—PreScouter—Custom Intelligence from a Global Network of Experts*. PreScouter. www.prescouter.com/2021/03/the-applications-of-embeddedsensors-in-the-manufacturing-value-chain/

Carrà, M. (2022). *L'ultima trovata di Esselunga: Un esoscheletro robot per ridurre gli sforzi fisici dei lavoratori*. Forbes Italia. https://forbes.it/2022/02/11/lultima-trovata-diesselunga-un-esoscheletro-robot-per-ridurre-gli-sforzi-fisici-dei-lavoratori/

Cirillo, V., Fanti, L., Mina, A., & Ricci, A. (2021). *Digitalizing Firms: Skills, Work Organization and the Adoption of New Enabling Technologies*. LEM Working Paper Series. Pisa, Italy: Sant'Anna School of Advanced Studies.

Costantino, F., Falegnami, A., Fedele, L., Bernabei, M., Stabile, S., & Bentivenga, R. (2021). New and emerging hazards for health and safety within digitalized manufacturing systems. *Sustainability*, *13*(19), 10948. https://doi.org/10.3390/su131910948

Directorate-General for Internal Market, Industry, Entrepreneurship and SMEs. (2019). *A Vision for the European Industry Until 2030: Final Report of the*

Industry 2030 High Level Industrial Roundtable. Publications Office of the European Union. https://data.europa.eu/doi/10.2873/34695

Directorate-General for Research and Innovation, Breque, M., De Nul, L., & Petridis, A. (2021). *Industry 5.0: Towards a Sustainable, Human Centric and Resilient European Industry*. Publications Office of the European Union. https://data.europa.eu/doi/10.2777/308407

Directorate-General for Research and Innovation, & Müller, J. (2020). *Enabling Technologies for Industry 5.0: Results of a Workshop with Europe's Technology Leaders*. Publications Office of the European Union. https://data.europa.eu/doi/10.2777/082634

Directorate-General for Research and Innovation, & Vanderborght, B. (2020). *Unlocking the Potential of Industrial Human–Robot Collaboration: A Vision on Industrial Collaborative Robots for Economy and Society*. Publications Office of the European Union. https://data.europa.eu/doi/10.2777/568116

Directorate-General for the Internal Market, Industry, Entrepreneurship and SMEs. (2017). *PMI definition user guide*. https://op.europa.eu/it/publication-detail/-/publication/79c0ce87-f4dc-11e6-8a35-01aa75ed71a1/language-it.

Dom Nicastro. (2021). *Real-world examples of artificial intelligence (AI) in the workplace*. CMSWire.Com. www.cmswire.com/digital-workplace/8-examples-ofartificial-intelligence-ai-in-the-workplace/

European Commission. (2020). *European Skills Agenda—Employment, Social Affairs & Inclusion—European Commission*. https://ec.europa.eu/social/main.jsp?catId=1223

(2019). *Regolamento (UE) 2019/2152*, EUR-Lex, GU L 327. http://data.europa.eu/eli/reg/2019/2152/oj/ita

Faccio, M., Granata, I., Menini, A., Milanese, M., Rossato, C., Bottin, M., Minto, R., Pluchino, P., Gamberini, L., Boschetti, G., & Rosati, G. (2022). Human factors in cobot era: A review of modern production systems features. *Journal of Intelligent Manufacturing*. https://doi.org/10.1007/s10845-022-01953-w

Graus, E., Özgül, P., & Steens, S. (2021). *Artificial Intelligence: Shaping the Future of Work with Insights from Firm-Level Evidence*. ROA External Reports.

International Data Corporation. (2021). *Organizations Are Forecast to Spend Nearly $656 Billion on Future of Work Technologies in 2021, According to New IDC Spending Guide*. IDC: The Premier Global Market Intelligence Company. www.idc.com/getdoc.jsp?containerId=prUS48040921

International Telecommunication Union. (2022). *Smart Sustainable Cities*. ITU. www.itu.int:443/en/ITU-T/ssc/Pages/info-ssc.aspx

ISTAT. (2020). Integrazione tra registro esteso delle principali variabili economiche delle imprese "Frame SBS" e l'indagine campionaria sulle tecnologie dell'informazione edella *comunicazione "ICT"*. 10. www.istat.it/it/files/2020/03/Principali-risultati-nota-metodologica.pdf

ISTAT. (2021). Misure di produttività-Anni 1995–2020. www.istat.it/comunicato-stampa/misure-di-produttivita-anni-1995-2021/#:~:text=La%20produttivit%C3%A0%20del%20capitale%20(rapporto,%2C%2C%25%20nel%202020

ISTAT. (2022). Imprese e ICT—Anno 2021. www.istat.it/it/archivio/265333

Keidanren. (2018). *Society 5.0—Co-Creating the Future.* European Economic and Social Committee. www.eesc.europa.eu/en/news-media/presentations/society-50-cocreating-future

Kumar, A., Bhatia, A., Kashyap, A., & Kumar, M. (2023). "LSTM Network: A Deep Learning Approach and Applications." *Advanced Applications of NLP and Deep Learning in Social Media Data.* IGI Global. 130–150.

Kumari, R. (2021). Top 10 Companies That Uses Big Data. Analytics Steps. www.analyticssteps.com/blogs/companies-uses-big-data

Marr, B. (2022). The Best Examples of Digital Twins Everyone Should Know About. Forbes. www.forbes.com/sites/bernardmarr/2022/06/20/the-best-examples-of-digitaltwins-everyone-should-know-about/

Microsoft Teams. (2021). *Connect Guru with Microsoft Teams for a Collaborative Wiki.* Techcommunity.microsoft.com. https://techcommunity.microsoft.com/t5/microsoftteams-blog/connect-guru-with-microsoft-teams-for-a-collaborative-wiki/ba-p/2051229

Kumar, M., Ali Khan, S., Bhatia, A., Sharma, V., & Jain, P. (2023) "A Conceptual Introduction of Machine Learning Algorithms," *2023 1st International Conference on Intelligent Computing and Research Trends (ICRT), Roorkee, India*, pp. 1–7. https://doi.org/10.1109/ICRT57042.2023.10146676

Østergaard, E. H. (2017). Factory Automation: Welcome to Industry 5.0—ISA. Isa. Org. www.isa.org/intech-home/2018/march-april/features/welcome-to-industry-5-0

Paschek, D., Mocan, A., & Draghici, A. (s.d.). (2020). Industry 5.0—The Expected Impact of Next Industrial Revolution. Reskilling Workers for Industry 4.0. McKinsey. www.mckinsey.com/businessfunctions/operations/our-insights/building-the-vital-skills-for-the-future-of-work-inoperations

Rossato, C., Pluchino, P., Cellini, N., Jacucci, G., Spagnolli, A., & Gamberini, L. (2021). Facing with collaborative robots: The subjective experience in senior and younger workers. *Cyberpsychology, Behavior, and Social Networking*, 24(5), 349–356. https://doi.org/10.1089/cyber.2020.0180

StartUs Insights. (2019). 4 Top Industrial Energy Efficiency Solutions Impacting the Industry. StartUs Insights. www.startus-insights.com/resources/

Statista. (2020). Industry 4.0 Technology Impact Organizations Worldwide 2020. Statista. www.statista.com/statistics/1200006/industry-40-technology-greatest-impactorganizations-worldwide/

Chapter 10

Managing Industry 5.0
The Next Frontier for Artificial Intelligence and Machine Learning Algorithms

Varsha Sahni and Ekta Bhaggi

10.1 INTRODUCTION

Technology has driven industries through various stages of evolution, from the mechanization of Industry 1.0 to the digital revolution of Industry 4.0. Researchers have been working on enabling computers to learn independently for more than 50 years, starting with the creation of computers in the 1950s that needed to be operated by humans. This development is a milestone for computer science, business, and human civilization. Computers have advanced to the point that they can, in a sense, do new tasks independently. Artificial intelligence (AI) of the future will interact and adapt to humans by using their natural language, gestures, and emotions. As more intelligent terminals become widely used and interconnected, people will not only reside in actual physical space but also in a digitally virtualized network. AI refers to the theory, methods, and technologies that enable machines, particularly computers, to analyze, replicate, exploit, and study human thought processes and behavior.

AI is the study of aspects of human activity, building an intelligent system, making computers perform jobs that humans were previously only able to perform, and using computer hardware and software to mimic the fundamental theories, methods, and strategies of human behavior. AI is being utilized increasingly frequently in particular industrialization and commercialization initiatives, indicating new tendencies in development. Big data and deep learning (DL) have become the standard for AI development. Robots can already learn and think like humans and do more complicated jobs because of artificial neural networks or ANNs. AI has progressively moved into the research and development (R&D) and manufacturing phases of technology, which were started by experimental research. Commercial speech and image recognition products, natural language processing (NLP), and predictive analysis have matured [1–5].

AI is also responsible for the rapid advancement of technology in the social and economic spheres of life, which has changed industry and society. Throughout history, technological progress has led to distinct social stages within society. The development of tools marked the beginning of Society 1.0, also known as the hunting society. This was followed by Society 2.0, the agrarian society, which was marked by the rise of agriculture. The Industrial Revolution marked the beginning of Society 3.0, i.e., the industrial society. Finally, we have Society 4.0, the data society, where data plays a crucial role in every aspect of life and determines the existing social structure. Industry 5.0 was created to prioritize sustainability and social justice, alongside digitalization and AI-driven technology (Industry 4.0).

This combination of AI with Society 5.0 is a great boom to society. In the complex realm of AI and machine learning (ML), there are numerous algorithms, each with distinct characteristics and applications. ML trains machines to manage data more effectively. Sometimes, even after analyzing the data, it can be difficult for us to comprehend the information that has been collected. In such cases, ML can be used to help us. A concise overview of popular AI and ML algorithms is presented in the next sections.

10.2 SUPERVISED LEARNING

Supervised learning is the foundation of cutting-edge ML methods and is essential to developing Society 5.0, a future in which technological progress and human well-being coexist harmoniously. This type of learning algorithm is vital in many applications that improve our daily lives in the current paradigm of society. Supervised learning uses labeled datasets to train models to make precise predictions and classifications. Supervised learning is an ML task that involves learning a function to translate input to output using example input–output pairs. This type of learning requires help from outside algorithms using labeled training data, which is a collection of training instances [6–10]. The training and test datasets are separated, and the output variable from the training dataset needs to be categorized or forecasted. All algorithms use the training dataset to identify patterns, which they then apply to the test dataset to make predictions or classify data. The workflow of supervised learning is shown in Figure 10.1.

The most well-known supervised ML algorithms have been covered in this chapter.

10.2.1 Decision Tree

Decision trees are crucial tools in the context of Society 5.0, as they help us navigate the complexities of our data-driven and networked society. Decision tree algorithms, a type of supervised learning, are essential for decision-making across various fields. Decision trees provide a simple and easy-to-understand framework for addressing a wide range of problems, from resource allocation strategies in smart cities to personalized recommendation systems that enhance our digital experiences. These algorithms

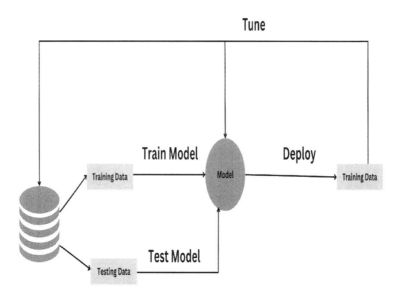

Figure 10.1 Supervised learning workflow.

play a critical role in Society 5.0 by optimizing decision paths, which in turn leads to more effective and efficient solutions to complex problems. Decision trees show how human needs and algorithms can combine to improve outcomes and create a sustainable future. They can guide urban planning or assist in health care.

10.2.2 Naïve Bayes

Within the framework of Society 5.0, Naïve Bayes algorithms show themselves to be adaptable instruments that utilize probabilistic reasoning to tackle a wide range of issues in our globalized society. Naïve Bayes algorithms are widely used in recommendation systems, fraud detection systems, and sentiment analysis applications. They are particularly good at predicting outcomes based on probabilities and dependencies. They are especially well-suited for real-time decision-making in a variety of disciplines because of their efficiency and simplicity. Naïve Bayes models can help with diagnostic procedures in the health care industry and help identify unusual patterns that point to possible security risks in the cybersecurity industry. Naïve Bayes algorithms highlight the interdependence of data-driven insights and human-centered applications as Society 5.0 develops, providing answers that support the main objective of promoting a more knowledgeable, flexible, and responsive society.

10.2.3 Support Vector Machine

In Society 5.0, technology is completely woven into the fabric of human experience, and Support Vector Machines (SVMs) are integral to this. These ML techniques are used in many different socioeconomic fields and are especially good at classification and regression tasks. SVMs use their ability to handle complicated and multidimensional data to help with disease classification and individualized treatment recommendations in the health care industry. SVMs help with resource allocation and urban planning in the context of smart cities under Society 5.0, maximizing efficiency through predictive modeling. Furthermore, by spotting patterns suggestive of malicious activity, SVMs play a crucial role in cybersecurity by aiding in the detection and prevention of cyber threats. SVMs are a great example of how advanced technology and human-centric requirements can work together to create a harmonious coexistence. Intelligent algorithms can enhance decision-making and promote societal well-being as we navigate the terrain of Society 5.0 [11].

10.3 UNSUPERVISED LEARNING

Unsupervised learning is an important tool for innovation in the current era of Society 5.0. It enables the seamless integration of technology into various aspects of our globalized society. By analyzing large amounts of unlabeled data, unsupervised learning algorithms can detect hidden patterns and structures with greater ease than they would be able to with annotated datasets. These algorithms are crucial in industries like banking, where they help prevent fraud by identifying anomalies, and in smart cities, where they assist in resource allocation and urban planning using clustering techniques. Unsupervised learning is an ML type with no teacher or correct answers. The algorithms are responsible for finding and displaying interesting patterns within the data. Unsupervised learning algorithms learn a few features from the data, but they can recognize the class of the data when it is introduced using previously learned features. The primary applications of unsupervised learning are feature reduction and clustering [12]. The workflow of unsupervised learning is shown in Figure 10.2.

10.4 NEURAL NETWORKS WORKFLOW

The most well-known unsupervised ML algorithms are covered in this chapter.

10.4.1 Principal Component Analysis

Principal component analysis (PCA) is a key technique for finding important patterns and reducing the dimensionality of complex datasets in the dynamic

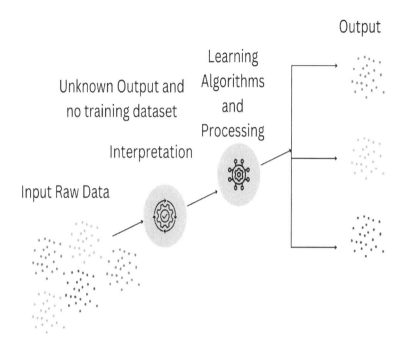

Figure 10.2 Unsupervised learning workflow.

environment of Society 5.0. This unsupervised learning technique is essential for unraveling complex relationships and uncovering latent structures within large datasets. PCA helps identify important characteristics for disease diagnosis and treatment coordination in industries such as health care. PCA helps to effectively analyze urban data in the context of smart cities in the context of Society 5.0, supporting decision-makers to make optimal use of available resources and infrastructure design. PCA, a feature extraction and data compression tool, is a prime example of how cutting-edge analytical techniques can meet the changing needs of a technologically advanced society. In Society 5.0, PCA provides actionable insights for decision-makers and data scientists, resulting in a more knowledgeable, effective, and flexible social environment [13].

PCA is a statistical technique that helps reduce the dimensionality of a dataset by transforming a set of variables that may be correlated into a set of values of linearly uncorrelated variables, or principal components, using an orthogonal transformation. This method is useful to speed up computations and provide an explanation of a set of variables' variance–covariance structure using linear combinations. PCA is often used as a method of dimensionality reduction.

10.4.2 K-Means Clustering

K-means clustering is an essential unsupervised learning method in the context of Society 5.0, providing answers and insights into the intricacies of our globalized society. This method is particularly good at dividing up different datasets into different clusters according to similarities, which makes it easier to see underlying patterns and structures in massive amounts of data. K-means clustering is essential for urban planning in smart city contexts. It helps planners find patterns in space, improve traffic flow, and allocate resources more effectively. This algorithm helps categorize patients in the health care industry, which enables individualized treatment programs and focused treatments. K-means clustering is a potent tool that will be useful as Society 5.0 develops. It promotes effectiveness and well-informed decision-making across several fields and is an example of how cutting-edge data-driven approaches may coexist peacefully with the demands of an intelligent and connected society.

One of the most straightforward unsupervised learning algorithms for resolving the well-known clustering problem is K-means. The process uses a set number of clusters to classify a given dataset easily and straightforwardly. Determining k centers—one for each cluster—is the basic notion. These centers need to be positioned cleverly as different locations yield varied outcomes. Placing them as far apart as feasible is hence the preferred option. The next action is to connect each point in a given dataset with the closest center. The first step is finished, and an early group age is finished when no points are outstanding. As the barycenter of the clusters produced by the previous step, we now need to recalculate k new centroids.

10.5 SEMI-SUPERVISED LEARNING

Semi-supervised learning is a strategic method that effectively combines the advantages of supervised and unsupervised learning to address the complex problems of network society in the dynamic context of Society 5.0. This new paradigm is particularly useful when labeled data is difficult to obtain, as it takes advantage of a larger pool of unlabeled data and a smaller number of annotated examples. This methodology is important for a variety of applications such as smart cities and health care, helping with resource efficiency and predictive maintenance. It also helps stratify patient risk and individualize treatment. Semi-supervised learning serves as a link in Society 5.0, bringing together the flexibility and investigative potential of unsupervised approaches with the accuracy of supervised procedures. This eventually improves decision-making processes in a variety of industries within our technologically sophisticated society.

Semi-supervised ML involves combining both supervised and unsupervised ML techniques. This technique can be particularly useful in ML and data mining applications where obtaining labeled data is a time-consuming process. With semi-supervised learning, existing unlabeled data can be put to good use [14–16].

The most popular supervised ML techniques involve using labeled data to train an ML algorithm. In contrast, semi-supervised learning techniques use both labeled and unlabeled data to create a better model. In this way, semi-supervised learning can be seen as a "middle ground" between supervised and unsupervised learning.

There are various semi-supervised learning methods available, and in the following discussion, we will explore a few of them.

10.5.1 Transductive Support Vector Machines

In semi-supervised learning, transductive support vector machines (TSVMs) are a popular method for handling partially labeled data. There has been mystery surrounding it as its generalization-based underpinning is not well understood. To maximize the margin between the labeled and unlabeled data, it is utilized to label the unlabeled data. It is an NP-hard problem to find an exact solution using TSVM. NP-hard problems are a type of computational problems that are believed to be difficult to solve. NP stands for "nondeterministic polynomial time" [17].

10.5.2 Generative Models

A model that generates data is called a generative model. It models the entire set of data, or the class as well as the features. All algorithms that model $P(x,y)$ are generative as they can create data points using this probability distribution. For each component, one labeled example is sufficient to verify the distribution of the mixture.

10.5.3 Self-Training

A classifier gets taught with some labeled data when it engages in self-training. Then, unlabeled data is given to the classifier. In the training set, the predicted labels and the unlabeled points are combined. After that, the same process is carried out once again. The term "self-training" emerges from the fact that the classifier is self-learning.

10.6 REINFORCEMENT LEARNING

Reinforcement learning is a transformational force in the era of Society 5.0, influencing the intricate dynamics of our interconnected society and how intelligent systems interact with each other. Inspired by behavioral psychology, this ML paradigm enables agents to discover the best behaviors for them through trial and error in a specific setting. Reinforcement learning is used in many different applications to support adaptive decision-making and ongoing development, from self-governing systems in smart cities to customized suggestions in e-commerce. It contributes to individualized care plans and treatment optimization in the medical field. Reinforcement learning becomes a pivotal component in the development of intelligent, self-improving systems that negotiate the complexities of human–technology interaction as Society 5.0 takes shape. This helps to create a society in which AI not only enhances human capabilities but also cleverly responds to changing circumstances. A branch of ML called reinforcement learning studies how software agents should behave in a given setting to maximize a concept known as cumulative reward. Along with supervised learning and unsupervised learning, reinforcement learning is one of the three fundamental paradigms in ML.

10.7 MULTITASKING LEARNING

The development of multitasking AI approaches is a major advancement toward more adaptable and versatile intelligent systems in the era of Society 5.0. This paradigm shift acknowledges the increasing complexity of problems in our globalized world and calls for AI systems that simultaneously handle multiple tasks. Multitasking AI uses sophisticated algorithms to transition between different activities with ease, making decisions more efficiently in real time. Whether it is used in smart cities to manage various urban activities or in health care to diagnose and prescribe treatments at the same time, it demonstrates the agility required to handle the complex issues facing Society 5.0.

Multitask learning (MTL) is a subfield of ML that aims to solve multiple problems at once by leveraging commonalities across several tasks [18]. This approach can improve learning efficiency by serving as a regularizer. In MTL, the knowledge contained in all tasks is utilized to enhance the learning of a particular model. Unlike conventional DL approaches that focus on solving a single task using a single model, MTL can solve multiple related tasks or a subset of them. Although these tasks may not be identical, they can be related to one another [19].

10.8 ENSEMBLE LEARNING

As a fundamental component of Society 5.0, ensemble learning offers a cooperative and synergistic method of tackling difficult problems. Ensemble learning techniques leverage the power of merging many models to improve overall forecast robustness and accuracy in this era of linked systems. When used in smart cities for efficient resource allocation or in health care for thorough diagnosis, ensemble learning takes advantage of model variety to help decision-makers come to more informed conclusions as a group. Ensemble techniques like Random Forests and Boosting algorithms are collaborative and meet the complex needs of Society 5.0, where a multitude of interrelated issues require flexible and all-encompassing solutions [20].

10.8.1 Boosting

A family of algorithms known as "boosting" is used to transform weak learners into strong learners. In ensemble learning, boosting is a technique used to reduce variation and bias. The foundation of boosting is the query "Can a set of weak learners create a single strong learner?" put out by Kearns and Valiant. A classifier is considered a poor learner if it has an arbitrarily high correlation with the correct classification, whereas a strong learner does not have this correlation.

10.8.2 Bagging

When an ML algorithm's accuracy and stability need to be improved, bagging or bootstrap aggregating is used. It can be used in regression and classification. Additionally, bagging reduces variance and aids in managing overfitting.

10.9 DEEP LEARNING (NEURAL NETWORK)

In the context of Society 5.0, the importance of DL, especially in the form of neural networks, is ushering in an era of transformation where AI reaches unprecedented heights. Inspired by the complex structure of the human brain, neural networks have revolutionized the way machines perceive, learn, and make decisions. This paradigm shift is particularly evident in applications ranging from health care, where neural networks improve diagnostic accuracy and drug development, to smart cities, where they optimize traffic management and energy distribution. In Society 5.0, the depth and complexity of neural networks will enable machines to understand complex patterns in large datasets, enabling advanced NLP, image recognition, and predictive analytics. DL through neural networks is proving to be fundamental as we live in this intelligent society, facilitating progress to meet the evolving needs of a highly connected and technologically sophisticated world. It can be used in regression and classification. Additionally, bagging reduces variance and aids in managing overfitting [21].

With a collection of algorithms designed to emulate the workings of the human brain, a neural network aims to identify underlying relationships in a given set of data. This definition of neural networks includes both artificial and organic neural network systems. The neural network produces

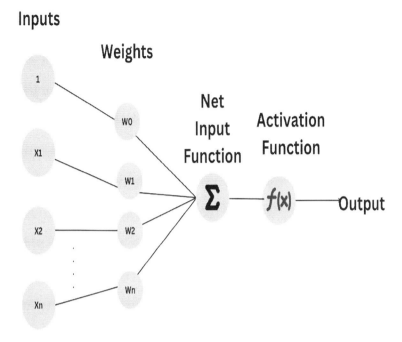

Figure 10.3 Neural networks workflow.

the best result by adapting to changing input without requiring output criteria redesign. AI is the source of the notion of neural networks, which is quickly gaining traction in the trading system development space [22]. The behavior of an ANN is the same. It functions in three levels. Input is received by the input layer and processed by the hidden layer by using the activation function. An activation function in a neural network is a mathematical operation that calculates the output or activation of a neuron by applying it to the weighted sum of its input values. This function brings nonlinearity to the network and allows it to represent complex relationships and patterns in the data. The computed output is finally sent via the output layer.

The working of neural network learning is shown in Figure 10.3.

10.9.1 Supervised Neural Network

The inputs and outputs of a supervised neural network are predetermined. The actual and expected results of the neural network are compared. The parameters are adjusted according to the error and fed back to the neural network. Feedforward neural networks use supervised neural networks.

10.9.2 Unsupervised Neural Network

The input and output of the neural network are unknown to it beforehand. The network's primary function is to classify the data based on certain commonalities. The neural network groups the inputs after determining their correlation.

10.9.3 Reinforced Neural Network

Goal-oriented activities include, for instance, learning how to accomplish a difficult target (goal), maximizing points scored in a game over several rounds, or maximizing along a specific dimension over several steps. The system is referred to as a reinforcement learning system. Even without experience, people can perform superhuman under the correct situations. Reinforcement happens when an algorithm, similar to how young children are rewarded for excellent conduct, punishes itself when it makes a mistaken decision and rewards itself when it makes the right one [23].

10.10 INSTANCE-BASED LEARNING (K-NEAREST NEIGHBOR)

In Society 5.0, instance-based learning (IBL) solves problems flexibly, aligning with personalization and adaptation. Unlike traditional rule-based or model-based approaches, IBL relies on using examples or stored instances to make decisions and forecasts. IBL thrives in Society 5.0, where context-aware systems and tailored experiences are essential. This approach is well-suited to applications such as smart cities that customize urban planning strategies to meet specific community needs, as well as personalized health care that recommends treatments based on the medical records of individual patients. By using saved instances and experience-based learning, IBL is a human-focused and adaptable approach that addresses the diverse and evolving requirements of our globalized world [24]. A family of classification and regression techniques known as "instance-based learning" generates a class label or prediction based on how similar a query is to its nearest neighbor(s) in the training set. IBL algorithms, in clear contrast to other techniques like decision trees and neural networks, do not abstract from particular instances. Instead, they merely store all the data and use the query's nearest neighbor(s) to determine the answer when a query is made. A straightforward supervised ML technique for solving regression and classification issues is the k-nearest neighbors (KNN) method. Its main disadvantage is that, although it is simple to use and comprehend, it becomes noticeably slower as the volume of data being used increases [25].

10.11 CONCLUSION

Industry 5.0 management is currently facing a crucial turning point where the integration of ML and AI algorithms has the potential to revolutionize innovation and industrial processes. The study of advanced algorithms, ranging from supervised learning and neural networks to ensemble approaches and beyond, highlights the dynamic synergy between technology and industrial evolution during this disruptive era. These algorithms work together to improve efficiency, decision-making, and flexibility, creating unprecedented opportunities for increased productivity and expansion.

REFERENCES

1. Akundi, A., Euresti, D., Luna, S., Ankobiah, W., Lopes, A., & Edinbarough, I. (2022). State of Industry 5.0—Analysis and identification of current research trends. Applied System Innovation, 5(1), 27.
2. Chander, B., Pal, S., De, D., & Buyya, R. (2022). Artificial intelligence-based Internet of Things for Industry 5.0. Artificial intelligence-based Internet of things systems (pp. 3–5). Springer International Publishing.
3. Cui, M., & Zhang, D. Y. (2021). Artificial intelligence and computational pathology. Laboratory Investigation, 101(4), 412–422.
4. Ekta & Varsha. (2024). 239 Enhancing rating and learning through clustering in artificial intelligence. In Artificial intelligence and Society 5.0: Issues, opportunities, and challenges (pp. 239–245). CRC Press.
5. Huang, S., Wang, B., Li, X., Zheng, P., Mourtzis, D., & Wang, L. (2022). Industry 5.0 and Society 5.0—Comparison, complementation and co-evolution. Journal of Manufacturing Systems, 64, 424–428.
6. Jiang, C., Zhang, H., Ren, Y., Han, Z., Chen, K. C., & Hanzo, L. (2016). Machine learning paradigms for next-generation wireless networks. IEEE Wireless Communications, 24(2), 98–105.
7. Kallem, S. R. (2012). Artificial intelligence algorithms. IOSR J Computer Engineering (IOSRJCE), 6(3), 1–8.
8. Blanton, M. J. A. M. (Ed.). (2009) *Algorithms and Theory of Computation Handbook*, Second Edition. Boca Raton FL, USA: CRC Press.
9. Liew, A. (2007). Understanding data, information, knowledge, and their inter-relationships. Journal of Knowledge Management Practice, 8(2), 1–16.
10. Lu, Y. (2019). Artificial intelligence: A survey on evolution, models, applications and future trends. Journal of Management Analytics, 6(1), 1–29.
11. Mahesh, B. (2020). Machine learning algorithms review. International Journal of Science and Research (IJSR) [Internet], 9(1), 381–386.
12. Morales, E. F., & Escalante, H. J. (2022). A brief introduction to supervised, unsupervised, and reinforcement learning. In Biosignal processing and classification using computational learning and intelligence (pp. 111–129). Academic Press.
13. Nahavandi, S. (2019). Industry 5.0—A human-centric solution. Sustainability, 11(16), 4371.

14. Nitzberg, M., & Zysman, J. (2022). Algorithms, data, and platforms: The diverse challenges of governing AI. Journal of European Public Policy, 29(11), 1753–1778.
15. Raja Santhi, A., & Muthuswamy, P. (2023). Industry 5.0 or Industry 4.0S? Introduction to Industry 4.0 and a peek into the prospective Industry 5.0 technologies. International Journal on Interactive Design and Manufacturing (IJIDeM), 17(2), 947–979.
16. Rane, N. (2023). ChatGPT and Similar Generative Artificial Intelligence (AI) for Smart Industry: Role, challenges, and opportunities for industry 4.0, industry 5.0 and society 5.0. Challenges and Opportunities for Industry, 4.
17. Rashidi, H. H., Tran, N. K., Betts, E. V., Howell, L. P., & Green, R. (2019). Artificial intelligence and machine learning in pathology: The present landscape of supervised methods. Academic Pathology, 6, 2374289519873088.
18. Rožanec, J. M., Novalija, I., Zajec, P., Kenda, K., Tavakoli Ghinani, H., Suh, S., ... & Soldatos, J. (2023). Human-centric artificial intelligence architecture for industry 5.0 applications. International Journal of Production Research, 61(20), 6847–6872.
19. Talib, M. A., Majzoub, S., Nasir, Q., & Jamal, D. (2021). A systematic literature review on hardware implementation of artificial intelligence algorithms. The Journal of Supercomputing, 77, 1897–1938.
20. Thomasian, N. M., Eickhoff, C., & Adashi, E. Y. (2021). Advancing health equity with artificial intelligence. Journal of Public Health Policy, 42, 602–611.
21. Torres-García, A. A., Garcia, C. A. R., Villasenor-Pineda, L., & Mendoza-Montoya, O. (Eds.). (2021). Biosignal processing and classification using computational learning and intelligence: Principles, algorithms, and applications. Academic Press.
22. Wahl, B., Cossy-Gantner, A., Germann, S., & Schwalbe, N. R. (2018). Artificial intelligence (AI) and global health: How can AI contribute to health in resource-poor settings? BMJ Global Health, 3(4), 16.
23. Xu, X., Lu, Y., Vogel-Heuser, B., & Wang, L. (2021). Industry 4.0 and Industry 5.0—Inception, conception and perception. Journal of Manufacturing Systems, 61, 530–535.
24. Yıkılmaz, I. (2020). New era: The transformation from an information society to a super-smart society (society 5.0). In G. Mert, E. Şen, & O. Yılmaz (Eds.), Data, information and knowledge management (pp. 85–112). Nobel Bilimsel Eserler.
25. Zhang, B., Zhu, J., & Su, H. (2023). Toward the third-generation artificial intelligence. Science China Information Sciences, 66(2), 121101.

Chapter 11

Industry 5.0
Security Based on Optimal Thresholding

Archika Jain and Devendra Somwanshi

11.1 INTRODUCTION

This chapter includes an introduction to watermarking, its classification, characteristics, applications, and requirements It also includes a comparison with steganography and cryptography, details of wavelet, wavelet transform, applications of wavelet, advantages of wavelet, discrete wavelet transform (DWT), segmentation, picture segmentation, application of picture segmentation, classification of segmentation, thresholding, purpose of thresholding, thresholding algorithm, thresholding methods and drawbacks of thresholding.

11.1.1 Brief about Watermarking

Hide the data or extract the information from digital multimedia by using the watermarking technique. For example, extract the watermark from audio, video, digital pictures, and documents [7].

Digital watermarking is the most well-known watermarking technology, which is gaining popularity, especially for adding undetectable identifying markings like author or copyright information. Digital watermarking is a technique for concealing information into a signal so that it cannot be extracted by unauthorized individuals. Digital watermarking embeds a signal into the content without degrading its quality, as well as inserting watermark information into other forms of media known as cover work. Computerized watermarks are embedded in the data, making it impossible for third parties to claim ownership. While some watermarks are visible, the majority of them are not [7]. Figure 11.1 shows real picture and watermark picture.

11.1.2 Watermarking Principle

Figure 11.2 shows the principle of watermarking. There are two types of algorithms in watermarking. The first algorithm is the embedding algorithm, in which real pictures and watermark pictures are used as input data.

Figure 11.1 Real Picture and Watermark Picture.

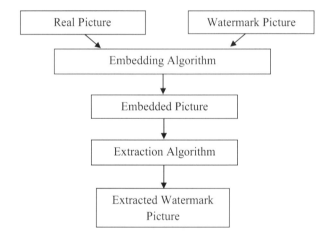

Figure 11.2 Watermarking System Block Diagram.

After this, an embedding algorithm is applied and a watermarked picture is produced. Then for detecting the watermarked picture, an extraction algorithm has to be applied.

A watermarking system is usually divided into two distinct steps.

11.1.2.1 Embedding

Figure 11.3 shows the embedding process. In embedding, an algorithm embeds the watermark picture into the real picture and produces a watermarked picture.

Figure 11.3 Embedding Process.

Figure 11.4 Extraction Process.

11.1.2.2 *Extraction*

Figure 11.4 shows the extraction process. In extraction, an algorithm extracts the watermark picture from the real picture.

11.2 LITERATURE REVIEW

This study comprises evaluating, synthesizing, summarizing, and framing the issue statement through a literature study. The Review Process Adopted, Categorical Review, Common Findings, Strengths and Weaknesses, and Issue-by-Issue Solution Approaches with Comparative Analysis are all covered in this chapter.

In the literature review, more than 40 research papers were reviewed that were published in various research journals in the period of the year 2001 to 2023 in the area of Picture Security. The literature review process contains more than 40 research papers that include a five-stage analysis, findings, and outcomes have been discussed. This method made it simple to classify the literature, analyze, synthesize, summarize, and formulate the problem statement and objectives.

11.2.1 Review Process Adopted

A literature study is required to have an understanding of the research topic and the challenges that have been solved and will be resolved later. Table 11.1 shows the details of research paper reviewed.

Table 11.2 shows the comprehensive assessment of more than 40 research articles published in the field of "Visible Watermarking" from 2001 to 2023 was conducted.

Table 11.1 Details of Research Paper Reviewed

S.NO.	Name of Issue	Total Paper Reviewed	Number of Paper Reviewed		
			Conf.	Journal	Transaction
1	Visible Watermarking	40	38	1	1

Table 11.2 Categorical Review of Research Paper

Authors	Year	Approach	Input Parameters	Results
Joo et al.	2001	Watermarking techniques applied on multimedia	Gray pictures	Inserting a watermark to protect the copyright of media.
Nagmode et al.	2005	High-capacity data hiding scheme	Binary pictures	Although the DPC (double processing case) had the highest embedding capacity, the picture's visual quality was not up to par.
Potdar et al.	2005	DWT based non-blind watermarking technique	Gray pictures	Show watermarking against content protection, copyright management, content authentication, and tamper detection.
Tripathi et al.	2006	Novel DCT and DWT-based watermarking technique	Digital pictures	This technique would not be threatened by picture cropping and JPEG. And achieve maximum PSNR at 64×64 watermark picture.
Ni et al.	2008	Robust lossless data hiding technique	Gray pictures	Show high visual quality of gray pictures and provide data embedding capacity.
Li et al.	2008	A recoverable picture for privacy protection	Color pictures	Reduce the amount of privacy data required, allowing all of the privacy data to be integrated into the privacy-protected picture without the need for data compression.

Table 11.2 (Cont.)

Authors	Year	Approach	Input Parameters	Results
Na-Li et al.	2008	A robust algorithm of digital picture watermarking based on discrete wavelet transform	Gray pictures	Proposed algorithm is invisible and robust against common picture processing and cropping operation.
Fu and Wang	2008	For digital picture authentication, a basic watermarking approach is used	Gray pictures	The suggested approach is safe, quick, and resistant to the quantization attack. The real picture is not required for the watermark extraction process in particular.
Yang et al.	2009	Digital picture watermarking technique using iterative blending	Gray-scale pictures	Robust against several attacks, such as cutting, median filtering, rotation and JPEG compressing.
Sun et al.	2009	A new watermarking method for convert communication and copyright protection	Gray-scale pictures	Hidden information was invisible, and the algorithm was robust to general picture processing operations.
Cika et al.	2009	Picture watermarking scheme based on DWT and BCH	Gray-scale pictures	The proposed watermarking scheme is resistant to common attacks like Gaussian filtering, JPEG and JPEG200 compression, median filtering.
Yeh et al.	2010	Watermarking technique based on discrete wavelet transform	Color pictures	Take less time and give high PSNR and NC values for color pictures.

(continued)

Table 11.2 (Cont.)

Authors	Year	Approach	Input Parameters	Results
Gunjal and Manthlkar	2010	Strongly robust digital picture watermarking scheme based	Gray pictures	The correlation factor for different attacks like noise addition, filtering, rotation, and compression ranges from 0.90 to 0.95. The PSNR with weighting factor 0.02 is up to 48.53 db.
Chandra and Pandey	2010	Visible watermarking technique	Gray pictures	Gives the improving watermark robustness
Cao et al.	2010	A new watermarking method based on the DWT and Fresnel diffraction transforms has been developed	Gray pictures	Some picture processing techniques, such as JPEG lossy compression, median filtering, Gaussian smoothing, and random cropping attacks, are resistant to this strategy. The hiding of the watermark picture using Fresnel diffraction transforms is good, and the watermark's resilience is strengthened.
Ayangar and Talbar	2010	Watermarking scheme based on DWT and SVD	Gray pictures	It is robust against various attacks including geometric attacks.
Sujatha and Sathik	2010	A novel DWT-based blind watermarking scheme	Gray scale pictures	Watermark is robust against those attacks such as adding noises, filtering, intensity adjustment, and histogram equalization.
Salama et al.	2011	A robust digital picture watermarking technique	Binary picture	This technique improved the robustness of the other watermarking techniques and keeps the watermarked picture imperceptible.

Table 11.2 (Cont.)

Authors	Year	Approach	Input Parameters	Results
Arya et al.	2011	LWT and SVD-based watermarking technique	Gray scale pictures	2 level LWT gives the higher PSNR and MSSIM values than 1 level LWT.
Dejun et al.	2011	Robust digital picture watermarking scheme based on SVD and DWT	Gray-scale pictures	Robustness against picture processing operations.
Deb et al.	2012	Combined DWT and DCT-based watermarking technique	Gray-scale pictures	Shows the correlation between the real watermark and the extracted watermark is more than 0.9.
Qianli and Yanhong	2012	Digital picture watermarking algorithm based on DWT and DCT	Gray pictures	Watermarking is robust to the common signal processing techniques including JPEG compressing, noise, low pass filtering, and cutting.
Zhang et al.	2012	Watermarking technique based on wavelet transform	Color pictures	Watermark picture has good invisibility and strong robustness for common signal processing and attack.
Aniyan and Deepa	2013	DCT based blind watermarking	Color pictures	Hardware implementation shows a low-cost, high-performance watermarking system.
Umaamaheshvari and Thanushkodi	2013	Robust picture watermarking based on block-based error correction code	Gray pictures	Provide the weighted addition rule embedding and extraction scheme which was improving the efficiency of the proposed system.
Pathak and Dehariya	2014	SVD–DWT based approach	Gray pictures	In this, watermarks inserted in the lowest frequencies are resistant to a attacks, and watermarks set in to uppermost frequencies are resistant to another attacks.

(continued)

Table 11.2 (Cont.)

Authors	Year	Approach	Input Parameters	Results
Khorramdin et al.	2014	Improving reversible picture watermarking technique based on interpolation method	Gray-scale pictures	Has higher picture fidelity compared to previous schemes.
Ibrahim et al.	2014	Non-blind picture watermarking technique based on interlacing	Color pictures	Excellent robustness against different attacks
Chalamala et al.	2014	Picture watermarking method based on contour-let transform	Gray scale pictures	Contour-let based method is more robust than wavelet-based method even under several attacks.
Kakkirala et al.	2014	Blind picture watermarking technique	Gray-scale pictures	Extracts watermark without cover picture and also proved that this method is robust against different signal and non-signal processing attacks.

11.3 THEORETICAL ASPECTS OF PROPOSED WORK

11.3.1 OTSU

OTSU's approach, named after Nobuyuki OTSU, is used in computer vision and picture processing to automatically execute clustering-based picture thresholding or the reduction of a gray level picture to a binary picture. Following a bi-modal histogram, the technique assumes that the picture has two classes of pixels: foreground pixels and background pixels. It determines the best threshold for dividing the two classes so that their intra-class variance is minimized and their inter-class variance is maximized as the sum of pairwise squared distances is constant.

11.3.2 OTSU Algorithm

Step 1: Read real picture.
Step 2: Initialize variables *gry*, *b*, *prob*, and *meantot*.
Where
gry = gray picture *b* = Picture histogram
prob = probability of real picture
meantot = total mean of real picture
Step 3: Compute probability and total mean of real pictures.

$$prob = b / length(a(:)) \quad \quad (3.1)$$

$$meantot = gry * prob \quad \quad (3.2)$$

Step 4: Set up initial weight $w0$ and $w1$.

$$w0 = sum(prob(1:m)) \quad \quad (3.3)$$

And $\therefore w1 = 1 - w0 \quad \quad (3.4)$

Step 5: The pixels are divided into two classes, cls0 with gray levels [1, ..., *m*] and cls2 with gray levels [*m* + 1, ..., 256].
Step 6: Compute mean of two classes and total variance.

$$meancls0 : (gry(1:m) * prob(1:m)) / w0 \quad \quad (3.5)$$

$$meancls1 : (gry(m+1:256) = prob(m+1:256)) / w1 \quad \quad (3.6)$$

$$tot\,var = ((gry - (meantot * onces(1,256))) \wedge 2) * prob \quad \quad (3.7)$$

Step 7: Compute variance of two classes and between-class variance.

$$var\,cls0 = ((gry(1:m) - (meancls0 * ones(1,m))) \wedge 2) * prob(1:m) \quad (3.8)$$

$$var\,cls1 = ((gry(m+1:256) - (meancls1 * ones(1,256-m))) \wedge 2) \\ * prob(m+1:256) \quad \quad (3.9)$$

$$var\,btw = sum(prob(1:m)) * sum(prob(m+1:256)) \\ * ((meancls1 - meancls0) \wedge 2) \quad \quad (3.10)$$

Step 8: Calculate maximum thresh value.

$$max\,thresh(m) = var\,btw / tot\,var \quad \quad (3.11)$$

11.3.3 Design and Implementation of Proposed Work

11.3.3.1 An OTSU's Process Flow Algorithm by Using Discrete Wavelet Transform

A detailed summary process of the OTSU-DWT process is shown below.

176 Artificial Intelligence and Communication Techniques

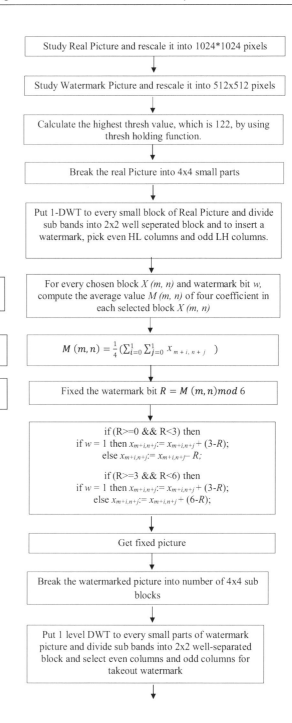

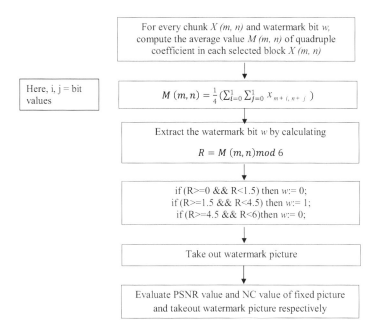

11.4 EXPERIMENTAL RESULTS AND ANALYSIS

11.4.1 Experimental Variations

For performance evaluation of PSNR and NC values based on different techniques like OTSU and DWT and following picture file format was used.

- TIFF picture file format

The following algorithms were used for the experimentation.

- Basic DWT algorithm
- OTSU-DWT algorithm

Experimental analysis was carried out on picture files which are described as follows:

- **Scenario 1:** Performance analysis of an OTSU-DWT algorithm
- **Scenario 2:** Comparison of PSNR and NC values based on an OTSU-DWT algorithms

Table 11.3 Performance Analysis of an OTSU-DWT Algorithms

S. No.	Group	PSNR Values of Real Picture	PSNR Value of Fixed Picture	Improve PSNR after Fixed Watermark	NC Value of Takeout Watermark Picture
1	Group 1	45.50	47.23	3.80%	0.80
2	Group 2	20.40	26.65	30.63%	0.56
3	Group 3	79.20	87.03	9.88%	0.47
4	Group 4	85.35	87.03	1.96%	0.73
5	Group 5	69.89	71.28	1.98%	0.79
6	Group 6	22.79	26.78	17.50%	0.41
7	Group 7	35.00	37.18	6.22%	0.88
8	Group 8	42.15	45.91	8.92%	0.67
9	Group 9	33.25	39.47	18.70%	0.65
10	Group 10	35.00	37.14	6.11%	0.64
11	Group 11	24.56	27.31	11.19%	0.56

11.4.2 Performance Results

Table 11.3 shows the performance analysis of PSNR values of embedded pictures and NC values of removed watermark pictures are calculated using an OTSU-DWT algorithm.

By analyzing Figure 11.5, it is seen that there is significant increase in PSNR values after embedding watermark rather than decrease.

Figure 11.6 shows the percentage improvement of PSNR values after applying OTSU-DWT algorithms. It is seen that maximum percentage improvement, i.e. 30.63% is for Group 2 and minimum percentage improvement, i.e. 1.96% for Group 4.

From Figure 11.7, it is clearly visible that the NC value of the takeout watermark picture is 0.88 highest for Group 7, and lowest 0.41 for Group 6, indicating that there is much quality degradation from real watermark to extracted one.

11.4.3 Comparison of PSNR and NC Values That Based on an OTSU-DWT Algorithms

Table 11.4 shows the performance analysis of PSNR and NC values for different tiff pictures that is based on OTSU-DWT algorithms.

The PSNR value of chilli.tiff is the greatest, as shown in Figure 11.8. The PSNR value of the inserted picture of chilli.tiff is also greater after using the OTSU-DWT algorithms.

We get highest PSNR value for monkey.tiff and the lowest for chilli.tiff, as shown in Figure 11.9.

Table 11.5 shows the performance comparison of the NC values. From Figure 11.10 it is clearly visible that NC value of counts.tiff is highest and after using an OTSU-DWT algorithms, a takeaway watermark image was created.

As demonstrated in Figure 11.11, the NC value is greatest for jpr.tiff and lowest for counts.tiff.

Industry 5.0: Security Based on Optimal Thresholding 179

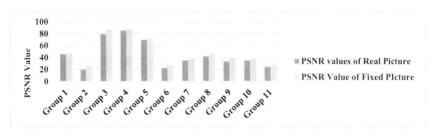

Figure 11.5 PSNR Values of Real and Fixed Pictures While OTSU-DWT Algorithms Used.

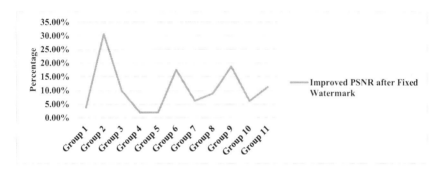

Figure 11.6 Percentage Improvement of PSNR Values While Using OTSU-DWT Algorithm.

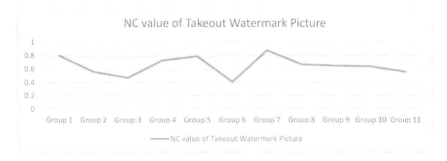

Figure 11.7 After Using OTSU-DWT Algorithms, NC Values of Takeout Watermark Pictures.

Table 11.4 Performance Analysis of PSNR and NC Values That Based on an OTSU-DWT Algorithms

Real Pictures	PSNR Values of Real Picture	Basic DWT Algorithm: PSNR Values of Fixed Pictures	OTSU-DWT Algorithm: PSNR Values of Fixed Pictures	Basic DWT: Improve PSNR after Fixed Watermark	OTSU-DWT: Improve PSNR after Fixed Watermark
home.tiff	44.40	45.60	46.25	3.56%	2.10%
money.tiff	19.50	20.57	25.58	9.10%	28.65%
wise.tiff	77.10	78.45	85.46	2.20%	8.89%
chilli.tiff	83.47	83.55	85.86	1.09%	0.86%
leaf.tiff	70.90	71.05	73.78	1.18%	2.18%
woman.tiff	20.79	23.70	25.78	8.40%	15.50%
ghar.tiff	37.00	37.18	39.18	0.61%	7.22%
shena.tiff	44.15	44.04	46.91	3.15%	8.90%
taj.tiff	32.25	32.79	37.47	3.63%	16.70%
wow.tiff	33.00	35.01	35.14	3.74%	5.11%
sugar.tiff	25.56	26.33	28.31	4.13%	12.19%

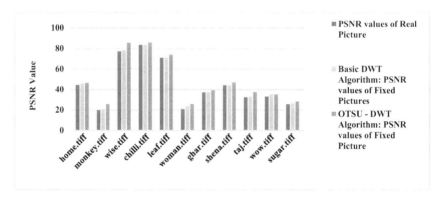

Figure 11.8 Basic DWT and OTSU-DWT Are Compared.

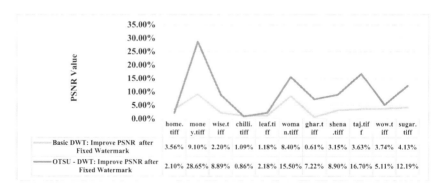

Figure 11.9 Performance Enhancement of PSNR Values.

Table 11.5 NC Values Performance Comparison

Watermark Pictures	Basic DWT Algorithm: NC Values of Takeout Watermark Picture	OTSU-DWT Algorithm: NC Values of Takeout Watermark	Basic DWT Algorithm: Decrement in NC Values	OTSU-DWT Algorithm: Decrement in NC Values
ship.tiff	0.68	0.70	20%	18%
pira.tiff	0.44	0.46	56%	43%
banna.tiff	0.44	0.48	58%	54%
car.tiff	0.75	0.74	29%	29%
jahaj.tiff	0.80	0.80	22%	22%
jpr.tiff	0.40	0.42	61%	60%
counts.tiff	0.90	0.90	13%	13%
suits.tiff	0.74	0.77	46%	43%
vehicle.tiff	0.65	0.66	37%	36%
sweet.tiff	0.65	0.69	38%	37%
campas.tiff	0.53	0.58	49%	45%

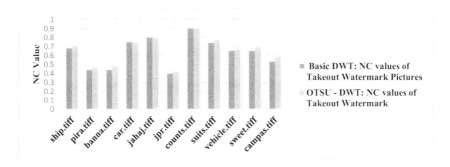

Figure 11.10 Comparative Analysis of NC Values.

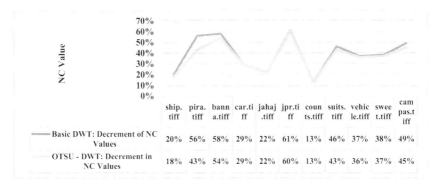

Figure 11.11 Comparison of NC Values for Both the Algorithms.

11.5 CONCLUSION AND FUTURE SCOPE

We have two algorithms in the field of Industry 5.0 in which we compare our PSNR and NC values. PSNR value is highest for monkey.tiff and lowest for chill.tiff, as well as NC values, is highest for jpr.tiff and lowest for counts. tiff after applying an OTSU-DWT algorithm. In the future, the amount of DWT utilized may be adjusted, and the improvement in performance can be tracked. More watermarking techniques can be applied to a variety of photos to see how well they work. There are a few additional criteria that might be examined to perform a thorough study.

REFERENCES

1. Aniyan, A.; Deepa, J., "Hardware implementation of a robust watermarking technique for digital pictures," 2013 IEEE Recent Advances in Intelligent Computational Systems (RAICS), pp. 293, 298. IEEE, Trivandrum, India, 19–21 Dec. 2013.
2. Ayangar, V.R.; Talbar, S.N., "A novel DWT-SVD based watermarking scheme," 2010 International Conference on Multimedia Computing and Information Technology (MCIT), pp. 105, 108. IEEE, Sharjah, United Arab Emirates, 2–4 March 2010.
3. Chalamala, S.R.; Kakkirala, K.R.; Mallikarjuna, R.G.B., "Analysis of wavelet and contourlet transform based picture watermarking techniques," 2014 IEEE International Advance Computing Conference (IACC), pp. 1122, 1126. IEEE, Gurgaon, India, 21–22 Feb. 2014.
4. Chandra, M.; Pandey, S., "A DWT domain visible watermarking techniques for digital pictures," 2010 International Conference on Electronics and Information Engineering (ICEIE), vol. 2, pp. V2-421, V2-427. IEEE, Kyoto, Japan, 1–3 Aug. 2010.
5. Cao, C.; Wang, R.; Huang, M.; Chen, R., "A new watermarking method based on DWT and Fresnel diffraction transforms," 2010 IEEE International Conference on Information Theory and Information Security (ICITIS), pp. 430, 433. Springer, China, 17–19 Dec. 2010.
6. Deb, K.; Al-Seraj, M.S.; Hoque, M.M.; Sarkar, M.I.H., "Combined DWT-DCT based digital picture watermarking technique for copyright protection," 2012 7th International Conference on Electrical & Computer Engineering (ICECE), pp. 458, 461, 20–22 Dec. 2012.
7. Li, G.; Ito, Y.; Yu, X.; Nitta, N.; Babaguchi, N., "A discrete wavelet transform based recoverable picture processing for privacy protection," 15th IEEE International Conference on Picture Processing, 2008. ICIP 2008, pp. 1372, 1375, 12–15 Oct. 2008.
8. Gunjal, B.L.; Manthlkar, R.R., "Discrete wavelet transform based strongly robust watermarking scheme for information hiding in digital pictures," 2010 3rd International Conference on Emerging Trends in Engineering and Technology (ICETET), pp.124, 129, 19–21 Nov. 2010.
9. Ibrahim, M.M.; Kader, N.S.A.; Zorkany, M., "A robust picture watermarking technique based on picture interlacing," *2014 31st National Radio Science Conference (NRSC)*, pp. 92, 98, 28–30 Apr. 2014.

10. Pathak, Y.; Dehariya, S., "A more secure transmission of medical pictures by two label DWT and SVD based watermarking technique," 2014 International Conference on Advances in Engineering and Technology Research (ICAETR), pp. 1, 5. INFOCOMP Journal of Computer Science, 1–2 Aug. 2014.
11. Pin Yeh, J.; Wellu, C.; Jen Lin, H.; Hsuan wu, H., "Watermarking technique based on DWT associated with embedding rule," *International Journal of Circuits*, vol. 4, no. 2, pp. 542–552, 2010.
12. Potdar, V.M.; Song, H.; Chang, E., "A survey of digital picture watermarking techniques," 2005 3rd IEEE International Conference on Industrial Informatics, 2005. INDIN '05, pp. 709, 716. IEEE, Perth, WA, Australia, 10–12 Aug. 2005.
13. Shahabaz; Somwanshi, D.K.; Yadav, A.K.; Roy, R., "Medical images texture analysis: A review," *2017 International Conference on Computer, Communications and Electronics (Comptelix)*, pp. 436–441. IEEE, 2017.
14. Sujatha, S.S.; Sathik, M.M., "Feature based blind approach for robust watermarking," *2010 IEEE International Conference on Communication Control and Computing Technologies (ICCCCT)*, pp. 608, 611. IEEE, Nagercoil, India, 7–9 Oct. 2010.
15. Tripathi, S.; Jain, R.C.; Gayatri, V., "Novel DCT and DWT based watermarking techniques for digital pictures," 18th International Conference on Pattern Recognition, 2006. ICPR 2006, vol. 4, pp. 358, 361. IEEE, Hong Kong, China, 2006.
16. Umaamaheshvari, A.; Thanushkodi, K., "Robust picture watermarking based on block based error correction code," 2013 International Conference on Current Trends in Engineering and Technology (ICCTET), pp. 34, 40. IEEE, Coimbatore, India, 3–3 July 2013.
17. Qianli, Y.; Yanhong, C., "A digital picture watermarking algorithm based on discrete wavelet transform and discrete cosine transform," 2012 International Symposium on Information Technology in Medicine and Education (ITME), vol. 2, pp. 1102, 1105. IEEE, Hokodate, Hokkaido, 3–5 Aug. 2012.
18. Zhang, Y.; Wang, J.; Chen, X., "Watermarking technique based on wavelet transform for color pictures," 2012 24th Chinese Control and Decision Conference (CCDC), pp. 1909, 1913. IEEE, Taiyuan, 23–25 May 2012.
19. Fu, Y.-G.; Wang, H.-R., "A novel discrete wavelet transform based digital watermarking scheme," 2nd International Conference on Anti-Counterfeiting, Security and Identification, 2008. ASID 2008, pp. 55, 58. IEEE, Guiyang, China, 20–23 Aug. 2008.
20. Na Li; Xiaoshi Zheng; Yanling Zhao; Huimin Wu; Shifeng Li, "Robust Algorithm 8 of Digital Image Watermarking Based on Discrete Wavelet Transform," *Electronic 9 Commerce and Security*, pp. 942, 945, Aug. 2008.
21. Yang, Shun-liao; Zhang, Zheng-bing., "Digital Image Watermarking Using Iterative 12 Blending Based on Wavelet Technique," *Multimedia Information Networking and 13 Security*, 2009. MINES '09. vol. 2, pp. 83, 86, 14 Nov. 2009.
22. Sun, Q.D.; Zuo, J.C., "A New Watermarking and Information Hiding Technique 16 Based on Discrete Wavelet Transform," *Information Science and Engineering (ICISE)*, pp. 1067, 1070, 26–28 Dec. 2009.

23. Cika, P., "Watermarking Scheme Based on Discrete Wavelet Transform and Error 19 Correction Codes," *Systems, Signals and Image Processing*, 2009. pp. 1, 4, 18–20 June 2009.
24. Salama, A.; Atta, R.; Rizk, R.; Wanes, F., "A robust digital image watermarking 22 technique based on wavelet transform," *System Engineering and Technology (ICSET)*, pp. 100, 105, 27–28 June 2011.
25. Arya, M.S.; Siddavatam, R.; Ghrera, S.P., "A hybrid semi-blind digital image 25 watermarking technique using lifting wavelet transform — Singular value 26 decomposition," *Electro/Information Technology (EIT)*, pp. 1, 6, 15–17 May 2011.
26. Khorramdin, M.; Amini, M.; Torabi, N.; Mahdavi, M., "Improving reversible image 7 watermarking using additive interpolation technique," *Telecommunications (IST)*, pp. 961, 965, 9–11 Sept. 2014.
27. Kakkirala, K.R.; Chalamala, S.R., "Block based robust blind image watermarking 10 using discrete wavelet transform," *Signal Processing & its Applications (CSPA)*, pp. 58, 61, 7–9 March 2014.

Chapter 12

Industry 5.0

Revolutionizing Energy Management through Smart Grid Integration and Sustainable Solutions

Sonal I. Shirke, Payal Bansal, and Sushil Jain

12.1 INTRODUCTION

Energy management strategies are essential for both residential and commercial systems to reduce energy expenses and raise revenue. According to recent assessments, there is an increasing need for power globally; by 2040, it is expected to expand by 40%. The most recent market research conducted by the European Union (EU) makes it clear that the costs associated with producing and distributing electricity are now a major concern. The creation of a grid architecture that can accommodate dispersed and renewable energy sources (RES) for power generation is the answer to this problem, and it should also be included into the antiquated, rigid, and overburdened centralized electrical systems (Mourtzis et al., 2022).

As a result, information is constantly exchanged between energy suppliers and customers in modern grids as energy is distributed (Mourtzis et al., 2022). The clean, sustainable, and intelligent modern electrical power grid driven by technology is anticipated to be the most significant human achievement of the twenty-first century, ensuring a sustainable future. The term "smart grid" refers to this intelligent electrical power grid. Becoming customer-centered rather than producer-centered is one of the basic ideas at the heart of the smart grid. In fact, this idea aligns with Industry 5.0's first fundamental pillar, which is people-centeredness (Khan et al., 2023).

Aiming to deliver a sustainable, affordable, and efficient energy supply, the smart grid offers new features and opportunities by connecting producers, consumers, and smart meters. According to the paradigm, consumers are encouraged to join the grid and become "prosumers" by producing, distributing, or selling energy. People then become an integral component of the grid's operation and have the ability to decide what is best for their own energy consumption. Through mitigating the adverse effects of the existing power infrastructure, contemporary energy management aims to improve the global environment. This is achieved by establishing optimal integration of RES and conceptualizing a shift away from energy production that is centered on fossil fuels. Renewable energy is produced by consumers using wind turbines and solar panels, which they subsequently provide to other grid users. The grid facilitates data and energy exchange among its stakeholders, which results in meaningful information that can be employed to enhance the grid's operation and distribution of energy (Mourtzis et al., 2023).

Among the many useful features of the smart grid is energy demand response (DR) support, which attempts to cut expenses by offering guidance on device usage during times of peak demand and elevated prices. To decrease peak hours and increase stability, it integrates effective load handling. Another part of it is decentralized energy production, which uses RES to allow people or other stakeholders to add to the power grid (Mourtzis et al., 2023). Most crucially, for effective deployment, accurate and timely power forecasting is a prerequisite for proper dynamic power generation distribution. Smart grids with artificial intelligence (AI) capabilities are necessary to accomplish this goal. They enhance the effectiveness of controlling variable energy sources, like solar power, by offering dynamic, real-time oversight and administration of electricity generation and consumption, thereby revolutionizing power systems. The field of renewable energy has greatly benefited from the application of AI. These contributions include managing supply volatility, forecasting weather and grid stability and reliability, operating energy storage systems, and creating and overseeing markets. Ultimately, this comprehensive approach contributes to closing the environmental gap between energy production and consumption, paving the way for Industry 5.0.

The integration of Industry 5.0 concepts with smart grid technology has resulted in advanced energy management systems (EMSs) that improve efficiency and change the way that sustainable electricity sources are integrated. However, digitization has also brought forth problems, most notably electricity theft, which puts the security of the smart grid network at jeopardy. Efforts are under motion to enhance the efficacy of electricity theft detection in these networks. This project perfectly conforms to Industry 5.0 requirements, with a focus on robustness in particular. Resilience in the context of smart grids seeks to guarantee continuous, high-quality power even in the face of environmental, physical, and cyber threats that may cause power systems to fail. The smart grid's digitization is consistent with Industry 5.0's goal of fusing human intellect with technological prowess. At present, the grid is undergoing a metamorphosis, progressing from a mere electrical current conduit to a robust and environment-responsive system.

12.2 ENERGY MANAGEMENT

12.2.1 What Is Energy Management?

Global pressure sometimes requires innovating infrastructure. Agreeing to sign the Paris Agreement at COP21, many countries have promised to work toward a society with a minimal carbon footprint. As a result, the nation is dedicated to moving toward renewable energy. Countries must reduce the cost of renewable energy, encourage energy conservation and better supply and demand management in densely populated areas, and offer energy at reduced costs to a sparsely populated consumer population in regional cities along with their suburbs to create a society free of carbon emissions as can be seen in Figure 12.1. Should they neglect these responsibilities, energy costs would increase and the power grid will become unstable. Failure to address these challenges effectively could lead to increased energy prices and destabilization of the power grid. Such outcomes would not only inconvenience consumers but also weaken the competitiveness of businesses, impeding the country's economic growth and compromising its overall productivity (Hitachi-UTokyo Laboratory, 2020).

Figure 12.2 illustrates the symbiotic relationship between the historical development of industrial revolutions and the evolution of energy. Originating in the late eighteenth century, the first industrial revolution

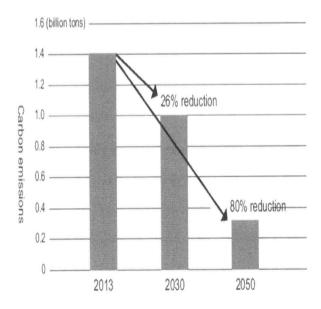

Figure 12.1 Carbon reduction targets (Hitachi-UTokyo Laboratory, 2020).

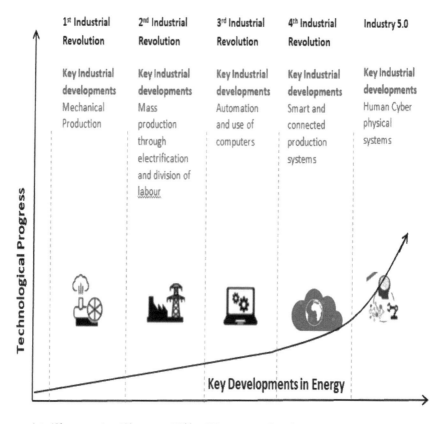

Figure 12.2 Important Energy Developments Alongside the Industrial Revolution (Mourtzis et al., 2023).

symbolized the dawn of mechanized production, displacing manual labor with machinery. Subsequently, the second industrial revolution, sparked around a century later, was fueled via the extensive use of electricity in production processes and the creation of global electricity networks. The third industrial revolution further advanced by prioritizing the combination of computer and automation technologies, optimizing production efficiency. Presently, the fourth industrial revolution, commonly referred to as Industry 4.0, leverages intelligent, networked systems to improve efficiency and flexibility across industries. This networked environment of devices, systems, and machines within and beyond industrial settings has ushered in a new era of heightened intelligence.

By recognizing the common threads between the principles of sustainable energy transition and the tenets of Industry 4.0, a pathway toward Industry 5.0 can be forged. Industry 5.0 envisions a future where sustainable energy practices are seamlessly integrated into the fabric of industrial operations, fostering a harmonious coexistence between technological advancement and ecological responsibility (Mourtzis et al., 2023).

12.2.2 Challenges in Energy Management

You can use the following subdomains to illustrate the main difficulties with energy management (Mourtzis et al., 2023):

1. Prosumer: Prosumers have the ability to produce power and engage in the market. Opportunities for overcoming obstacles include the use of battery energy storage systems (BESS) and DR, the progress of mathematical models of electrical appliances and the creation of sophisticated technologies for real-time monitoring, and the implementation of home energy management systems (HEMS).
2. Aggregation and community: Aggregators coordinating with community energy markets have been identified by current research; however, further focus is required to create hardware and software utilizing cutting-edge algorithms for DR integration and optimization, and assess if it would be feasible to move to take part in community electricity markets, and develop business and asset management models using DR programs and optimization.
3. Market regulation: Further study is required to examine the connection between prosumer optimization models, DR programs, and market regulation to enhance market regulation. Using DR as supplementary services, assessing how DR affects price regulation, looking into the legal ramifications of selling excess energy, and creating public policies to maintain market equilibrium are important factors to take into account. The development of experimental projects, market modeling research, and the assessment of aggregating agent opportunities can all benefit from mathematical formulations. Depending on the area, several economic models may be used and small-scale consumers are encouraged to participate in DR programs when they are permitted.

12.3 INDUSTRY 5.0 IN ENERGY MANAGEMENT

According to Friedman and Hendry, the Industry 5.0 forces information technologists, industry practitioners, and philosophers to concentrate on taking human considerations into account when integrating technologies into industrial processes (Maddikunta et al., 2022).

As can be seen from Figure 12.3, there are essentially three primary concepts that comprise Industry 5.0 (Zafar et al., 2023):

1. Human-centered approach: This value prioritizes production according to the needs and interests of people.
2. Sustainability: To save the environment and create circular processes that recycle, reuse, and repurpose materials while boosting output and efficiency, Industry 5.0 takes the lead.
3. Developing a solid plan to protect vital infrastructure in the event of a catastrophe is a necessary component of building resilience.

The EMS is a strategic component of Industry 5.0, which is revolutionizing industry operations. These are the links between Industry 5.0 and energy management power forecasts (Khan et al., 2023) and the Figure 12.3 depicts the same:

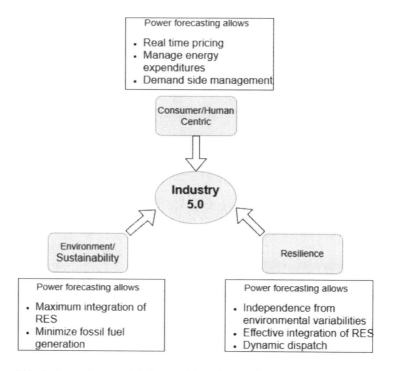

Figure 12.3 Basic attributes of Industry 5.0 and its relation to energy management (Mourtzis et al., 2023).

- Consumer-centric: A basic tenet of the smart grid, utilized for energy management, is the emphasis on the customer rather than the producer. This idea supports Industry 5.0's first fundamental pillar, which is people-centeredness. The fundamental tenet of conventional power networks is that consumers should consume power passively, having no control over how it is produced. The smart grid's prosumers, or producers and consumers, will have greater control over the energy produced and used. Real-time dynamic pricing will be used to bill electricity consumers based on their consumption of clean or dirty energy. When the utility is able to predict power reliably and confidently notify users in advance, it can put this principle into practice. Efficient power forecasting methods and methodologies utilizing machine learning and AI must be linked into the smart grid to give consumers this flexibility. The suggested research project tackles the problem of accurate power forecasting, which makes it related to the deployment of smart grid in Industry 5.0.
- Integration of renewable energy: It is anticipated that the current power systems, which contribute significantly to the generation of greenhouse gases and global warming, will be replaced with more environmentally friendly and sustainable systems of energy. Because of the power of AI and machine learning. Power forecasting powered by AI has the potential to be extremely helpful in reaching this goal. Industry 5.0 uses the smart grid to incorporate renewable energy into industrial processes. Manufacturers can employ renewable energy generation and the real-time monitoring and management of dispersed energy resources to replace traditional energy sources.
- Grid resilience and stability: Resilience is one of the concepts included in standards of Industry 5.0. The goal of resilience in a smart grid is to guarantee uninterrupted, high-quality power in the face of various physical, cyber, and environmental risks that could interfere with the power infrastructure. The power flow is becoming more and more reliant on both short- and long-term environmental circumstances as RES are being incorporated into electricity networks across the globe. Variations in the surrounding conditions will alter the RES's output and may have a significant impact on the power system's supply and demand balance. Power grid operators will be able to anticipate and address power imbalances and maintain the grid's resilience in these circumstances with the use of accurate and effective power forecasting. By integrating blockchain technology into energy transfers, the process will become more resistant to cyberattacks.

Numerous technologies facilitate the diverse uses of Industry 5.0. Some of the technologies that enable Industry 5.0 are cobots, digital twins, AI, Internet of Everything, 6G, blockchain, and more. Blockchain and AI are the technologies that enable smart energy management. In the forthcoming sections, we will discuss about their involvement and contribution in energy management.

Industrial revolutions aim to raise the standard of living in society and make life easier for consumers in addition to directly addressing and improving industry demands in a productive manner. Consequently, any industrial revolution should always be accompanied by economic expansion. Here, the population's ever-increasing demand for electricity is met under the best circumstances possible by addressing the constraints of the twenty-first century, which include the nonrenewable nature of conventional sources, the rising cost of fuel, and the growing demand for electricity.

Countries with limited energy resources or those that import raw materials must be able to respond to these challenges with wisdom. In view of this, the idea of "smart cities" is growing in popularity. As the primary user of energy sources in metropolitan areas, the electrical industry has taken up the majority of the focus. This is prompting urban planners to shift the focus of the city's intelligence toward smart energy cities.

There have been a number of modifications to the manner that electricity is generated. Centralized energy was replaced by decentralized energy, and distributed energy was created by integrating RES (Tabaa et al., 2020).

Sustainable development in the energy sector is made possible in large part by the digital revolution. Smart energy management is made possible by the use of digital technologies like the AI, Internet of Things (IoT) and big data analytics, which are being used to monitor and control environmental effects, optimize resource use, and encourage sustainable energy production and consumption (Yin et al., 2023).

The most important aspect of human development in the twenty-first century, ensuring a future driven by technology, is anticipated to be the development of the clean, sustainable, and intelligent contemporary electrical power grid (Khan et al., 2023).

12.4 MICROGRID

A microgrid is where the idea for a smart grid first emerged. A microgrid is a small power grid that can operate entirely on its own or in conjunction with other local power systems. A microgrid is any isolated, small-scale power plant with independent production, limits, and energy storage capabilities. Microgrids are widely used to provide backup power to the main power grid during peak hours. Usually, they are run on RES like solar and wind power, or generators. Microgrids can run independently in "island mode" and separate from the networked macrogrid when needed by technological or financial circumstances. As power outages can result in significant revenue losses and protracted restart times for numerous manufacturing operations, industrial and commercial microgrids are designed to guarantee the security and dependability of the power supply. Consequently, the circular economy should be considered when designing industrial microgrids (Pandey et al., 2023).

Hence comes the concept of smart grid.

12.5 SMART GRID

12.5.1 Why Smart Grid?

Energy production, distribution, and transmission are all parts of the energy production process. Although they are not typically included in the production of power, energy trading and storage are crucial to the development of renewable energy. While renewable energy is more sustainable than conventional resources in many ways, it is not as reliable in terms of power delivery. Moreover, harnessing renewable energy is challenging due to its inaccessible nature. The unpredictability of electricity supplies and inconsistencies in the market are causing backlash against renewable energy. Specifically, more efficient operation of distribution networks for resources used in the renewable energy production is required.

Industry 5.0's technology prowess can help renewable energy systems by providing strong flexibility and increasing energy transparency. Industry 5.0 has the potential to enhance the accessibility of intermittent renewable energy in several industries, such as biomass energy generation, hydropower plants, solar power, and wind power. In this case, distribution network component management requires a smart distribution grid. In particular, smart grids and microgrids can facilitate energy distribution, which in turn can lead to a rise in the usage of renewable energy. Smart energy production control systems, emergency DR mechanisms, smart consumption, transmission line fault detection, and power loss management are a few uses for integrating Industry 5.0 technology in electricity distribution (Pandey et al., 2023).

When power systems are operated in a decentralized market, the distribution system will be significantly impacted. With dispersed generation growing, it is essential to implement market structures that support local energy production and consumption. The local electricity market helps maintain a balance between supply and demand for energy locally and reduces the need for more expensive infrastructure development.

To encourage more retailers and owners of decentralized energy resources to engage in the market, governments in certain nations have created electricity markets. Transactions can be carried out directly between participants without the need for any authority agents, saving participants' time and money. This phenomenon is referred to as peer-to-peer (P2P) energy trading, which operates within the framework of microgrids within the distribution system. This model relies on the exchange of information not only between prosumers but also among electrical lines. Prosumers, individuals who both produce and consume energy, are typically equipped with RES for electricity generation. They engage in energy trading with consumers, forming interconnected networks for energy exchange. Thus, the microgrid is created.

Energy policies are now being implemented by numerous nations to promote photovoltaic (PV) energy or renewable energy self-consumption from the prosumers' standpoint. In fact, India has announced to have an over-the-counter (OTC) energy market where buyers and sellers can directly transact the power and discuss the contract and price. This will happen for both conventional (coal, gas, hydropower) and renewable (solar, wind).

And all this is incorporated in a smart grid.

12.5.2 What Is a Smart Grid?

Modern digital technology is used by the smart grid, an electricity distribution system, to enable two-way communication among stakeholders. Through continuous observation, analysis, and regulation of the electrical current, useful data can be obtained to build a more intelligent, flexible, and reliable grid that can predict energy consumption and expenses (Mourtzis et al., 2023).

From a technical standpoint, smart grid's three main systems are as follows (Fang et al., 2012):

1. Smart infrastructure system: This is the energy, communication, and information infrastructure that underpins the smart grid and facilitates three essential operations: (1) Sophisticated information management, monitoring, and metering; (2) cutting-edge technology for communication; and (3) enhanced production, distribution, and use of electricity.
2. Smart management system: The smart management system is the name of the smart grid subsystem that provides advanced control and management features.
3. Smart protection system: The smart protection system is the smart grid subsystem that provides failure protection, advanced grid reliability analysis, and security and privacy protection services.

The smart grid uses a number of significant components to accomplish its objectives (Zafar et al., 2023). Among these are:

- Energy DR support, which seeks to cut expenses by offering guidance on how to use devices during times of peak demand and higher rates.
- Effective load management shortens peak hours and increase stability.
- Decentralized energy production makes use of RES to allow individuals or other stakeholders to contribute to the power system.

- The self-healing and resilience of the grid reduce errors and disturbances. The grid quickly identifies and isolates errors, reroutes power flows, and more effectively restores service because of automated monitoring and control technologies. This shortens the duration of outages, improving the electricity grid's reliability.

12.5.3 Components of a Smart Grid

All of the physical equipment connected to the electrical grid is part of the smart grid. The items in question are devices, which provide the following attributes related to energy: Advanced metering, home/industrial automation, and intelligent grid control and management.

12.5.3.1 Customer Domain Components

- Meter: This is the name for the point-of-sale instrument that is used to measure electricity usage across different systems and domains. Keep in mind that the meter can connect with the provider, distribution, and operation domains, among other domains. A meter can function as a stand-alone instrument that connects other domains in this way.
- Equipment and appliances for customers: A device or instrument designed to do a specific job, especially a household electrical appliance like a refrigerator or TV. It can be an electronic machine or appliance with the potential to be controlled and observed.
- Distributed energy resource and storage: On a customer site, energy-generating resources like solar, wind, and others are used to generate and store energy, which is then interfaced with the Home Area Network (HAN) controller to carry out energy-related tasks.

The term "electric vehicle" refers to a car with an electric motor that is primarily powered by a rechargeable battery, which can be charged via a gasoline-powered alternator or by plugging into the grid; the term "customer premise display" refers to a device that lets customers view usage and cost information inside their house or workplace.

- Gateway for the home area network: This is an apparatus that acts as a bridge between the devices in the client domain and the distribution, operations, service provider, and customer domains.

12.5.3.2 Distribution Domain Components

- Distribution data collector: A device that collects data from multiple sources and changes or converts it into different formats.

- Automation field devices: Multifunctional installation tools that satisfy a wide range of planning, control, and measurement needs, as well as system performance reporting for utility staff
- Remote Terminal Unit (RTU) and Intelligent Electronic Devices (IED): These gadgets gather information from power equipment and sensors, and they can also give feedback control directives. For example, they can trip circuit breakers in the event of anomalies in current, voltage or frequency, or they can adjust voltage levels to keep the system stable by raising or lowering voltage levels.
- Distribution sensor: An apparatus that gauges a physical attribute and transforms it into a signal that an observer can interpret to determine the state of the electrical distribution network.

12.5.3.3 Service Provider Domain Components

- Computer and networking equipment: Utility companies and third-party providers use these devices to run their regular business operations, which include billing, aggregation, retail, and other applications.

12.5.4 Challenges for Smart Grid Implementation

To fully reap the rewards of smart grid implementation, both industrialized and developing nations must solve the following major problems and issues (Mourtzis et al., 2022):

1. Technology development: A completely new, highly dependable, and resilient communication infrastructure needs to be built, either independently of or connected with the current World Wide Web, to make the grid even smart. Thus, developing and implementing cutting-edge parts, such as flexible AC transmission system devices, energy-efficient storage devices, smart appliances, smart meters, and high-voltage DC transmission devices is essential.
2. Quality power for every home: The current grid will need to be updated and expanded to guarantee the maintenance of a high-quality supply for every household. To further reduce the supply gap during peak hours and peak energy costs, distributed renewable energy generation and the ability to shift loads from off-peak to peak periods should be encouraged.
3. Diminishment of losses in transmission and distribution: To comply with international standards, transmission and distribution losses should be kept to a minimum. The primary variables influencing the transmission and distribution loss include financial losses, a fall in collection efficiency, and technical losses brought on by a weak grid.

4. Interoperability and cybersecurity: Interoperability standards are developed by the United States. These standards include advanced metering infrastructure and smart grid end-to-end security, building automation, a revenue metering information model, application-level EMS interfaces, inter-control center communications, substation automation and protection, phasor measurement unit communications, and information security for power system control operations.
5. Customer support: One of the biggest challenges to implementing the smart grid is the general lack of customer awareness of issues in the energy sector. Therefore, customer support for smart grid implementation will be crucial to lowering peak load usage and promoting distributed renewable generation of energy. The deployment of an intelligent grid will improve the reliability and quality of the electricity supply. New alternatives for utility users will also be advantageous, including the ability to save money by shifting loads from peak to off-peak hours, green electricity, and an easy-to-use interface. Nonetheless, customers need to support their utilities and be informed of emerging technologies to profit from the smart grid on a personal and a national level.

Further, we will be discussing about the two main technologies which pave the way for energy management in the smart grids.

12.6 BLOCKCHAIN IN SMART GRID

12.6.1 Why Blockchain?

It is anticipated that Industry 5.0's automated procedures would include human impact through the use of precise and accurate control and modeling systems. Industry 5.0 is anticipated to support numerous various systems, including smart grids. As open wireless networks are used for data sharing, security becomes essential to protecting industry systems. Hence, to secure the industrial perimeters and promote synergy, reliable and secure data flow is essential. Because of its intrinsic qualities of immutability, chronology, auditability, and timestamped ledger, blockchain is a favored option for enabling security in Industry 5.0 ecosystems (Verma et al., 2022). Thus blockchain contributes toward making the application resilient.

12.6.2 What is Blockchain?

Blockchain, a distinctive form of data structure, functions by executing and recording transactions within blocks. Each block contains a timestamp and a cryptographic hash link connecting it to the preceding block, ensuring the integrity and sequentiality of the recorded data. This structure guarantees that once data is added to the blockchain, it remains immutable and cannot be altered retroactively.

Conceptually, blockchain operates as a distributed database, devoid of a central authority, where multiple entities interact without mutual trust. In this decentralized environment, participants can rely on the collective integrity and consensus of the network rather than a single controlling entity. This decentralized nature fosters a system where transparency, security, and reliability are maintained through the collaborative efforts of its participants.

A chain of ledgers made up of many individual blocks is what is commonly referred to as a blockchain. In addition, because blocks cannot be modified without also modifying the entire chain of blocks, the system is more resilient and the attacker faces a challenge. Each of the preceding blocks must be altered to change the current block. Blockchain technology stands out as a highly secure approach for transmitting digital assets, currency, and contracts without relying on intermediaries.

Blockchain functions as a public ledger of all transactions during digital events between the involved parties. Mining is the practice of producing new blocks. All parties come to an agreement to check each block, ensuring the integrity and reliability of the system.

The basic blockchain setup is shown in Figure 12.4. The first block in any blockchain network is called as the genesis block. As seen in figure, each block is uniquely identified by its information and hash. Every block also includes the hash of the earlier block as shown in it. This serves as an additional layer of security, because changing one block invalidates all subsequent blocks (Chitchyan & Murkin, 2018).

A central authority is not present with blockchain technology, therefore participants in the network directly execute and validate transactions. Participants in the network are also in charge of storing transaction data. Because third-party intermediaries are no longer necessary, transaction fees are greatly lowered, which is a major benefit of this technology.

Additionally, as each user has a copy of the list of transactions, the ledger's contents can be independently checked by other users on the network.

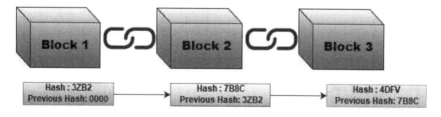

Figure 12.4 Blockchain representation.

Transparency is ensured by this. Furthermore, as every network participant contributes to the validation of the transactions by comparing their version of the ledger with others, there is no requirement for a trusted authority to provide security. Blockchain is difficult to tamper with without a sizable number of users doing so at the same time due to its decentralized structure. As previously indicated, encryption, namely hash functions, which are difficult mathematical procedures that permit only authorized users to decipher the transactions, also protects it.

Additionally, smart contracts, which are self-enforcing pieces of software embedded in the ledger and define the rules behind each transaction, enable transactions to be carried out automatically. These contracts introduce changes to the ledger based on previously defined terms of agreement and, as a result, speed up the process.

12.6.3 Blockchain in Energy Management

Numerous businesses are becoming more interested in blockchain, including the energy industry, which is one of the fields that blockchain technology has the most potential to disrupt. To manage new sources, however, to maintain supply consistency and safety, new solutions are needed. Blockchain technology seems to hold potential for solving some of the problems that face today's energy infrastructure. The decentralized architecture improves fault tolerance and communication speed. Blockchain technology is ideal for carrying out business operations in smart grids as it includes smart contracts (Verma et al., 2022). It has a major contribution in making the grid resilient.

In the context of smart grid technology, an authentication mechanism needs to meet the following important security requirements to be considered reliable. The requirements listed below must be fulfilled (Ayub et al., 2023):

- Mutual authentication: To ensure that only authorized smart meters may access electric services inside a certain smart grid, mutual authentication between smart meters and their service providers is necessary. This authentication process must be finished to maintain the integrity and security of the system.
- Anonymity: A smart meter's identity should be concealed to prevent unauthorized parties from associating it with a specific individual or location. This can be accomplished by using anonymous or pseudonymous identifiers, which hide the real identity of the smart meter.
- Untraceability: It is imperative to prevent other parties from keeping an eye on or tracking certain smart meters' activities and consumption trends. Using techniques like encryption, data aggregation, and secure communication protocols, smart meters may exchange anonymous and untraceable data.

- Resilience against attacks: The authentication system needs to demonstrate resistance to a wide range of recognized attacks, such as replay attacks, impersonation, physical capture, and other security risks. This resilience is essential for supporting the system's ability to bear a variety of security risks and continue to function in hostile environments.

Below is a brief summary of the blockchain's key energy sector applications toward meeting the above requirements of grid resilience.

12.6.3.1 Billing

Blockchain has the ability to enhance the present billing system as a database technology. As everyone uses energy in a different way, payment alternatives are ultimately not well suited to the needs of the consumer. As there is typically a cumulative amount to pay for consumed electricity on the energy bill, individual energy profiles are still barely known. Understanding usage trends and putting pay-as-you-go or micropayment systems in place might have a significant impact on both energy firms and consumers. Automated billing improves on the present, dated systems.

12.6.3.2 Data Management

The provision of clients with information and control over their energy use is a component of the development of smart grids. Depending on the infrastructure at hand, different communication methods are used for this. Blockchain would make this type of data transfer more dependable and secure, as well as guarantee transparency and make it more difficult to tamper with the records on either side. Additionally, it contains a lot of data, like source of energy, which were previously hard to track. It is important to note that having access to time-stamped data is extremely essential to energy utilities and gives them the chance to enhance their operations and provide better services.

12.6.3.3 Peer-to-peer Energy Trading

The installation of decentralized grids is supported by P2P energy trading, because it only involves two other network participants. One of the most promising blockchain applications for the energy sector to date may be this one. Due to the implementation of the smart contract infrastructure, Ethereum is the most widely used platform for P2P trading. As was already mentioned, there is no longer a need for third-party middlemen in blockchain applications. Energy can be freely and trustfully exchanged between network participants. It enables all energy consumers to participate in the market, giving them more control over their energy use and a wider range of energy source options. In contrast to centralized energy systems, where they have no choice but to buy their energy, they may also be able to negotiate a cheaper price. Prosumers who own generating assets can now transform into suppliers, utilizing their energy output to the fullest, and either selling or storing any excess. As blockchain is expected to ensure the capacity to trace the source of the energy, there is still the option to purchase green energy at a competitive price.

12.6.3.4 Grid Flexibility

Because RES are intermittent, the system needs to be more adaptable. The strain on the system can be significantly decreased if the extra electricity is stored or directed to the parts of the network that require it the most. Blockchain technology, for instance, allows consumers to interact with local producers, resulting in better energy management and source coordination. Another advantage of flexible grids is dynamic pricing, which is decided by all market players and is based on actual transmission costs. By coordinating a large number of transactions among numerous participants and reaching a just consensus, blockchain can speed up these procedures.

12.6.3.5 Cybersecurity

As the energy sector grows more digitalized, it is crucial to ensure that smart grids are secure. Due to the interconnectedness of smart grids, any security breach might have catastrophic consequences, including blackouts and power disruptions. Using blockchain technology can help to increase the cybersecurity of smart grids. Because blockchain is decentralized, it provides a transparent and secure means of transferring and storing data. Blockchain-powered smart grids can protect against cyberattacks, ensure data integrity, and protect user privacy. By implementing security measures like identity management, access controls, and threat detection systems, cybersecurity may prevent hostile assaults and data breaches in any industry. Furthermore, blockchain technology can establish a safe and transparent platform for data management and exchange among several stakeholders in the specific industry and its subdomains (Mourtzis et al., 2023).

12.6.3.6 Demand–response Management

The DR control unit and the customer's smart meter transmitting data securely using shared keys and mutual authentication are prerequisites for resilience. The collected data is saved on the blockchain for privacy and integrity purposes, enabling trustworthy and efficient DR management based on the data that has been stored (Ayub et al., 2023).

12.6.4 Case Studies

A smart grid framework has been devised and implemented in a Greek university campus as part of research study on the democratization of energy (Mourtzis et al., 2023). In the end, the provided study work aims to establish the groundwork for traditional consumers—including institutions of higher learning—to become prosumers. Blockchain makes consumer goals more appealing by expanding the potential for integrating local and retail market mechanisms. Blockchain can also create a transparent and safe platform for data management and exchange between many departments and stakeholders in the Smart Campus. The main obstacles and chances for future growth have been identified and are being showcased based on DR.

The authors of this chapter suggest a privacy-focused authentication method for consumer electronics-based smart grid networks that require safe, customer-centric DR management (Ayub et al., 2023). To improve resilience to such assaults and guarantee the integrity of DR data, their suggested protocol makes use of blockchain-based authentication. To confirm the session key security and validate the efficacy of our suggested strategy against several threats, they have carried out both formal and informal security evaluations. Comparing their protocol to other similar protocols, they found that theirs offers more functionality and better security at a cheaper computation and communication cost. They have also measured the computational time required to implement their protocol's blockchain component while accounting for the effects of different block addition and transaction counts.

In this work, a method to protect privacy during the energy trading process was proposed (Guan et al., 2021). To increase the security and dependability of the model that safeguards personal data, a model known as the Privacy Preserving Blockchain Energy Trading Scheme (PP-BCETS) was proposed. A credibility-based consensus process was employed to boost the effectiveness of the operation. According to the findings, privacy leakage is resolved without compromising system performance.

This work aims to improve the grid's resilience using an alternative method (Zafar et al., 2023). A major danger to the security and dependability of the system is the possibility of electricity theft. Using deep federated learning (FL) and a Convolutional Gated Recurrent Unit (ConvGRU) model, this work proposes a safe and dependable theft detection method for smart grid networks. To train a ConvGRU model on dispersed data sources while maintaining data privacy, the suggested technique makes use of FL. The investigation's findings show how well the deep FL-based ConvGRU model performs in precisely identifying power theft. The comparison study demonstrates the superior performance of the suggested method, which is demonstrated by the high scores attained for recall, accuracy, precision, and F1.

12.7 ARTIFICIAL INTELLIGENCE IN SMART GRID

12.7.1 Why AI?

The World Commission on Environment and Development proposed the strategic idea of "sustainable development" and defined it in a report titled "Our Common Future," which was published on April 27, 1987. The goal of this concept is to create balanced development in the three areas of economics, environment and society by "meeting current needs without compromising the ability and opportunities of the next generation to meet their own needs" (Tahir et al., 2023). Sustainable economic development has an internal driving force and technological help from technological innovation. Technological innovation modifies production models, increases the efficiency and rationality of resource allocation, decreases production time and costs, and forms a green industrial system. It also lessens the environmental harm that traditional industries sustain while promoting economic growth momentum. As was previously said, smart grid technologies allow for the efficient management of variable energy sources like solar power by providing real-time tracking and management of the production and use of electricity. AI technologies can be used for predicting and as inference to enhance energy flow management and increase its predictability and efficiency (Rojek et al., 2023).

12.7.2 AI in Energy Management

Artificial intelligence has shown to be a valuable tool for computer and machine replication of human intelligence (Mourtzis et al., 2022). Similar to this, AI is helpful in the energy sector as it makes it possible to manage the enormous volumes of data generated within a smart grid and keeps up with its growing complexity. AI is very useful for monitoring, operating, maintaining, and storing electrical energy produced in the renewable energy industry. As a result, the following can be used to summarize the main characteristics of AI:

- Energy generation taking volatility in the supply into account
- Stability and dependability of the grid
- Demand on the grid and weather forecasting
- Demand-side management of the grid and energy storage operations
- Market design and administration

Artificial intelligence in smart grids utilizing RES is leveraged across four primary domains (Rojek et al., 2023):

1. Smart management, which includes PV grid autonomy: AI has the potential to improve PV grid autonomy and smart management by

optimizing grid control and operation. To make informed decisions regarding when and how power is generated, stored, and delivered, machine learning algorithms are able to analyze real-time data from PV panels, weather forecasts, energy demands, and other pertinent aspects. As a result, the grid can more effectively balance supply and demand, improve energy efficiency, and react dynamically to changing circumstances.
2. AI has the potential to enhance PV grids' dependability and robustness by anticipating and averting malfunctions or breakdowns. By analyzing sensor and device data, predictive maintenance algorithms can find abnormalities and possible issues before they cause system outages. More responsiveness and less downtime can be achieved with AI-based issue detection and diagnostics, which also increases power supply reliability.
3. Human-centric design approach: Creating AI solutions that are easy to use, simple to comprehend, and consistent with human needs and behavior is the goal of the human-centric approach to AI design. When it comes to PV grids, this can mean developing user interfaces that let consumers or grid operators communicate with AI systems in a natural way. AI can also assist with demand forecasting, which enables grid operators to make well-informed decisions based on patterns of energy usage and human behavior.
4. Data security and privacy: Data security and privacy: AI in PV grids needs to provide sensitive data security and privacy. Data protection laws and strong security measures must be followed by AI systems that process data from PV panels, energy usage, and other sources. Safe data transfer, access control methods, and encryption are essential for preserving data privacy.

12.7.3 Challenges

The area of basic AI analysis of residential solar PV networks still has several unmet research needs, some of which are as follows:

- Integration of data in real time: There may still be work to be done in terms of efficiently integrating real-time data from multiple sources, such as weather forecasts, energy use trends, and PV system performance. Grid analysis may become more precise and adaptable if AI algorithms that can smoothly integrate these data streams are developed.
- User-friendly decision support: Experts or researchers should perform the majority of AI analysis for residential PV grids. There might be research needed in the development of user-friendly tools or interfaces that enable small enterprises and homeowners to quickly

understand AI-based insights and decide how best to optimize their PV systems.
- AI is capable of predicting maintenance requirements and detecting defects in PV systems. But there might be research needed in creating algorithms that can precisely forecast certain maintenance needs or simplify the diagnosis of system errors, enabling nonexperts to take the necessary action.
- Localized energy storage optimization: AI can maximize the utilization of PV energy storage; nevertheless, there can be a research gap in creating algorithms that consider regional energy laws, subsidies, and market dynamics. These algorithms could assist homeowners in determining when to take energy from the grid, sell extra energy back to the grid, or store it.
- Long-term performance evaluation: Short- to medium-term optimization is the main focus of many AI assessments. Assessing the long-term effectiveness of AI-guided strategies, taking into account variables like system degradation, technical advancements, and shifting energy landscapes, may still need more investigation.

12.7.4 Case Studies

As environmental concerns and the cost of fossil fuel energy continue to rise, PV systems are becoming more and more common in residential applications in Poland and throughout Europe (Rojek et al., 2023). The purpose of this study is to determine how reliable a home PV system's short-term performance is; these forecasts are typically generated 24 hours in advance. Based on an actual household profile (chosen energy storage installation), the authors conducted a comparative analysis of several energy management strategies using both conventional techniques and a range of AI tools. Their work is innovative and significant because it shows that even basic AI solutions can be useful in inference and forecasting to enhance energy flow management and increase its predictability and efficiency in the context of residential PV networks.

This study uses a hybrid deep learning model (DLM) to provide a novel method for improving PV power plant power output predictions under dynamic environmental conditions (Khan et al., 2023). To accurately capture the spatial and temporal relationships within meteorological data that are essential for making correct forecasts, the hybrid DLM combines the strengths of long short-term memory (LSTM) networks, convolutional neural networks (CNN), and bidirectional LSTMs (Bi-LSTM). With its exceptional accuracy and robustness in power production prediction under a range of weather scenarios, this hybrid technique demonstrates its promise for effective PV power plant management in the smart grid. These study results have great promise for real-world applications in PV power plant management in addition to their contribution to the field of renewable energy. Stakeholders may accelerate the shift to clean and sustainable energy sources by using this information to inform decision-making and maximize PV power generation efficiency.

The optimization of DR in smart energy systems (SESs) is the main topic of this study (Javed et al., 2023). Within the confines of this manuscript, the writers executed and showcased outcomes from the initial phase of DR optimization, and offered a conceptual framework for employing edge AI-driven environmental assessment in stage 2. Any SES could benefit from the adaptation of this novel two-stage DR optimization combo. This innovative approach ensures low-latency, secure communication across different industrial, and smart home IoT devices. The study intends to promote sustainable energy practices and increase the overall efficiency of SESs by concentrating on room-level optimization.

The goal of this endeavor is to reduce domestic electricity costs with smart energy management approaches based on AI (Mateen et al., 2023). To reduce expenses while allowing for a reasonable delay in appliance scheduling, the authors suggest a brand-new hybrid Genetic Flower Pollination Algorithm (GFPA) that draws inspiration from nature. To build a hybrid technique, their suggested GFPA algorithm incorporates parts of the Flower Pollination Algorithm (FPA) and Genetic Algorithm (GA). We assume a scalable town with 1, 10, 30, and 50 households, respectively, to evaluate the performance of the suggested approach. The suggested method determines the ideal scheduling scheme that maximizes user comfort (UC) and concurrently minimizes Electricity Cost (EC) and Peak to Average Ratio (PAR). They make the assumption that every home has the same appliances and uses the same amount of electricity. In terms of cost reduction, simulation results indicate that their proposed scheme, GFPA, outperforms unscheduled, GA, and FPA-based solutions when applying the Critical Peak Pricing (CPP) signal using various Operational Time Intervals (OTIs), achieving, on average, 98%, 36%, 23%, and 22%, respectively. These results indicate that their plan can be used in real-world scenarios to increase power grid sustainability and efficiency.

To control power consumption and lower costs in IoT-enabled smart residential buildings, this work presents a DR-based EMS based on AI techniques (Shreenidhi & Ramaiah, 2022). It does this by effectively scheduling household appliances to operate during off-peak hours rather than during peak hours. To predict the best scheduling pattern for the smart appliances, the authors use a Two-stage Deep Dilated Multi-Kernel Convolutional network (DDMKC)-Modified Elephant Herd Algorithm (MEHOA) for HEMS. This algorithm uses the customer's resource features and price-based DR programs as the input. To control the load, they divide the appliances into three categories: Critical appliances, power-flexible appliances, and time-flexible appliances. Next, to improve the user experience, consumers are grouped together. Subsequently, the DDMKC learning model is utilized to anticipate pricing signals. It acquires knowledge of price signals and produces future forecast prices for the purpose of optimal scheduling. The MEHOA scheme arranges the appliances' power usage according to a schedule that chooses which one should be powered on at which time. The suggested model is assessed using several criteria and contrasted with the current models. In comparison to other state-of-the-art methods, the outcome demonstrates that energy consumption is reduced and performance is enhanced.

12.8 SUMMARY

An emphasis on switching to RES is part of the worldwide movement toward a low-carbon society. However, to achieve this aim, a number of issues must be resolved, such as lowering the cost of renewable energy, encouraging energy efficiency, successfully regulating the supply and demand for energy, and guaranteeing that everyone has access to reasonably priced energy. If these issues are not resolved, consumers and businesses may suffer from higher energy costs and unstable power systems.

The development of industrial revolutions, from mechanization to Industry 4.0's incorporation of smart technology and, more recently, Industry 5.0, shows a path toward sustainable energy transitions.

Utilizing digital technology to facilitate bidirectional communication between stakeholders, the smart grid is an example of an updated electrical distribution network. To facilitate effective load managing, decentralized energy production, and grid resilience, key subsystems include smart infrastructure, management, and protection systems.

Numerous case studies illustrate how blockchain-based systems can improve grid resilience, secure DR management, protect privacy in energy trade, and identify power theft. Based on these findings, blockchain technology has the potential to solve important problems and boost smart grid networks' dependability and efficiency.

To optimize energy management and handle the intricacies of smart grid operations, AI is crucial. Numerous case studies demonstrate how AI may be used to improve power grid efficiency, encourage sustainability, and hasten the switch to sustainable energy sources.

In general, the shift to sustainable energy management in the twenty-first century and the resolution of global energy concerns hinge on the integration of smart technology, RES, and decentralized energy systems.

REFERENCES

Ayub, M. F., Li, X., Mahmood, K., Shamshad, S., Saleem, M. A., & Omar, M. (2023). Secure consumer-centric demand response management in resilient smart grid as Industry 5.0 application with blockchain-based authentication.

IEEE Transactions on Consumer Electronics, 70, 1. https://doi.org/10.1109/TCE.2023.3320974

Chitchyan, R., & Murkin, J. (2018). *Review of Blockchain Technology and Its Expectations: Case of the Energy Sector*. March. http://arxiv.org/abs/1803.03567

Fang, X., Misra, S., Xue, G., & Yang, D. (2012). Smart grid—The new and improved power grid: A survey. *IEEE Communications Surveys and Tutorials*, 14(4), 944–980. https://doi.org/10.1109/SURV.2011.101911.00087

Guan, Z., Lu, X., Yang, W., Wu, L., Wang, N., & Zhang, Z. (2021). Achieving efficient and privacy-preserving energy trading based on blockchain and ABE in smart grid. *Journal of Parallel and Distributed Computing*, 147, 34–45. https://doi.org/10.1016/j.jpdc.2020.08.012

Shreenidhi, H. S., & Ramaiah, N. S. (2022). A two-stage deep convolutional model for demand response energy management system in IoT-enabled smart grid. *Sustainable Energy, Grids and Networks*, 30, 100630. https://doi.org/10.1016/j.segan.2022.100630

Hitachi-UTokyo Laboratory. (2020). *Society 5.0: A People-Centric Super-Smart Society*. Spinger Open. https://library.oapen.org/bitstream/handle/20.500.12657/41719/2020_Book_Society50.pdf?sequenc#page=59

Javed, S., Tripathy, A., van Deventer, J, Mokayed, H., Paniagua, C., & Delsing, J. (2023). An approach towards demand response optimization at the edge in smart energy systems using local clouds. *Smart Energy*, 12, 100123. https://doi.org/10.1016/j.segy.2023.100123

Khan, U. A., Khan, N. M., & Zafar, M. H. (2023). Resource efficient PV power forecasting: Transductive transfer learning based hybrid deep learning model for smart grid in Industry 5.0. *Energy Conversion and Management: X*, 20, 100486. https://doi.org/10.1016/j.ecmx.2023.100486

Maddikunta, P. K. R., Pham, Q. V., Prabadevi, B, Deepa, N., Dev, K., Gadekallu, T. R., Ruby, R., & Liyanage, M. (2022). Industry 5.0: A survey on enabling technologies and potential applications. *Journal of Industrial Information Integration*, 26, 100257. https://doi.org/10.1016/j.jii.2021.100257

Mateen, A., Wasim, M., Ahad, A., Ashfaq, T., Iqbal, M., & Ali, A. (2023). Smart energy management system for minimizing electricity cost and peak to average ratio in residential areas with hybrid genetic flower pollination algorithm. *Alexandria Engineering Journal*, 77, 593–611. https://doi.org/10.1016/j.aej.2023.06.053

Mourtzis, D., Angelopoulos, J., & Panopoulos, N. (2022). Smart Grids as product-service systems in the framework of energy 5.0—A state-of-the-art review. *Green Manufacturing Open*, 1(1), 5. https://doi.org/10.20517/gmo.2022.12

Mourtzis, D., Angelopoulos, J., & Panopoulos, N. (2023). Personalized Services for Smart Grids in the framework of Society 5.0: A Smart University Campus Case Study: Smart Campus. *Technical Annals*, 1(2). https://doi.org/10.12681/ta.34199

Pandey, V., Sircar, A., Bist, N., Solanki, K., & Yadav, K. (2023). Accelerating the renewable energy sector through Industry 4.0: Optimization opportunities in the digital revolution. *International Journal of Innovation Studies*, 7(2), 171–188. https://doi.org/10.1016/j.ijis.2023.03.003

Rojek, I. Mikołajewski, D. Mroziński, A. & Macko, M. (2023). Machine learning- and artificial intelligence-derived prediction for home smart energy systems with PV installation and battery energy storage. *Energies*, *16*, 6613. https://doi.org/10.3390/en16186613

Tabaa, M., Monteiro, F., Bensag, H., & Dandache, A. (2020). Green Industrial Internet of Things from a smart industry perspectives. *Energy Reports*, *6*, 430–446. https://doi.org/10.1016/j.egyr.2020.09.022

Verma, A., Bhattacharya, P., Madhani, N., Trivedi, C., Bhushan, B., Tanwar, S., Sharma, G., Bokoro, P. N., & Sharma, R. (2022). Blockchain for Industry 5.0: Vision, opportunities, key enablers, and future directions. *IEEE Access*, *10*, 69160–69199. https://doi.org/10.1109/ACCESS.2022.3186892

Yin, S., Liu, L., & Mahmood, T. (2023). New trends in sustainable development for Industry 5.0: digital green innovation economy. *Green and Low-Carbon Economy*, 1–8. https://doi.org/10.47852/bonviewGLCE32021584

Zafar, M. H., Bukhari, S. M. S., Abou Houran, M., Moosavi, S. K. R., Mansoor, M., Al-Tawalbeh, N., & Sanfilippo, F. (2023). Step towards secure and reliable smart grids in Industry 5.0: A federated learning assisted hybrid deep learning model for electricity theft detection using smart meters. *Energy Reports*, *10*, 3001–3019. https://doi.org/10.1016/j.egyr.2023.09.100

Chapter 13

Dynamic Human–Artificial Intelligence Collaboration Framework for Adaptive Work Environments in Industry 5.0

Nisha Banerjee and Aniket Bhattacharyea

13.1 INTRODUCTION

As we stand in the new era of innovation in technology and Industry 5.0, the design of work environments is undergoing a huge transformation due to the integration of artificial intelligence (AI) into various areas of industrial processes which marks the advancement of automation in upcoming new generation projects. In their policy brief (Breque *et al.*, 2021), the European Commission unveiled "Industry 5.0," a vision for the future of European enterprise. Industry 5.0 focuses on sustainable production, maintaining a high level of robustness, and putting the welfare of industry workers at the forefront of the production process, so that industry may become a resilient source of prosperity. The integration of AI technologies along with human skills and capabilities ensures flexibility in task allocation to optimize productivity and adapt to evolving industry demands, leading to the design of a dynamic human–AI collaboration framework for adaptive work environments. In human–machine teams, task distribution should be planned to make the most of each actor's strengths, maintain equitable task distribution, and maintain the significance and manageability of human labor as discussed by Kaasinen *et al.* (2022). The main aim of the research paper is to redefine how industries operate in the era of intelligent systems, by developing innovative work environments that can effectively navigate the challenges as well as opportunities in the Industry 5.0 landscape. Human–machine teams can consist of human actors with varying skill sets, cooperative robots, and AI-based systems for oversight or support. This research paper focuses on the construction of a collaborative framework that not only acknowledges the symbiosis of human and AI but also actively cultivates an environment where these entities complement each other seamlessly by making real-time adjustments, continuous learning, and efficient allocation of tasks to achieve optimal performance, adapting to the evolving needs of the industry. By exploring the dynamic interplay between human intuition, creativity, and the analytical prowess of AI, this paper aims to emphasize the need for dynamic and innovative work environments by contributing to the blueprint of adaptive work environments poised to thrive amidst the challenges and opportunities presented by Industry 5.0.

13.2 LITERATURE REVIEW AND BACKGROUND STUDY

Several studies from various scholarly works within the context of Industry 5.0, provide vital insights into contributing to an overall understanding of the challenges and opportunities associated with dynamic human–AI collaboration frameworks in adaptive work environments. Research by Smith *et al.* (2019) explains the importance of establishing smooth and effective interaction frameworks that facilitate collaboration between humans and AI in Industry 5.0. This research discusses the design principles for user interfaces and decision support systems to enhance the user experience and promote synergy between AI-driven analytics and human intuition. Brown and Miller (2020) studied the concept of agility in Industry 5.0 and its implications for industry work environments. The literature review explores how adaptive frameworks allow quick response to changing conditions, thereby focusing on the need for a dynamic collaboration model to ensure flexible operation, allowing organizations to swiftly adapt to evolving circumstances and maintain a competitive edge. The research of Chen *et al.* (2021) delves into the importance of continuous learning in Industry 5.0 in the competitive market. Their study emphasizes the necessity of creating congenial environments where both human workers and AI systems can grow their skills collaboratively, such that the workforce is better able to handle evolving tasks and challenges. Studies by Gupta and Singh (2018) focus on task allocation optimization and reveal insights into algorithms and methodologies that enhance efficiency by intelligently distributing tasks based on the strengths of human workers and the capabilities of AI systems. The study aims to enhance overall performance by ensuring that tasks are allocated to the entities with less current level of workload, maximizing accuracy in results and minimizing bottlenecks. Works by Kim and Park (2019) investigate the broader organizational impact of integrating AI into work environments in Industry 5.0. The extant literature suggests that successful human–AI collaboration requires a cultural shift within organizations, fostering an environment that values both human creativity and the analytical capabilities of AI. The work by Thompson *et al.* (2020) explores how cognitive ergonomics principles can be applied to optimize the collaboration between humans and AI systems, thereby enhancing decision-making processes and boosting overall system performance. Investigating the importance of human-centric design in AI systems, another study (Wang *et al.*, 2019) advocates for designing AI technologies that align with human cognitive processes, preferences, and work patterns in the evolving landscape of Industry 5.0. Focusing on the adaptive capabilities of learning algorithms, Yang *et al.* (2021) explore how AI systems can dynamically adjust their behavior based on real-time human interactions, contributing to more responsive and effective collaboration. Addressing the crucial aspect of trust in collaborative environments, this research (Zhang et al., 2022) investigates the role of explainability in AI systems, examining how transparent decision-making processes enhance trust between human operators and AI algorithms. Stern and Becker (2019) point out that many future industrial workplaces present themselves as human-machine systems, where tasks are carried out via interactions with several workers, production machines, and cyber–physical assistance systems. O'Neill et al. (2020) argue that collaboration with agents and with robots is fundamentally different due to the embodiment of the robots. For example, the physical presence and design of a robot can influence the interactions with humans, the extent to which the robot engages and interests the human, and foster trust development. As robotics technology has advanced, robots have increasingly become capable of working together with humans to accomplish joint work, thus raising design challenges related to interdependencies and teamwork (Ma et al., 2018). Van Diggelen et al. (2018) suggest design patterns support the development of effective and resilient human-machine teamwork. They claim that traditional design methodologies do not sufficiently address the autonomous capabilities of agents, and thus often result in applications where the human becomes a supervisor rather than a teammate (Romero and Stahre, 2020). These summaries provide a glimpse into various dimensions of human–AI collaboration in Industry 5.0, offering insights into cognitive ergonomics, design principles, adaptive algorithms, trust-building mechanisms, cross-disciplinary models, and integration in cyber–physical systems (CPSs).

13.3 METHODOLOGY

Future industry floors are envisioned as dynamic teams made up of humans, cooperative robots, and autonomous agents that can respond quickly to shifting demands in the work environment. While humans are adept at coming up with original ideas and innovative solutions, AI is good at learning from historical data. An ensemble that is more capable than the sum of human and AI skills can be created by combining these two complementary points of view (Chakraborti et al., 2018). Introducing clever solutions offers a chance to reflect on the roles played by all parties involved, posing queries like: How should the members of the team communicate and work together? What are the new responsibilities for each team member? How are the works dynamically allocated in real time? Systems involving humans and machines should be designed with worker roles and the corresponding skill requirements in mind. This is consistent with the original sociotechnical design value system (Mumford et al., 2000) which states that the needs and rights of every employee should always come first, regardless of how the industry changes in terms of technology and organizational structures. Additionally, employees should have the flexibility to switch roles on the spot in the dynamic Industry 5.0 environment if they need to learn new skills or just want a change. The extensive body of research on collaborative cognitive systems (Hollnagel et al., 2005; Woods et al., 2006) offers a solid foundation for comprehending the design's primary focus, which is the system as a whole with many human and machine actors and shared objectives. To comprehend and enhance the dynamic nature of the system, wherein the skills and capacities of all the actors vary over time and the task allocation needs to be adjusted adaptably to the changing environment, complementing approaches and methods are necessary. A cognitive system can be thought of as an adaptable system that plans and adjusts its activities based on information about the environment and itself. This type of system operates with an eye on the future (Hollnagel et al., 1999). According to Jones et al. (2018) and Chacon et al. (2020), smart manufacturing systems are collaborative agent systems that can employ a wide range of agents, including machine, human, and organizational agents, in addition to hardware and software, to achieve shared objectives. According to Chacon et al. (2020), the Industry 4.0 paradigm shift from doing to thinking has made room for cognition to emerge as a fresh viewpoint on intelligent systems. A fundamental concept according to Gibson (1979) is that the environment shapes the system and its purpose, and that these factors must be taken into account when preserving adaptive behavior. The "intelligent environment," or an entire adaptive system, is then the focus of design (Norros et al., 2009). The roles that the human actors have personally established are dependent on their current skill set, but they are also flexible enough to shift roles within the team as they continue to grow. It is obvious that once it has been created and put into use, it begins to evolve on its own as intelligent machine actors and human actors alike pick up new abilities and skills. We suggest that three interesting approaches and methods can be added to a collaborative cognitive systems approach, based on our own experiences and the literature analysis:

- By employing empirical, evidence-based analyses to describe sociotechnical networks, the actor-network theory helps in understanding the functioning and evolution of a dynamic human–machine collaboration that can be aided by an understanding of actor-network theory.
- The concept of operations is a useful technique that offers a way to explain various players and the relationships among them. It better enables co-design and development activities with important stakeholders as well as the dynamic character of the entire system.
- Ethical questions about human responsibilities, task distribution, and machine-based decision making are brought up by human–machine

collaborations. The steadily changing joint system may bring up new ethical problems that were not considered in the original plan.

In this section, we break down the concepts of operations, actor-network theory, and ethically conscious design into smaller sections and examine the specific benefits that these approaches and techniques potentially offer to the creation and use of human–machine teams. We also point up possible dangers and difficulties.

13.3.1 Actor-Network Theory

Humans and machines will form new teams and relationships as a result of Industry 5.0. As Latour and Wooglar (2013) illustrated in their seminal study, namely "Laboratory Life: The Construction of Scientific Facts," actor-networks are typically analyzed empirically using an ethnographical method, which highlights the intricacy of industry work. Employees must communicate with a variety of actors in many digital workplaces, including other people, different machines, and emerging technology. Digitalized networks allow for three different forms of interaction: machine-machine interaction, social human-to-human communication, and human–machine interaction. When dealing with technology devices, the interaction is referred to as human–computer interaction (HCI) in other sorts of digitalized networks where at least one actor is non-human. It can be challenging to determine the results of various activities in the network processes or even to examine causal relationships. Cobots and other AI-based actors may be able to make decisions and carry them out in the real world, but they are neither legal entities in society nor autonomous actors (Niemelä *et al.*, 2007; Law, 1992). It is typically a descriptive framework meant to highlight the relations between the actors. However, neither social connections nor the causality of actions is explained by actor-network theory (ANT). The problem with ANT is that it becomes challenging to determine human–machine teams if fieldwork is poorly done, descriptions are lacking, or analyses are too hazy. Creating harmonious and collaborative human–machine teams can be facilitated by using the findings of actor-network studies. After that, there is a chance the ANT method would not be able to generate relevant data for the initial stages of design.

13.3.2 Concept of Operations

IEEE standard 1362 defines a ConOps document as a user-oriented document that describes a system's operational characteristics from the viewpoint of the end user. The issue with ANT is that it can be difficult to assess human–machine teams when there is inadequate fieldwork, inadequate description, or imprecise or superficial analysis. There is a possibility that the ANT approach will not be able to produce pertinent data for the preliminary stages of design after that. A ConOps paper typically includes the main parts of the system, stakeholders, tasks, and an explanation of how it works (Tommila *et al.*, 2013). ConOps documents usually comprise written descriptions, but they may also contain unofficial photos meant to highlight key features of the proposed system, like its objectives, and operating protocols. Laarni and Koskinen (2017) suggested a classification scheme for robot swarm management complexity. The most complicated scenario has teams of humans from numerous units working together with swarms from various classes. This scenario also considers machine–machine interactions in autonomous or semi-autonomous swarms, as well as human teamwork. Although an effective approach for co-designing human–machine systems with relevant stakeholders and for shedding light on design decisions, collaborative robotics, human–machine teams have some risks and limitations. These can be observed in action at the industrial infrastructure level in a particular ConOps. As noted by Mostashari *et al.* (2012), ConOps may be perceived as burdensome rather than beneficial due to the extensive documentation requirements. Therefore, persuading the design team and other stakeholders of the significance of ConOps may prove to be difficult. As ConOps development is frequently a drawn-out and difficult process, estimating the resources needed might be difficult.

13.3.3 Ethically Aware Design

When deciding how to divide up work between humans and machines, for instance, it's crucial to take into account values like respecting workers' autonomy, privacy, and dignity. One should also think about how to make the most of both human and machine strengths and develop work practices that enhance or maintain the sense of meaning at work. Applying the prior work by Kaasinen *et al.* (2018) outline five ethical standards to promote ethics-aware design in the context of Operator 4.0 solutions. The ethical concepts of privacy, autonomy, integration, dignity, dependability, and inclusion serve as the foundation for the guidelines, which provide one guiding sentence for each theme. Despite the perception that ethics is a crucial topic of design, designers may be discouraged from taking ethical considerations into account because there are no practical tools or integrations with the design process. Four fundamental concepts serve as the foundation of an ethical impact assessment methodology (Wright, 2011): beneficence, justice, autonomy respect, and non-maleficence. Each of these principles has multiple subsections and sample questions for a designer to help them think through how technology affects their respective tenets. Three complimentary strategies to assist in designing Industry 5.0 human–machine teams as collaborative cognitive systems are provided in this section. Table 13.1 presents a summary and comparison of the approaches with respect to their

Table 13.1 Comparison of the approaches

Basis	Actor Network Theory	Concept of Operations	Ethical Awareness
Nature of the approach	Theory	Activity or tool based	Ethical values integrated into the design
Method of collection of data	Observational values from different networks	stakeholder workshops	Stakeholder interview within the design team
Main questions	How do humans and machines work together? How does the dynamic communication take place? Who are the actors?	Who are the stakeholders? How are the task allocated? What are the other user requirements?	What ethical principles should be integrated into the design?
Benefits	Humans and machines studied without any bias for a smooth cooperation	Relationship between stakeholder and the design of this system is seen from the end users' viewpoint for better understanding	Identification and consideration of ethical aspects are integrated and supported
Risk and challenges	Analysis can be hazy and lacking clarity	Drafting is a tough job for allocating resources	Too much focus on risk of misuse as a result forgetting the main advantages
Result	Connection between the human machine team formation	Responsibility and task equally divided	Ethical guidelines to be followed during the design
Type of research	Empirical and qualitative	Mainly focus on system engineering	Research on the design issues

objectives, outcomes, advantages, and drawbacks. Table 13.1 presents an overview and contrast of the three methods.

13.3.4 A Conceptual Framework for the Design of a Cognitive Assistant Support

According to the literature analysis and research, another great solution to make a dynamic and adaptive work environment consisting of human–machine teams could be the use of a CPS. CPSs that are centered on human–machine interaction are able to transition from a physical interaction paradigm to a cognitive one. However, as these systems' information and communication capacities grow, their complexity increases beyond what can be comprehended by the industry's present conventional user interfaces. Consequently, in order to keep the system operating within stable parameters, the operator would require assistance. In addition, during field operation, the operator might obtain the industry's work plan, necessitating the requirement for extra information that requires access to the situation- and task-oriented, location-independent data (Hollnagel et al., 2010) Autonomy of the human in human–machine collaboration is, in general, considered superior to that of a machine (Simmler and Frischknecht, 2021). With the development of highly complex and dynamic human–machine systems, comes the idea of forming independent(active) machine agents for the application in dynamic and uncertain environments to support and assist human workers (Krupitzer et al., 2020), which are, despite the technological enhancement, a significant factor in the design of manufacturing systems. Besides acting as a mere assistant for human operators, machine entities can actively collaborate with humans on joint activities, sharing and dividing tasks according to their respective strengths and capabilities. Figure 13.1 shows a design is proposed so as to make a predictive decision-making system that understands the load on the human and accordingly divides the task helping in better work allocation for faster results.

According to the ethics by design method (Niemelä et al., 2014), one of the most crucial first steps is to encourage ethical thinking from the design team and other pertinent stakeholders as early as possible in the process. Numerous advantages come with this approach: it promotes commitment to the stated ideals or principles, engages multiple ethical viewpoints on the design, and enables proactive ethical thinking. When developing user experience goals for design, ethics can be taken into consideration and explored as one of the themes in user research or stakeholder workshops. Additionally, specific design exercises that match the organization's design process might be used to address ethical issues. In actuality, for instance, imagining future conditions and going over them with employees, specialists, or the design teams may assist in determining the design's possible ethical ramifications and may direct the design process further (see, e.g., Heikkilä et al., 2018). Figure 13.2 shows the conceptual model framework for designing the idea of a cognitive assistant support so as to allow dynamically task allocation in the Industry 5.0.

Characterization of human–machine collaboration can be done by the nature of the collaboration, i.e., the distribution of tasks between agents according to their abilities and capacities. In this way, we follow definitions of the fundamental characteristics of human–machine

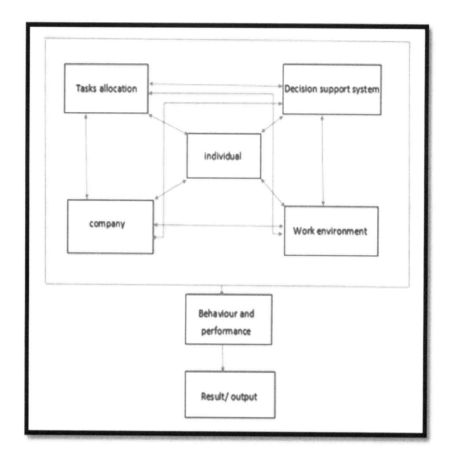

Figure 13.1 A design is proposed to understand work overload and thus allocate tasks in real time.

collaboration (Simmler and Frischknecht, 2021). Symbiotic relationships between humans and machines are discussed in human–machine collaboration (Lu *et al.*, 2021; Wang *et al.*, 2019; Gerber *et al.*, 2020). The symbiotic partnership may also facilitate a depth of collaboration, in which both agents are pursuing similar goals, benefit mutually, and become smarter (Jarrahi, 2018).

They form a partnership of agents which is capable of solving problems the individual members alone would not be able to tackle (Wang *et al.*, 2019). Requirements of Human Machine Systems (HMS) systems are shortly described in the following, based on Lu *et al.* (2021), Gerber *et al.* (2020), and Wang *et al.* (2019):

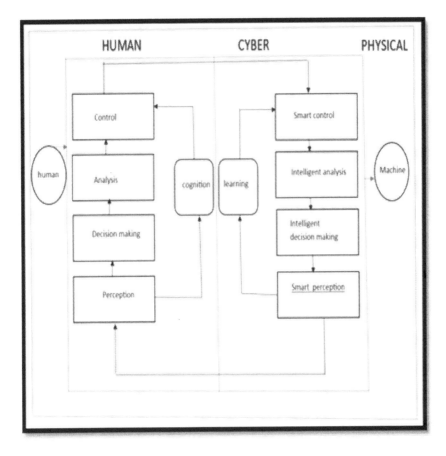

Figure 13.2 Model of a CPS for human–robot workplace according to joint cognitive system.

- Balanced **autonomy**, all agents are inherently autonomous, i.e., in an equal position, they form together a team.
- Accurate **context-awareness**, all agents are context-aware, i.e., their actions and decisions are grounded on the actual physical and cognitive circumstances.
- Transparent **representation**, all agents apply at least partially shared representations of the environment to formulate common goals, take roles, execute plans, and solve tasks dynamically adjusted in real time.
- Effective **communication**, all agents continuously engage with each other.
- **Social wellness**, the ability to detect and respond to human distress or fatigue in physical and mental performance to assure well-being.

- Dynamic **adaptability**, the performance of a symbiotic system improves over time, by adapting to new situations, changing conditions, and learning from failures and successes based on feedback from the environment.
- Natural **human-centricity**, the ability to focus on the human's needs

A cognitive assistant's assistance to their human colleagues in a work team is crucial: Keeping team members informed about significant developments in both internal and external information while avoiding taking their focus away from the main duties: keeping an eye on intra- and inter-team interactions; monitoring work completion and progress toward the team's objectives; giving prompt performance evaluation and direction on fixing team mistakes. The proposed conceptual framework for the creation of cognitive assistants will adhere to a thorough operational definition of cognition, as illustrated in Figure 13.3, which uses two iterative loops of three

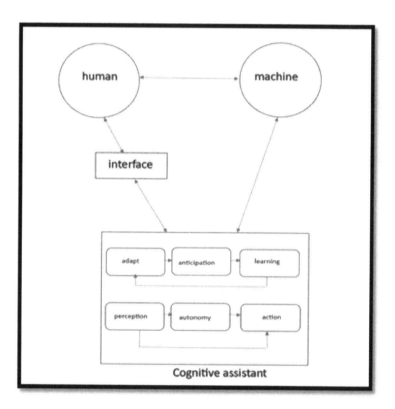

Figure 13.3 Joint cognition system using cognitive assistant.

essential attributes: anticipation, adaptation, and learning; and action, perception, and autonomy.

Goal-oriented behavior, autonomy, interaction through cooperation and communication, intention reading, interpretation of expected and unexpected events, prediction of the outcome of one's own and other people's actions, action selection and evaluation, adaptation to changing circumstances, learning from experience, and monitoring and correcting one's own performance are some of the attributes that can explain the variety of skills that the cognitive assistant should possess. It also involves choosing the machine learning (ML) algorithm, which ranges from rule-based learning to deep learning. The operator's needs must be specified when choosing the level of assistance, and factors like interpretability, accuracy, and explainability must be taken into account. A reinforcement learning algorithm is combined with artificial neural networks to create a ML model. Figure 13.4 shows the collaboration between humans and machines for dynamic learning in Industry 5.0.

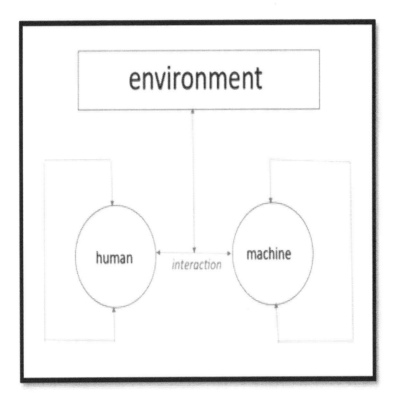

Figure 13.4 Collaborative human and machine learning.

Given that the operator may not always possess prior AI expertise, the ML professional can evaluate the precision of the employed ML technique and provide best practices to enhance explainability. Learning of AI agents is the subject of ML research, currently encompassing a wide range of learning algorithms, (semi-/un-) supervised learning, reinforcement learning, active learning, or deep learning (Russell and Norvig, 2021). Trial-and-error is one of the most fundamental learning techniques humans use. Evidently, interaction of agents in learning scenarios can support achieving agent-specific learning goals. Similarly, collaboration between agents supports achieving mutual learning goals and maximizes global payoffs. A concept of reciprocal learning between humans and machines in the context of human–machine collaboration in manufacturing has been defined by Ansari et al. (2018) as: "a bidirectional process involving reciprocal exchange, dependence, action or influence within human and machine collaboration on performing shared tasks, which results in creating new meaning or concept, enriching the existing ones or improving skills and abilities in association with each group of learners." Any human and any intelligent machine agent, regardless of the type of interaction (if any) or task, comprise of basic characteristics that affect learning namely, a knowledge base (e.g., previous experience, knowledge about the environment, skills, and competence), actuators (e.g., ways to act and interact), sensors(e.g., ways to receive feedback and information), the general ability to adapt or learn (e.g., change own behavior), a perception of the environment (e.g., interpretation of reality), and autonomous cognition or inference engine (e.g., reasoning, problem-solving), respectively. The characteristics affect the behavior of the agent and are required to learn.

13.4 DISCUSSION OF RESULTS AND ANALYSIS

We have researched various strategies and techniques that could facilitate a human-centered early design stage and the ongoing evolution of Industry 5.0 human–machine teams as productive, adaptable, and continuously developing joint cognitive systems on the manufacturing floor. A combined cognitive system constantly co-evolves as both human and machine actors learn, thereby acquiring new talents and skills after it has been initially built and put into use. Furthermore, the system needs to be flexible and resilient to adjust to shifting demands and expectations due to modifications in the production environment. Industry 5.0 visions depict human-centered, human-resilient, and sustainable factories of the future, where intelligent robots and human operators collaborate in teams. To adapt to dynamic changes in the environment, these teams must possess resilience. The ability of team members to switch between roles and the dynamic job distribution both need to represent resilience. This is significant for the creation of human–machine teams because, in these teams, humans and machines have separate roles and responsibilities while working toward shared objectives. Human operators are not just "using" the machines. Joint cognitive systems are also characterized by the dynamic and adaptable division of labor between humans and machines. Designing automated systems where the ability to keep control of a situation in the face of disruptive influences from the environment or the process itself is crucial has been done using the joint cognitive systems method. Just having techniques for the design of a joint cognitive system is insufficient; techniques that facilitate the monitoring and observation of a joint cognitive system are also required in order to comprehend the current state of the system, the abilities and skills of various human and machine actors, and the distribution of tasks among the actors. Additionally, co-design techniques and resources are required to enable pertinent parties to take part in the joint human–machine system design process. The cognitive systems engineering (CSE) method is suggested by Jones *et al.* (2018) and Chacon *et al.* (2020) for creating these multi-agent-based manufacturing systems. In order to understand how people utilize artifacts and how people and artifacts collaborate to develop and organize collaborative cognitive systems, CSE focuses on cognitive functions and examines how humans manage complexity. According to Jones *et al.*, there are three types of actors in Industry 4.0/5.0 manufacturing systems: technological, organizational, and human. According to Chacon *et al.* (2020), these agents are built on a synergistic fusion of many technologies, including multi-agent systems (MAS), AI, and the Internet of Things (IoT). Joint cognitive systems are those that are comprised of different human and machine agents in manufacturing systems. The foundation of CSE is modeling, including cognitive task modeling, which is frequently a very time-consuming procedure. In model-based design, human actors are frequently oversimplified (Gräßler *et al.*, 2021). The modeling of human performance may then be hindered if it fails to appropriately account for the heterogeneity observed in individual traits, abilities, and preferences. Van Diggelen *et al.* (2018) have identified another design challenge: conventional design approaches fail to adequately handle the autonomous capacities of computer agents, leading to systems in which humans function more as supervisors than as teammates. Given that the entire joint cognitive system is dynamically changing over time, it is critical to comprehend the entity, various players, and their respective functions. We suggest actor-network theory (ANT) as an antiquated but effective strategy for that reason. Actor-network analyses, which seek to characterize actor-to-actor interaction and bring network processes to light, necessitate ethnographic fieldwork combined with observation and other evidence-based analytical techniques. The impartial ANT method looks out for all agents, human and non-human alike. This is crucial because, in human–machine teams, the capabilities that are offered should be used in the same way regardless of the actor supplying the skill. One major problem with function allocation in joint cognitive system design is that the ideal function allocation is dependent on the operating environment. Multiple levels of autonomy are offered in dynamic function allocation, and a decision-making process is used to transition between these levels. Dynamic function scheduling is a subset of dynamic function allocation, wherein a specific function is reallocated according to the agents' schedule. For example, Hildebrandt et al. (2003) and Fairly et al. (1997) developed the ConOps technique for assigning roles and stakeholder requirements to the various components of the proposed system. It can be used to categorize and explain how a group of autonomous or semi-autonomous robots interacts with human operators. The fact that the suggested approaches and procedures need a fair amount of labor from human factor specialists, other members of the design team, and industrial stakeholders in order to be applied effectively presents a challenge. Designing sustainable human–machine collaborations requires careful consideration of ethical issues. As noted by Pacaux et al. (2020) techno-centered design frequently requires highly skilled human operators. Teams should continue to value human labor, and the design should not assume that human operators will handle every scenario in which machine intelligence is not able to do its function. Mutual understanding is necessary for human–machine collaboration to be successful, and this can easily result in machines watching people and their actions. Equitable task distribution between humans and machines is necessary. This highlights, among many other things, the necessity of giving ethical considerations, and careful thought at the outset and all the way through the design process. Proactive ethical thinking is supported by ethically conscious design, which integrates the viewpoints of many stakeholder groups and labor groups. Additionally, ethically conscious design gives the entire design team the means to dedicate themselves to sustainable design. Hence, a dynamic adaptive work environment can be created in Industry 5.0 by integrating AI with human entities as a team.

13.5 CONCLUSIONS AND FUTURE SCOPE

Intelligent assistance systems enable decisions to be made more quickly and efficiently; the qualification of manual tasks declines; the operator has real-time access to all the information required to make decisions; co-working with machines in the workspace requires less effort and attention; Thus, the following effects of this technological evolution are felt by the operator, among other things: less manual tasks qualify; the operator has access to all the information they need to make decisions in real-time; intelligent assistance systems facilitate quick decision-making; co-working in the workspace between machines and people requires less effort and attention. In conclusion, HMS systems possess the capability of perception, learning, communication, and decision-making for context-aware human–machine collaboration. We intend to put the concepts discussed in this paper into practice in the future. This will give a better knowledge of the real design difficulties and the practicality of the techniques and methodologies put forward here. A particular area of attention will be the monitoring and guidance of a joint human–machine system's ongoing co-evolution.

REFERENCES

Anderson, E., & Anderson, M. (2022). Ethical Considerations in Human–AI Collaboration: Transparency, Accountability, and Fairness. Ethics and Information Technology: 24(1), 45–63.

Ansari, F., Erol, S., & Sihn, W. (2018). Rethinking Human-Machine Learning in Industry 4.0: How Does the Paradigm Shift Treat the Role of Human Learning? Procedia Manufacturing, 23, 117–122.

Becerik-Gerber, B., Lucas, G., Aryal, A., Awada, M., Bergés, M., Billington, S. L., ... & Zhao, J. (2022). Ten Questions Concerning Human-Building Interaction Research for Improving the Quality of Life. Building and Environment, 226, 109681.

Breque, M., de Nul, L., & Petridis, A. (2021). Industry 5.0: Towards a Sustainable, Human-Centric and Resilient European Industry. European Commission, Directorate-General for Research and Innovation: Luxembourg.

Brown, R., & Miller, C. (2020). Agility in Industry 5.0: Adapting Work Environments for the Future. International Journal of Industrial Engineering, 12(2), 78–92.

Chacón, A., Angulo, C., & Ponsa, P. (2020). Developing Cognitive Advisor Agents for Operators in Industry 4.0. In New Trends in the Use of Artificial Intelligence for the Industry 4.0. Intech Open: The Hague, The Netherlands, 127.

Chakraborti, T., & Kambhampati, S. (2018). Algorithms for the Greater Good! On Mental Modeling and Acceptable Symbiosis in Human–AI Collaboration. arXiv:1801.09854.

Chen, L., Wang, Q., & Liu, X. (2021). Continuous Learning and Skill Adaptation in Industry 5.0: A Human–AI Collaborative Perspective. Journal of Advanced Manufacturing Technology: 37(4), 511–528.

Fairley, R.E., & Thayer, R.H. (1997). The Concept of Operations: The Bridge from Operational Requirements to Technical Specifications. Annals of Software Engineering: 3, 417–432.

Gibson, J.J. (1979). The ecological approach to visual perception. In The Theory of Affordances. Houghton Mifflin.

Gräßler, I., Wiechel, D., & Roesmann, D. (2021). Integrating Human Factors in the Model Based Development of Cyber-Physical Production Systems. Procedia CIRP, 100, 518–523.

Gupta, A., & Singh, S. (2018). Optimizing Task Allocation in Human–AI Collaborative Environments for Improved Efficiency. IEEE Transactions on Automation Science and Engineering, 15(1), 124–138.

Heikkilä, P., Honka, A., Mach, S., Schmalfuß, F., Kaasinen, E., & Väänänen, K. (2018). Quantified Factory Worker-Expert Evaluation and Ethical Considerations of Wearable Self-Tracking Devices. In Proceedings of the 22nd International Academic Mindtrek Conference, Tampere, Finland, 202–211.

Hildebrandt, M., & Harrison, M. (2003). Putting Time (Back) into Dynamic Function Allocation. In Proceedings of the Human Factors and Ergonomics Society Annual Meeting, Denver, CO, USA. Los Angeles, CA: SAGE Publications.

Hollnagel, E. (2010). Prolegomenon to Cognitive Task Design. In Handbook of Cognitive Task Design. CRC Press, 3–15.

Hollnagel, E., & Woods, D.D. (1999). Cognitive systems engineering: New wine in new bottles. International Journal of Human–Computer Studies: 51, 339–356.

Hollnagel, E., & Woods, D.D. (2005) Joint Cognitive Systems: Foundations of Cognitive Systems Engineering. CRC Press: Boca Raton, FL, USA, 2005. 11, 43–56.

Jarrahi, M.H. (2018). Human–AI Symbiosis in Organizational Decision Making. Artificial Intelligence and the Future of Work Business Horizons: 61(4), 577–586.

Jones, A.T., Romero, D., & Wuest, T. (2018) Modeling agents as joint cognitive systems in smart manufacturing systems. Manufacturing Letters: 17, 6–8.

Kaasinen, E., Anttila, A.-H., & Heikkilä, P. (2022). New industrial work-personalised job roles, smooth human–machine teamwork and support for well-being at work. In Human–Technology Interaction—Shaping the Future of Industrial User Interfaces; Röcker, C., Büttner, S., Eds. Springer Nature: Berlin/Heidelberg, Germany.

Kaasinen, E., Liinasuo, M., Schmalfuß, F., Koskinen, H., Aromaa, S., Heikkilä, P., Honka, A., Mach, S., & Malm, T. (2018). A Worker-Centric Design and Evaluation Framework for Operator 4.0 Solutions That Support Work Well-Being. In Proceedings of the IFIP Working Conference on Human–Work Interaction Design, Espoo, Finland . Springer: Berlin/Heidelberg, Germany.

Kim, Y., & Park, J. (2019). Impact of Human–AI Collaboration on Organizational Culture in Industry 5.0. Journal of Organizational Change Management: 32(4), 489–505.

Krupitzer, C., Müller, S., Lesch, V., Züfle, M., Edinger, J., Lemken, A., Schäfer, D., Kounev, S., & Becker, C. (2020). A Survey on Human–Machine Interaction in Industry 4.0. arXiv preprint arXiv:2002.01025.

Laarni, J., Koskinen, H., & Väätänen, A. (2017). Concept of Operations Development for Autonomous and Semi-Autonomous Swarm of Robotic Vehicles. In Proceedings of the Companion of the ACM/IEEE International Conference on Human–Robot Interaction, Vienna, Austria, 179–180.

Latour, B., & Woolgar, S. (2013). Laboratory Life. Princeton University Press: Princeton, NJ, USA.

Law, J. (1992). Notes on the Theory of the Actor-Network: Ordering, Strategy, and Heterogeneity. Systems Practice: 5, 379–393.

Lu, Y., Adrados, J.S., Chand, S.S., & Wang, L. (2021). Humans Are Not Machines—Anthropocentric Human–Machine Symbiosis for Ultra-Flexible Smart Manufacturing. Engineering, 7(6), 734–737.

Ma, L.M., Fong, T., Micire, M.J., Kim, Y.K., & Feigh, K. (2018). Human-Robot Teaming: Concepts and Components for Design. Field and Service Robotics. Springer: Berlin/Heidelberg, Germany.

Mostashari, A., McComb, S.A., Kennedy, D.M., Cloutier, R., & Korfiatis, P. (2012). Developing a Stakeholder-Assisted Agile CONOPS Development Process. Systems Engineering: 15, 1–13.

Mumford, E. (2000). Socio-Technical Design: An Unfulfilled Promise or a Future Opportunity? Springer: Boston, MA, USA, 33–46.

Niemelä, M., Kaasinen, E., & Ikonen, V. (2007). Ethics by Design—An Experience-Based Proposal for Introducing Ethics to R&D of Emerging ICTs. In ETHICOMP 2014-Liberty and Security in an Age of ICTs. CERNA: Paris, France.

Niemelä, J., Saarno, T., Bansal, T., Storbjörk, S., & Bratteteig, H. (2014). The Concept of Ecosystem Services in Adaptive Urban Planning and Design: A Framework for Supporting Innovation. Landscape and Urban Planning, 136, 132–142.

Norros, L., & Salo, L. (2009). Design of Joint Systems: A Theoretical Challenge for Cognitive Systems Engineering. Cognition, Technology & Work: 11, 43–56.

O'Neill, T., McNeese, N., Barron, A., & Schelble, B. (2020). Human–Autonomy Teaming: A Rview and Analysis of the Empirical Literature. Human Factors: 64(5).

Pacaux-Lemoine, M.P. (2020). *Human-Machine Cooperation: Adaptability of Shared Functions between Humans and Machines-Design and Evaluation Aspects* (Doctoral dissertation, Université Polytechnique Hauts-de-France).

Romero, D., & Stahre, J. (2021). Towards The Resilient Operator 5.0: The Future of Work in Smart Resilient Manufacturing Systems.

Romero, D., Stahre, J., & Taisch, M. (2020). The Operator 4.0: Towards Socially Sustainable Factories of the Future. Computers & Industrial Engineering: 139, 106128.

Shiga, J. (2007). Translations: Artifacts from an Actor-Network Perspective. Artifact—Journal of Design Practice: 1, 40–55.

Simmler, M., & Frischknecht, R. (2021). A Taxonomy of Human–Machine Collaboration: Capturing Automation and Technical Autonomy. AI & Society: 36(1), 239–250.

Smith, J., Johnson, M., & Williams, A. (2019). Designing Effective Human–AI Interaction Frameworks for Industry 5.0. Journal of Human–Computer Interaction: 25(3), 321–335.

Stern, H., & Becker, T. (2019) Concept and Evaluation of a Method for the Integration of Human Factors into Human-Oriented Work Design in Cyber-Physical Production Systems. Sustainability: 11, 4508.

Thompson, R., & Davis, P. (2020). Enhancing Cognitive Ergonomics in Human–AI Collaboration: A Review. International Journal of Human–Computer Interaction: 28(4), 567–582.

Tommila, T., Laarni, J., & Savioja, P. (2013). Concept of Operations (ConOps) in the Design of Nuclear Power Plant Instrumentation & Control Systems. In Proceedings of the 31st European Conference on Cognitive Ergonomics, Toulouse, France, A Working Report of the SAREMAN Project.

Van Diggelen, J., Neerincx, M., Peeters, M., & Schraagen, J.M. (2018). Developing Effective and Resilient Human-Agent Teamwork Using Team Design Patterns. IEEE Intelligent Systems: 34, 15–24.

Wang, H., & Zhang, L. (2019). Towards Human-Centric AI Design in Industry 5.0. Journal of Artificial Intelligence Research: 24(2), 211–228.

Woods, D.D., & Hollnagel, E. (2006). Joint Cognitive Systems: Patterns in Cognitive Systems Engineering. CRC Press: Boca Raton, FL, USA.

Wright, D. (2011). A Framework for the Ethical Impact Assessment of Information Technology. Ethics and Information Technology: 13, 199–226.

Yang, J., & Li, Y. (2021). Adaptive Learning Algorithms for Human–AI Collaboration in Dynamic Environments. Journal of Intelligent Systems: 35(3), 401–416.

Zhang, Q., & Liu, Y. (2022). Building Trust in Human–AI Collaboration: The Role of Explainability. Information and Management: 40(1), 89–105.

Chapter 14

An Overview of Artificial Intelligence Algorithms for Future Generation

Ritu and Bebesh Tripathy

14.1 INTRODUCTION

Algorithms for artificial intelligence (AI) have become the cornerstone of revolutionary technical developments that have the potential to reshape industry, the economy, and society in the future. AI research and development have reached previously unheard-of heights because of the unrelenting pursuit of intelligent systems that are able to see, reason, and learn from data. This has ushered in a new era of invention and discovery. We must investigate the significant consequences of AI algorithms for future generations as well as the numerous opportunities and problems they bring, as we are on the verge of a technological revolution. The origins of AI algorithms may be traced to the groundbreaking research of visionaries like Alan Turing, John McCarthy, and Marvin Minsky, whose contributions established the groundwork for the field's current period. The development of AI algorithms, from neural network designs to symbolic reasoning systems, is a reflection of the convergence of several paradigms and approaches targeted at simulating human intellect and problem-solving skills. We start by looking at the basic ideas and methods that AI algorithms are based on, explaining their benefits, drawbacks, and range of uses in different fields and businesses. We explore the diverse range of AI algorithms that are pushing the boundaries of innovation and discovery, from traditional algorithms like decision trees and support vector machines to state-of-the-art deep learning architectures and reinforcement learning strategies. Additionally, we highlight new directions and paradigms that are changing the field of AI research and development, such as neuromorphic computing, quantum computing-based algorithms, explainable AI, and federated learning. By dissecting these new paradigms' complexities, we aim to uncover new opportunities and challenges on the horizon, poised to redefine the boundaries of AI innovation and deployment in the years to come.

We also negotiate the ethical and societal ramifications of AI algorithms, addressing concerns about algorithmic bias, accountability, transparency, and privacy that highlight the need for responsible AI research and use. We want to steer toward a future where AI algorithms act as catalysts for good social change, empowerment, and inclusion, promoting a more sustainable and fair society for future generations by critically analyzing and reflecting on current events. However, the widespread use of AI algorithms also brings up important existential, social, and ethical issues that call for serious thought and contemplation. Concerns about algorithmic bias, data privacy, employment displacement, and autonomous weapon systems highlight the significance of developing and using AI responsibly. It is our responsibility as guardians of this revolutionary technology to wisely and honorably negotiate these moral and social dilemmas. Taking these factors into account, this chapter critically analyzes the ethical and legal frameworks required to guarantee the responsible use of AI algorithms in addition to highlighting the possible advantages and uses of these systems. By encouraging multidisciplinary cooperation, openness, and responsibility, we may work to fully utilize AI's promise to improve humankind while reducing dangers and unexpected repercussions. The potential and ramifications of AI algorithms are both thrilling and terrifying as we look into the future generations. By adopting a progressive outlook and a dedication to moral guardianship, we may steer clear of the current and create a future in which AI technologies enhance our quality of life, unleash human potential, and cultivate a fairer and more sustainable society for future generations. The important aspect of machine learning is shown in Figure 14.1.

Big data and cloud computing have accelerated the development of AI algorithms, allowing for previously unheard-of levels of complexity, scalability, and sophistication in AI systems. These days, AI algorithms underpin a wide range of applications that are essential to our everyday lives, including recommendation systems, driverless cars, virtual assistants, and many more. AI will have a significant and wide-ranging influence on future generations as it continues to penetrate every aspect of civilization. AI algorithms have the potential to revolutionize education by providing individualized learning experiences that cater to the requirements, preferences, and learning styles of each individual learner. Intelligent tutoring systems, instructional chatbots, and adaptive learning platforms are just a few instances of how AI might completely transform how we learn, grow as people, and interact with material in the digital era. Furthermore, AI algorithms provide creative solutions for ecosystem management, climate modeling, and resource optimization in the context of environmental sustainability. AI-driven technologies have the potential to reduce environmental degradation, protect biodiversity, and advance sustainable development for future generations. Examples of these technologies include smart energy grids, precision agriculture, wildlife protection, and pollution monitoring. Nevertheless, despite the potential and promise of AI algorithms, there are still a lot of unknowns and difficulties. When

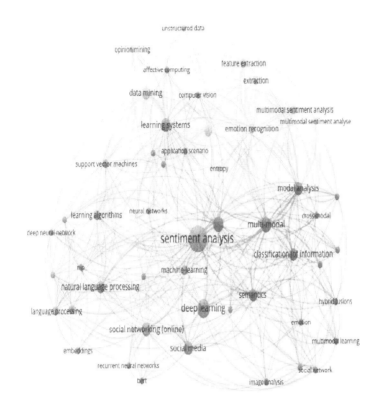

Figure 14.1 Some important aspects of machine learning.

talking about the societal ramifications of AI, concerns about algorithmic prejudice, the loss of jobs, and the consolidation of power in the hands of a few tech giants are common. In addition, moral conundrums involving the use of AI in surveillance technologies, autonomous weaponry, and predictive policing highlight the necessity of strong moral frameworks and government control to protect human rights and dignity. It is essential to take a multidisciplinary approach to managing these challenges, using ideas from computer science, ethics, sociology, psychology, and law. Through cultivating an environment of conscientious innovation and comprehensive governance, we can use the revolutionary capabilities of AI algorithms while maintaining the essential values of equity, openness, and accountability.

14.2 LITERATURE REVIEW

This landmark study highlights the concepts, designs, and uses of neural networks while offering a thorough introduction to deep learning. It acts as a fundamental resource for comprehending the developments in deep learning algorithms and their game-changing effects in a variety of fields [1]. The AlphaGo project is a groundbreaking example of AI, showing how deep neural networks and reinforcement learning may enable AI programs to win challenging games. This research demonstrates how AI may address problems that were previously thought to be beyond the capabilities of machines [2]. Neural network theory originated from Rosenblatt's work on the perceptron, which also served as a model for later advances in machine learning and AI. By introducing the idea of a perceptron as a neural network building block, the article opens the door for further developments in information processing and pattern recognition [3]. This work demonstrates dermatologist-level accuracy in skin cancer classification, highlighting the promise of deep-learning algorithms in medical diagnostics. The study highlights how AI is revolutionizing healthcare and provides new opportunities for early detection and treatment of diseases [4]. This research emphasizes the advancements in AI made possible using deep neural networks and reinforcement learning, much as the AlphaGo project. AI systems show their aptitude for strategic decision-making and intricate problem-solving by winning the age-old game of Go [5]. This thorough analysis emphasizes the significance of ethical issues in AI development and use while offering insights into the changing field of AI ethics standards. In order to resolve moral conundrums and social issues, the article emphasizes the necessity of responsible AI governance [6]. In mapping the current discourse on computational ethics, this article examines topics such as algorithmic bias, accountability, and transparency. It provides critical insightful information about the moral dilemmas raised by AI algorithms and the necessity of multidisciplinary cooperation in order to successfully resolve these dilemmas [7].

The Logic Theory Machine, an early AI system that mimics human cognitive processes through information processing, was introduced in this groundbreaking study. The study establishes the foundation for further advancements in the field of AI and advances our knowledge of the early attempts in the subject [8]. The renowned Turing Test, which serves as a standard for evaluating AI, is introduced in Alan Turing's groundbreaking study, which also investigates the idea of machine intelligence. The study sparks thought-provoking conversations on the nature of intelligence and the future powers of technology [9]. In this publication, McCulloch and Pitts propose a logical calculus that mimics the information processing of the nervous system, laying the groundwork for neural network theory. It presents basic ideas that serve as a basis for later advancements in artificial neural networks [10]. An important method for training neural networks, the backpropagation algorithm, is presented in this seminal publication. Backpropagation changed the area of deep learning and opened the door for many other developments in AI by making it possible to optimize neural network parameters efficiently [11]. This survey offers a thorough introduction to reinforcement learning, highlighting important ideas, methods, and uses in self-governing decision-making. For scholars and practitioners interested in comprehending the foundations and difficulties of reinforcement learning, it is an invaluable resource [12]. The transformer model, which emphasizes attention mechanisms above recurrent or convolutional structures, revolutionizes natural language processing and is introduced in the "Attention is All You Need" article. This innovative approach sets new benchmarks in language creation and interpretation by greatly increasing the efficacy and efficiency of sequence-to-sequence activities [13]. This study shows how deep neural networks may be used for medical picture processing, attaining skin cancer classification accuracy comparable to that of dermatologists. The study emphasizes how AI is revolutionizing healthcare by creating new avenues for early diagnosis and individualized therapy [14]. This study, which focuses on applying deep learning to electronic health records (EHRs), demonstrates the precision and scalability of AI algorithms in the analysis of massive amounts of medical data. AI-driven methods improve patient risk assessment, clinical decision support, and illness prediction by utilizing EHRs. This improves patient care and outcomes. Every one of these references offers a different perspective on the creation, use, and consequences of AI algorithms across a range of fields. When taken as a whole, these methods demonstrate the interdisciplinary character of AI research and its significant social influence [15].

14.3 ABOUT EVOLUTIONARY ALGORITHM

A class of optimization algorithms known as evolutionary algorithms is motivated by natural selection and biological evolution. Using a process comparable to natural selection, reproduction, mutation, and recombination, these algorithms work on a population of potential solutions, gradually improving them. Better-performing solutions have a higher chance of surviving and procreating, and this idea is reflected in the development of solutions. The idea of genetic representation, in which possible solutions are represented as strings of characters or parameters, is one of the fundamental ideas of evolutionary algorithms. Genetic operators like crossover and mutation may be applied to these representations, allowing the population to become more diverse and exploratory. Evolutionary algorithms look for the best or most efficient solutions by repeatedly applying these operations. Evolutionary algorithms have an advantage over standard optimization approaches in that they can explore high-dimensional, complicated search spaces. Evolutionary algorithms are capable of navigating difficult terrain by using stochastic variation operators to maintain a varied population of solutions. This allows them to find unique and counterintuitive solutions that gradient-based approaches would overlook. Moreover, evolutionary algorithms demonstrate resilience and flexibility when addressing ambiguous or noisy optimization tasks. Because evolutionary algorithms are population-based, they can withstand variations and disturbances in the search space, which makes them appropriate for uncertain and variable real-world optimization applications. Numerous fields, including engineering design, financial modeling, robotics, and bioinformatics, use evolutionary algorithms. Evolutionary algorithms are used in engineering for system-level optimization in complicated engineering systems, design synthesis, and parameter optimization. Evolutionary algorithms are used in finance for algorithmic trading, risk management, and portfolio optimization.

Evolutionary algorithms have also been combined with other computational methods, such as deep learning and machine learning, in recent years to provide several hybrid optimization strategies that are more scalable and perform better. These hybrid algorithms create potent optimization frameworks for addressing challenging issues in data-driven domains by fusing the local refinement and generalization skills of machine learning models with the global search capabilities of evolutionary algorithms. Natural Evolution as Inspiration: The principles of natural evolution, in which individuals within a population compete and reproduce based on their suitability to the environment, serve as a model for evolutionary algorithms. Evolutionary algorithms preserve a population of potential solutions to an issue and iteratively enhance these answers across generations, much like natural selection does. Evolutionary algorithms keep track of a population of solutions, as opposed to conventional optimization methods that work with a single solution. By avoiding local optima and enabling exploration of the solution space, this population of potential solutions to the optimization issue is diversified. In evolutionary algorithms, solutions are sometimes shown as people or chromosomes, where each chromosome contains a possible solution to the input issue. Depending on the type of issue being tackled, these representations might be binary strings, real-valued vectors, permutations, trees, or other structures. Every member of the population has their fitness assessed using a predetermined objective function or fitness metric. This function measures a solution's performance in relation to the goals and limitations of the challenge. Selection processes are used by evolutionary algorithms to decide which individuals will procreate and become members of the following generation. Similar to the process of natural selection, individuals with higher fitness values are more likely to be chosen for reproduction.

Genetic operators like crossover and mutation are used by evolutionary algorithms to generate offspring solutions from a subset of people. While

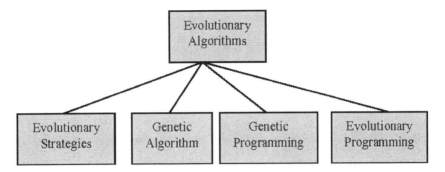

Figure 14.2 Hierarchical structure for evolutionary algorithms.

mutation adds random changes to preserve population diversity, crossover combines genetic material from two parent solutions to create new child solutions. The population changes in the direction of better answers as the algorithm advances over several generations. The method eventually converges to optimum or nearly optimal solutions as a result of the population's increased prevalence of individuals with higher fitness scores. It is expected that evolutionary algorithms in the future will benefit from developments in distributed and parallel computing systems. By making use of parallelism, these algorithms can speed up the search over vast solution spaces by evaluating several candidate solutions at once. To improve their robustness and performance, evolutionary algorithms can be coupled with various optimization strategies and metaheuristics. To more successfully address challenging optimization issues, hybrid approaches combine evolutionary algorithms with machine learning algorithms, constraint fulfillment strategies, or local search methods. Numerous fields, including engineering design, robotics, financial modeling, bioinformatics, data mining, and more, have seen success using evolutionary algorithms. It is anticipated that further iterations of evolutionary algorithms would broaden their scope of application to tackle novel issues in domains including sustainable development, transportation, healthcare, and renewable energy. The hierarchical structure for evolutionary algorithms is shown in Figure 14.2.

14.4 DEEP LEARNING AND NEURAL NETWORKS

Deep learning and neural networks represent a powerful paradigm in machine learning, inspired by the structure and function of the human brain. The neural network, a computational model made up of linked layers of artificial neurons that receive input data and learn to derive hierarchical representations, is the fundamental component of deep learning. Deep learning algorithms automatically discover hierarchical features straight from raw data, in contrast to conventional machine learning techniques that rely on manually created features. This allows for more efficient and scalable learning. Large-scale datasets and technological advancements in computing technology have made it easier to train intricate neural network topologies with millions of parameters, which has contributed to the success of deep learning. Deep neural networks (DNNs) have shown outstanding performance over a wide range of tasks, including image recognition, speech recognition, natural language processing, and reinforcement learning. The idea of backpropagation, a technique for training neural networks by repeatedly changing the network's parameters to minimize a predetermined loss function, is the foundation of deep learning. By calculating the gradient of the loss function in relation to the network parameters, backpropagation enables effective optimization using gradient descent or its variations. DNNs with numerous levels of abstraction can now be trained, which has made it possible to create sophisticated models that can recognize complex patterns and correlations in data. The use of convolutional neural networks (CNNs) for image and video processing tasks is one of the distinguishing characteristics of deep learning. CNNs use pooling layers for spatial sub-sampling and fully connected layers for classification or regression. CNN architectures, such as AlexNet, VGG, and ResNet, have achieved state-of-the-art performance on benchmark datasets like ImageNet, demonstrating the effectiveness of deep learning in computer vision applications and also for spatial feature extraction from input pictures, and convolutional layers for features.

Apart from image processing, sequential data modeling applications like voice recognition, machine translation, and time series prediction have been transformed by recurrent neural networks (RNNs) and their derivatives, such as long short-term memory (LSTM) networks. RNNs are able to learn from previous observations and predict future states because they keep hidden states over time steps, which allows them to grasp temporal relationships in sequential data. The success of deep learning has inspired significant developments in adjacent domains, including generative adversarial networks (GANs) for picture creation, reinforcement learning for sequential decision-making, and attention mechanisms for sequence-to-sequence tasks. These advancements have expanded the potential uses of deep learning and stimulated innovation in a variety of fields, including robotics, autonomous driving, healthcare, and finance. Deep learning, as a subset of machine learning, relies on the concept of neural networks. Neural networks are trained through two main stages: backpropagation and forward propagation. Predictions are formed and input data is sent via the network during forward propagation. The network's mistakes are then computed during backpropagation, and gradients are propagated backward through the network to update the weights and reduce the discrepancy between the expected and actual outputs. Numerous neural network topologies designed for distinct tasks and data kinds are included in deep learning. By using convolutional layers to capture spatial patterns and features, CNNs perform very well in image processing tasks. Sequential data processing is the focus of RNNs, with topologies such as LSTM networks being especially useful for capturing long-range relationships in sequences.

14.5 ETHICAL AND SOCIETAL IMPLICATIONS

Deep learning algorithms and other AI systems are not immune to bias, which can provide unfair or biased results. Unspoken presumptions in algorithm design, biased training data, or ingrained social preconceptions in the data can all be sources of bias. Ensuring equitable treatment for various groups necessitates addressing prejudice and encouraging fairness in AI systems. To learn and make predictions, deep learning algorithms frequently rely on enormous volumes of data, including personal information. Concerns over mass monitoring and privacy violations are rising as AI systems become more commonplace in day-to-day life. Clear regulations, strong data protection measures, and open data handling procedures are necessary to preserve people's privacy rights when utilizing AI technology. Transparency and accountability are hampered by deep learning algorithms' opacity. It can be difficult to analyze or explain the outputs of complex neural networks, which makes it tough to comprehend the reasoning behind AI-driven judgments. To promote trust and accountability in AI systems, tools for algorithmic transparency, auditing, and accountability must be established. Concerns over job displacement and economic disruption are raised by the broad deployment of AI, particularly automation driven by deep learning algorithms. AI has the potential to increase productivity and streamline procedures, but it may also result in the displacement of some sectors and jobs. To address the socioeconomic effects of AI, proactive measures including workforce retraining, lifelong learning, and inclusive economic policies are needed. Autonomous technologies powered by AI, like drones and self-driving automobiles, present moral dilemmas when it comes to making decisions under pressure.

14.6 CONCLUSION

In the rapidly advancing field of AI, where algorithms are always changing and expanding the scope of what is possible, it is critical that future generations grasp the foundations. This chapter's overview sheds light on the wide range of AI algorithms and their significant long-term ramifications. We have set out on a trip into the inner workings of AI, covering everything from the fundamentals of supervised and unsupervised learning to the complexities of reinforcement learning and evolutionary algorithms. Deep learning and neural networks have shown us their revolutionary power, thereby opening new avenues for image identification, natural language processing, and generative modeling. But immense power also entails great responsibility. The ethical and societal ramifications of AI algorithms must be considered in addition to our admiration for their promise. Bias, privacy issues, and employment loss highlight the necessity of careful planning and preventative steps to guarantee fair and responsible AI implementation.

REFERENCES

[1] LeCun, Y., Bengio, Y., & Hinton, G. (2015). Deep Learning. Nature, 521(7553), 436–444.

[2] Silver, D., Schrittwieser, J., Simonyan, K., et al. (2017). Mastering the Game of Go without Human Knowledge. Nature, 550(7676), 354–359.

[3] Rosenblatt, F. (1958). The Perceptron: A Probabilistic Model for Information Storage and Organization in the Brain. Psychological Review, 65(6), 386–408.

[4] Esteva, A., Kuprel, B., Novoa, R. A., et al. (2017). Dermatologist-level Classification of Skin Cancer with Deep Neural Networks. Nature, 542(7639), 115–118.

[5] Silver, D., Huang, A., Maddison, C. J., et al. (2016). Mastering the Game of Go with Deep Neural Networks and Tree Search. Nature, 529(7587), 484–489.

[6] Jobin, A., Ienca, M., & Vayena, E. (2019). The Global Landscape of AI Ethics Guidelines. Nature Machine Intelligence, 1(9), 389–399.

[7] Mittelstadt, B. D., Allo, P., Taddeo, M., et al. (2016). The Ethics of Algorithms: Mapping the Debate. Big Data & Society, 3(2), 2053951716679679.

[8] Newell, A., & Simon, H. A. (1956). The Logic Theory Machine—A Complex Information Processing System. IRE Transactions on Information Theory, 2(3), 61–79.

[9] Turing, A. M. (1950). Computing Machinery and Intelligence. Mind, 59(236), 433–460.

[10] McCulloch, W. S., & Pitts, W. (1943). A Logical Calculus of the Ideas Immanent in Nervous Activity. Bulletin of Mathematical Biophysics, 5(4), 115–133.

[11] Rumelhart, D. E., Hinton, G. E., & Williams, R. J. (1986). Learning Representations by Back-propagating Errors. Nature, 323(6088), 533–536.

[12] Kaelbling, L. P., Littman, M. L., & Moore, A. W. (1996). Reinforcement Learning: A Survey. Journal of Artificial Intelligence Research, 4, 237–285.

[13] Vaswani, A., Shazeer, N., Parmar, N., et al. (2017). Attention is All You Need. Advances in Neural Information Processing Systems, 30, 5998–6008.

[14] Esteva, A., Kuprel, B., Novoa, R. A., et al. (2017). Dermatologist-level Classification of Skin Cancer with Deep Neural Networks. Nature, 542(7639), 115–118.

[15] Rajkomar, A., Oren, E., Chen, K., et al. (2018). Scalable and accurate deep learning with electronic health records. npj Digital Medicine, 1(1), 18.

Chapter 15

Artificial Intelligence Algorithms
Conspectus and Vision

Arpita Tewari

15.1 INTRODUCTION

Artificial intelligence (AI) fulfills the new vision of Industry 5.0, which is defined based on the principles of human centrality, sustainability, and resiliency. Unlike the Fourth Industrial Revolution, which does not prioritize these goals, Industry 5.0 embraces AI to achieve them (Figure 15.1).

AI is undeniably a groundbreaking achievement in the field of computer science, poised to become an integral element of contemporary software in the forthcoming years and decades. This not only poses a potential risk but also offers a promising prospect. AI is set to be utilized in enhancing defensive and offensive cyber activities. Moreover, novel methods of cyber assault will be developed to exploit the specific vulnerabilities of AI technology. The significance of data will be magnified by the insatiable demand for extensive training data by AI, thereby reshaping our perception of data protection. It is imperative to establish cautious governance on a global scale to guarantee that this groundbreaking technology of Industrial Revolution 5.0 fosters widespread safety and prosperity [1].

15.1.1 Data Authority

NetApp, as the leading expert in hybrid cloud data management, recognizes the importance of data access, control, and governance. The NetApp data fabric offers a cohesive data management ecosystem that extends across various edge devices, data centers, and multiple hyper-scale clouds. The data fabric empowers organizations of various scales to expedite essential applications, enhance data transparency, optimize data security, and enhance operational flexibility [2].

NetApp AI solutions are based on the following key building blocks:

- **ONTAP software** enables AI and deep learning both on premises and in the hybrid cloud.

240 Artificial Intelligence and Communication Techniques

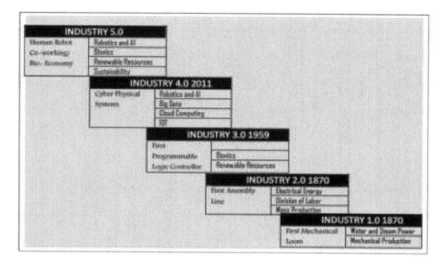

Figure 15.1 Progress of AI from IR1.0 to IR5.0.

- **AFF all-flash systems** accelerate AI and deep learning workloads and remove performance bottlenecks.
- ONTAP Select software enables efficient data collection at the edge, using Internet of Things (IoT) devices and aggregations points.
- Cloud Volumes can be used to rapidly prototype new projects and provide the ability to move AI data to and from the cloud.

15.1.2 Transformation of World through AI Techniques

AI is a revolutionary technology that is reshaping all facets of life. It is a versatile tool that empowers individuals to reconsider how information is integrated, data is analyzed, and the insights gained are leveraged to enhance decision-making processes. The purpose of this extensive analysis is to elucidate the concept of AI to a target audience consisting of policymakers, opinion leaders, and curious observers. Additionally, it aims to showcase the transformative impact of AI on the world and highlight the significant societal, economic, and governance-related questions it poses [3].

To optimize the advantages of AI, we propose a set of nine guidelines for progressing further:

Enumerated below are the primary contributions of this chapter:

- To delimit the extent of our investigation, it is imperative to consider the inherent qualities and attributes of diverse real-world

Artificial Intelligence Algorithms: Conspectus and Vision 241

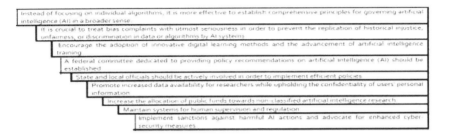

Figure 15.2 Progress Guidelines.

    data sets, as well as the proficiencies exhibited by different learning methodologies
- To offer a thorough perspective on AI algorithms that have the potential to improve the intelligence and functionalities of a data-centric application
- Exploring the suitability of machine learning (ML)-driven solutions across a range of practical fields
- To emphasize and succinctly outline the prospective avenues for research within the parameters of our study on intelligent data analysis and services (Figure 15.2).

15.2 TYPES OF REAL-WORLD DATA AND AI ALGORITHMS

15.2.1 Types of Real-World Data

The presence of data is commonly regarded as essential for developing an ML model or data-centric real-world systems. Data may exist in different formats, including structured, semi-structured, or unstructured. Additionally, "metadata" serves as a type of data that typically describes information about the data itself.

In the subsequent section, we will provide a concise overview of these various categories of data.

- **Structured:** It possesses a clearly defined framework and adheres to a data model, which adheres to a standardized sequence, exhibiting a high level of organization and effortless accessibility. Consequently, it is employed by either an entity or a computer program.

Structured data, such as names, dates, addresses, credit card numbers, stock information, and geo-location, are commonly stored in tabular formats within clearly defined systems like relational databases [4].

- **Unstructured:** Unstructured data lacks a predetermined format or structure, posing challenges in capturing, processing, and analyzing it. This type of data typically includes text and multimedia content such as sensor data, emails, blog entries, wikis, word processing documents, PDF files, audio files, videos, images, presentations, web pages, and various other forms of business documents.
- **Semi-structured:** Semi-structured data differs from structured data in that it is not stored in a relational database. Despite this distinction, semi-structured data possesses organizational properties that facilitate analysis. Examples of semi-structured data include HTML, XML, JSON documents, and NoSQL databases.
- **Metadata:** Data is not in its conventional form, but rather it is "data about data". The fundamental distinction between "data" and "metadata" lies in the fact that data represents the actual material that can categorize, quantify, or even record something in relation to an organization's data characteristics. Conversely, metadata provides a description of the pertinent data information, thereby imparting greater significance to data users. In the area of ML and data science, researchers use various widely used datasets for different purposes.

Various datasets are available for cybersecurity, smartphone usage, IoT, agriculture, e-commerce, health, and other application domains. Examples include NSL-KDD, UNSW-NB15, ISCX'12, CICDDoS2019, Bot-IoT for cybersecurity; phone call logs, SMS log, mobile application usages logs, mobile phone notification logs for smartphone datasets; IoT data; agriculture and e-commerce data; heart disease, diabetes mellitus, coronavirus disease 2019 (COVID-19) for health data, among others [5].

15.2.2 Type of AI Algorithms for IR5.0

Industry 5.0 signifies the merging of industrial manufacturing and cutting-edge technologies like AI, robotics, IoT, and big data analytics, with a significant focus on the collaboration between humans and machines. The utilization of AI algorithms is pivotal in facilitating and enhancing various processes within the realm of Industry 5.0. Below are a few noteworthy AI algorithms that hold relevance in the context of Industry 5.0 [6–8].

- **Predictive maintenance algorithms:** ML techniques can be employed to forecast equipment failures by analyzing sensor data, usage patterns, and historical maintenance records. This proactive approach aids in minimizing downtime, optimizing maintenance schedules, and prolonging the operational lifespan of machinery (Figure 15.3).
- **Quality control algorithms:** AI algorithms have the capability to analyze real-time sensor data and visual inspection images, allowing

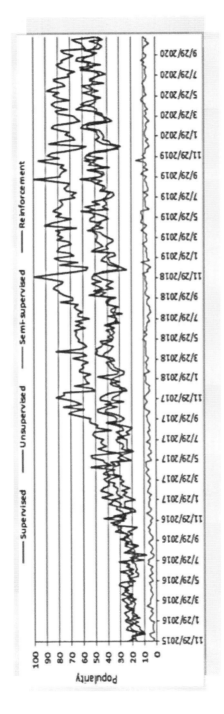

Figure 15.3 The Worldwide Popularity Score of Various Types of ML Algorithms.

them to effectively detect any defects or anomalies in products throughout the manufacturing process. This advanced technology empowers manufacturers to make proactive adjustments, ensuring that high-quality standards are consistently upheld and minimizing unnecessary waste.
- **Autonomous robotics algorithms:** Autonomous robots rely on algorithms to carry out a range of tasks, including material handling, assembly, and inspection, while minimizing the need for human involvement. These algorithms encompass path planning, obstacle avoidance, object recognition, and manipulation.
- **Supply chain optimization algorithms:** AI algorithms have the capability to enhance inventory management, demand forecasting, and logistics operations within the supply chain. By scrutinizing past data, market trends, and external influences, these algorithms can effectively optimize inventory levels, minimize transportation expenses, and enhance the efficiency of deliveries.
- **Human–robot collaboration algorithms:** Version 1: Industry 5.0 places significant emphasis on fostering collaboration between humans and robots within the workplace. Through the utilization of AI algorithms, robots are empowered to comprehend human gestures, intentions, and commands. This newfound ability enables them to actively participate in tasks such as assembly, maintenance, and inspection, thereby promoting a secure and highly efficient collaborative environment (Figure 15.3).
- **Customized production algorithms:** AI algorithms have the capability to examine customer preferences, market trends, and production capabilities to facilitate the mass customization of products. This includes adaptive manufacturing processes that can flexibly modify production parameters to fulfill specific customer needs.
- **Big Data analytics algorithms:** Industry 5.0 is responsible for the generation of extensive volumes of data derived from sensors, machines, and production processes. The utilization of AI algorithms in Big Data analytics facilitates the continuous monitoring, analysis, and optimization of manufacturing operations, thereby resulting in enhanced efficiency, quality, and productivity (Figure 15.3).
- **Bidirectional Encoder Representations from Transformers (BERT) variants:** Since BERT was introduced, numerous iterations and advancements have emerged, including RoBERTa, ALBERT, and ELECTRA. These models aim to overcome existing limitations and delve into novel architectures to augment language comprehension and generation tasks.
- **Meta-learning:** Meta-learning techniques empower models to acquire the ability to learn, through exposure to diverse tasks and the capacity to swiftly adjust to new tasks with minimal data. Strategies

like Model-Agnostic Meta-Learning (MAML) and Reptile have been introduced for scenarios involving learning from a small number of examples.
- **Graph neural networks (GNNs):** GNNs are increasingly recognized as effective instruments for examining and representing structured data, specifically graphs and networks. They find utility in various fields including social network scrutiny, recommendation engines, pharmaceutical exploration, and numerous other areas.
- **Continual learning:** Continuous learning algorithms allow models to learn gradually from a continuous flow of data while preventing catastrophic forgetting. Strategies like replay, regularization, and dynamic architectures are investigated to tackle the obstacles of life-long learning.

15.3 SPEEDY-GROWING AI ROLES OF 2024

The demand for AI expertise has seen a significant rise, highlighting the insatiable need for proficient professionals in this field. In recent years, the extraordinary progress of AI has revolutionized the operational dynamics of businesses, thereby giving rise to a multitude of employment opportunities. AI and/or ML engineers hold a pivotal position in the development and implementation of AI and ML systems. In the role of an AI/ML engineer, one would be entrusted with significant responsibilities, which encompass [9–11]:

- Designing ML models
- Preprocessing data
- Enhancing algorithms to efficiently solve complex business challenges
- **Data Scientist:** Data scientists possess advanced skills and expertise in extracting meaningful patterns, trends, and forecasts from data, enabling informed decision-making. This process is further improved by utilizing AI-powered data analytics. Their main objective is to uncover valuable insights from large datasets by leveraging cutting-edge technologies such as AI. The increasing need for data-driven insights has led to data scientists playing a crucial role in driving innovation, improving operational efficiency, and influencing the future of global industries.
- **AI Product Manager:** AI product managers play a crucial role in developing and implementing AI-powered products. They act as a bridge between technical experts and business decision-makers, converting complex AI features into practical product plans. With a deep understanding of AI technologies, market trends, and consumer needs, these individuals are able to identify opportunities where AI can deliver substantial benefits.

- **Natural Language Processing (NLP) Specialists:** In the realm of language-based data, specialists in NLP hold a significant position in transforming disorganized text into useful, actionable findings. These experts have the knowledge and skills required to create and execute algorithms and models that can understand, interpret, and extract valuable information from such unstructured data. Their pivotal contributions have played a crucial role in spearheading innovations like chatbots and virtual assistants, enhancing customer interactions and optimizing a range of language-related functions.
- **Computer Vision Engineers:** Computer vision engineers play a crucial role in various industries, driving advancements in fields like autonomous vehicles and medical imaging. Key competencies for this position encompass proficiency in programming languages such as Python and C++, familiarity with computer vision tools like OpenCV and TensorFlow, a strong grasp of image processing techniques, proficiency in deep learning algorithms, and the ability to optimize models effectively.

15.3.1 Research Interrogation for Most Important Advances in AI

The fundamental technologies encompass vision, speech recognition and synthesis, NLP (comprehension and generation), image and video synthesis, multi-agent systems, planning, decision-making, and the fusion of vision and motor skills for robotics. Moreover, innovative applications have surfaced across diverse fields such as gaming, medical diagnostics, logistical systems, autonomous vehicles, language translation, and interactive personal aids.

The sections that follow provide examples of many salient developments [12–13].

- **Language Processing:** ELMo, GPT, mT5, and BERT.3 are examples of neural network language models. These models have proven to be highly proficient in various language-related tasks such as machine translation, text classification, speech recognition, writing aids, and chatbots. Furthermore, their capabilities hold great potential for enhancing human–AI interactions in different languages and scenarios. The widespread presence of voice-control systems like Google Assistant, Siri, and Alexa can be attributed to the advancements in AI-powered voice recognition technology. An example of a conversational interface that leverages these advancements is Google Duplex.
- **Computer Vision and Image Processing:** Image-processing technology is utilized to create video-conference backgrounds and produce photo-realistic images commonly referred to as deep fakes. Algorithms operating on ImageNet, a vast repository of more than

14 million images employed for training and evaluating visual recognition systems, now perform their tasks at a speed that is 100 times quicker compared to only 3 years ago.

Real-time object detection systems, like YOLO (You Only Look Once), play a crucial role in video surveillance of crowds and are of great significance for mobile robots, including self-driving cars. These systems are designed to promptly identify significant objects as they appear in an image. Additionally, face recognition is another important application that benefits from such real-time object detection systems. Generative Adversarial Networks (GANs) have made it feasible to create visually realistic images and even videos. The integration of GAN technology, which enables image generation, and transformer technology, which facilitates text production, can be accomplished through diverse approaches.

- **Games:** The advancement of AI techniques has found a fruitful training ground and a platform for display in the development of algorithms for games and simulations involving adversarial scenarios. Notable examples include Atari video games, StarCraft II, Quake III, and Alpha Dogfight, which is a jet-fighter simulation sponsored by the US Defense Department. Additionally, classical games such as poker have also served as valuable arenas for the application of AI techniques.

The team behind AlphaGo, known as DeepMind, progressed to produce AlphaGoZero, which eliminated the reliance on human input by utilizing historical data from Go matches. AlphaGoZero autonomously formulated strategies and moves without external guidance. This concept was enhanced with the introduction of AlphaZero, a unified network structure capable of mastering Go, Shogi, and Chess at an expert level.

Bipedal and quadrupedal robots are making significant progress in terms of agility. For instance, Atlas, an advanced humanoid robot created by Boston Dynamics, has showcased remarkable capabilities such as jumping, running, backflipping, and navigating through uneven terrains—tasks that were previously deemed unattainable for robots. Similarly, Spot, a four-legged robot also developed by Boston Dynamics, excels in maneuvering challenging environments and is currently employed in construction sites for the delivery and monitoring of lightweight materials and tools. On the other hand, Cassie, a biped robot designed by Agility Robotics and Oregon State University, utilizes deep reinforcement learning to enhance its walking and running functionalities.

Throughout 2020, the progression of robotics was partially motivated by the imperative to uphold social distancing protocols amid the COVID-19 crisis. Notably, a collection of eateries in China unveiled a staff of 20 robots to assist in the cooking and serving of meals. Furthermore, early models of delivery robots were dispatched to specific locations like campuses to transport various items including books and food [14,15].

- **Mobility:** Autonomous vehicles, also known as self-driving cars, have emerged as a prominent field in the realm of deployed robotics. The development of self-driving cars necessitates the seamless integration of various cutting-edge technologies such as sensor fusion, AI planning and decision-making, vehicle dynamics prediction, real-time rerouting, inter-vehicle communication, and numerous others.

Autonomous vehicles are currently in operation in specific regions, such as Phoenix, Arizona, where driving and weather conditions are notably favorable. Additionally, outside Beijing, the presence of 5G connectivity enables remote drivers to intervene if necessary.

- **Health:** AI is becoming more prevalent in biomedical applications, especially in the areas of diagnosis, drug development, and fundamental life science investigations.

Various tools have been developed to identify a range of eye and skin disorders, as well as to detect cancers and assist in obtaining measurements required for clinical diagnosis. Certain systems now available are comparable to the diagnostic skills of experienced pathologists and radiologists, and can aid in reducing the burden of monotonous tasks, such as tallying the proliferation of cells in cancerous tissue.

Various technologies exist for digital medical transcription, analyzing electrocardiogram (ECG) systems, generating super-resolution images to expedite magnetic resonance imaging (MRI) processes, and formulating queries for clinicians to pose to pediatric patients.

- **Finance:** AI has seen a rise in its integration within the finance sector. Deep learning algorithms are now being utilized by numerous lenders to semi-automate the process of making lending decisions, and have revolutionized payment systems through innovations like credit scoring in platforms such as WeChat Pay.

Robo-advising, which refers to the automated provision of financial advice, is rapidly gaining popularity in the realm of personal finance. This technology is increasingly being adopted by individuals for investment purposes and overall financial planning. Financial institutions, on the other hand, are leveraging AI not only for fraud detection and cybersecurity enhancement but also for automating legal and compliance documentation, as well as identifying instances of money laundering.

- **Recommender systems:** In addition to an overall increase in online activities such as news, music, videos, e-commerce, and others, there have been significant advancements in AI technologies that drive recommender systems over the last 5 years. AI systems are now focusing more on comprehending the rationale behind recommending a specific item to a particular individual or query. For instance, Spotify utilizes audio analysis of music, while large language models like BERT are employed to enhance recommendations for news or social media content.

15.4 NEW HIGH-POWERED PARADIGM

AI research has diversified into emerging fields such as NLP, computer vision, and robotics, thus laying the foundation for the current AI revolution.

- **GPT: The Game Changer**

The rapid growth of AI can be mainly credited to the advancement of deep learning methods and the rise of extensive neural networks, like the Generative Pre-trained Transformer (GPT) series developed by OpenAI. GPT-3, which was introduced in 2020, serves as a notable illustration of the progression of AI, featuring 175 billion parameters and showcasing unparalleled abilities in comprehending and generating natural language. The achievements of GPT-3 and its forerunners underscore the promise of AI and have catalyzed additional exploration and advancement in the field. GPT-4, the most recent version, expands on the foundation laid by its predecessors and demonstrates heightened capabilities, propelling the frontiers of AI to new heights.

15.4.1 Key enabling technologies for Industry 5.0

Industry 4.0 relies on a wide range of primarily information and communications technology (ICT) technologies, whereas Industry 5.0 will predominantly rely on the "cognitive revolution" of these systems (Table 15.1). This revolution entails the evolution of current narrow domain knowledge capabilities to encompass much broader and context-aware cognition (Figure 15.4).

The ISA-95-defined automation pyramid outlines a model that progresses from field level to business level, with increasing complexity and abstraction. AI is generally employed in tasks where the objective can be clearly defined and training data can be acquired.

15.4.2 Industries Transformed by AI

AI has infiltrated every facet of our existence, completely transforming countless sectors in the process [16–20].

Table 15.1 Distinctness of Industry 4.0 and Industry 5.0

Industry 4.0	Industry 5.0
Emphasizing increased effectiveness via digital connectivity and artificial intelligence	Ensuring a framework that amalgamates competitiveness and suitability is crucial for the industry as it enables the industry to fully harness its potential as a fundamental driver of technological transformation.
The focus of technology is on the development of cyber physical entities	Highlights the significance of alternative forms of technology governance in promoting sustainability and enhancing resilience.
Aligned with improving business models to align with prevailing capital market dynamics and economics structures- ultimately aimed at reducing costs and maximizing profits for shareholders.	Digital devices empower workers promoting a technology driven approach that priorities the well-being of individuals.
The absence of emphasis on design and performance aspects is detrimental to achieving systemic transformation and the disconnection of resource and material utilization from adverse environmental, climate and social consequences	The scope of corporate responsibility is broadened to encompass their entire value chains thereby extending their accountability beyond their immediate operations.

- **Health care:** AI-driven diagnostic tools have enhanced the detection of diseases, leading to more precise and prompt diagnoses. Additionally, AI-facilitated drug discovery expedites the creation of novel treatments. Robotics and AI are crucial in precision surgery, improving patient results and shortening recovery periods.
- **Finance:** AI has become a crucial component of the financial industry, playing a significant role in various functions such as fraud detection and portfolio management. The emergence of robo-advisors and algorithmic trading platforms has democratized investing and made it more efficient. Additionally, AI-powered risk assessment models have enhanced the precision of credit and loan assessments.
- **Manufacturing:** AI is revolutionizing the fabricating industry through the integration of intelligent factories, where robotics and sophisticated automation systems enhance production processes, resulting in decreased inefficiencies and heightened productivity. Additionally, AI-driven predictive maintenance has emerged as an indispensable tool in minimizing equipment idle time and cutting down on operational costs.

Artificial Intelligence Algorithms: Conspectus and Vision 251

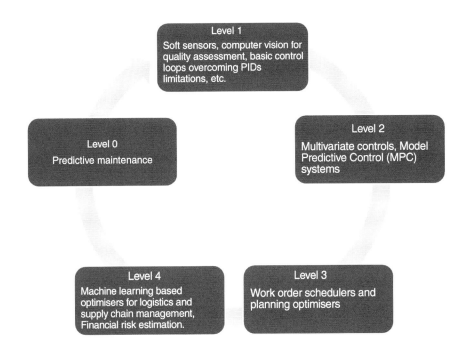

Figure 15.4 Different Levels of Cognitive Revolution.

- **Transportation:** Self-governing vehicles and traffic management systems energized by AI are transforming the field of transportation, offering the potential to decrease accidents and ease traffic congestion. Additionally, AI has been successfully implemented in logistics, enhancing the efficiency of supply chain operations and maximizing the efficiency of delivery routes.
- **Retail:** AI is revolutionizing the retail industry by facilitating customized suggestions, intelligent virtual assistants, and sophisticated inventory control. Additionally, AI-powered data analysis assists retailers in gaining a deeper comprehension of customer behavior and preferences, which enables businesses to empower their customers by allowing them to personalize and tailor their products and services to their specific needs and desires.
- **Agriculture:** AI-powered exactitude agriculture methods have led to enhanced crop productivity and efficient allocation of resources. By accurately detecting pests and diseases and determining the most favorable periods for planting and harvesting, AI technology is significantly contributing to the preservation of global food security.

15.5 THE FUTURE OF AI

AI is experiencing rapid growth, and the reasons behind this phenomenon are quite evident. Although the concept of AI has been in existence for several decades, it is only in recent times that we have witnessed the emergence of extensive AI programs. The remarkable progress in the field of AI can be primarily attributed to the advancements in big data and cloud computing technologies. The significance of big data in AI lies in its capacity to facilitate the training of AI systems with vast quantities of data. This process ensures that AI can acquire knowledge from past experiences and comprehend intricate scenarios. Moreover, cloud computing contributes significantly to the advancement of AI by providing online storage solutions for data. This enables convenient access to data from any location, enhancing the efficiency of AI systems. The potential of AI appears to be promising in the future. Although we are still in the process of discovering the optimal utilization of this technology, the advantages it offers are immense. With its ability to provide more precise predictions, enhance business operations, and develop even more intelligent robots, the future of AI seems to be filled with optimism.

15.5.1 Contemporary Provocation

AI algorithms are of utmost importance in the advancement of Industry 5.0 as they effectively streamline operations, boost efficiency, and foster harmonious interaction between humans and machines (Table 15.1). Nevertheless, they encounter various obstacles within this framework.

- **Data Quality and Availability:** Industry 5.0 heavily depends on the data obtained from sensors, machines, and diverse systems. The task of guaranteeing the quality, dependability, and accessibility of this data can pose significant challenges, particularly in settings characterized by a multitude of data sources, outdated systems, and isolated data repositories.
- **Interoperability and Integration:** The integration of AI algorithms into current infrastructure and systems within industrial environments may present complexities. Overcoming interoperability issues among various hardware and software platforms, along with ensuring compatibility with older systems, can pose notable obstacles.
- **Safety and Reliability:** AI algorithms utilized in industrial settings are required to comply with stringent safety regulations to uphold the dependability and authenticity of processes. The incorporation of resilience, error tolerance, and fail-safe protocols is essential in averting mishaps and reducing potential hazards.

- **Human–Machine Collaboration:** Industry 5.0 underscores the significance of human–machine collaboration in the workplace with AI algorithms operating in conjunction with human operators. Developing AI systems that can effectively engage with and enhance human workers, all while valuing their autonomy and expertise, necessitates a thorough examination of human factors and user experience.
- **Ethical and Societal Implications:** The implementation of AI algorithms in Industry 5.0 gives rise to ethical and societal issues concerning job displacement, privacy, bias, and fairness. It is crucial to uphold ethical AI standards, such as accountability, transparency, and fairness, to address possible risks and cultivate confidence in AI systems.

15.5.2 Growing Graph: The Usage of AI in Industry 5.0

Early Adoption (2010s): During the initial phases of Industry 5.0, the utilization of AI was predominantly confined to research and development initiatives, while pilot deployments were observed in specific sectors like automotive manufacturing, aerospace, and energy.

Expansion and Experimentation (Mid-2010s): With the advancement and increased availability of AI technologies, different sectors have started to explore the prevalent use of AI in a range of regions like predictive maintenance, quality assurance, supply chain management, and self-governing robotics.

Acceleration (Late 2010s to Early 2020s): The adoption of AI in Industry 5.0 experienced a notable surge during the late 2010s. This growth was primarily fueled by the widespread availability of data, advancements in deep learning and reinforcement learning techniques, and the emergence of Industry 4.0 initiatives.

Maturation and Integration (Mid to Late 2020s): In the latter half of the 2020s, AI technologies had emerged as indispensable elements within the ecosystems of Industry 5.0, experiencing extensive implementation across various industries and geographical areas. These AI algorithms were seamlessly incorporated into the industrial infrastructure, facilitating comprehensive automation, adaptable manufacturing processes, and astute decision-making on a large scale.

Collaborative robots, also known as cobots, integrated with AI functionalities have become a standard feature in manufacturing settings. These robots collaborate with human workers to efficiently carry out intricate tasks while ensuring accuracy and safety.

Transformation and Innovation (Late 2020s and Beyond): Moving forward, the employment of AI in Industry 5.0 is anticipated to persist in its evolution, leading to additional advancements and changes within various sectors. AI-driven technologies like augmented reality (AR), virtual reality (VR), digital twins, and edge computing are on the verge of assuming more prominent roles in improving productivity, optimizing resource allocation, and facilitating novel modes of interaction between humans and machines. Industries are poised to utilize the capabilities of AI to tackle the pressing issues of climate change, sustainability, and resilience. This will lead to the advancement of AI-based solutions for renewable energy, intelligent infrastructure, and circular economy endeavors.

15.5.3 The Subsequent of AI: Provocation and Occasion

Notwithstanding the remarkable advancements in AI technology, there remain obstacles and ethical dilemmas that need to be resolved as the field progresses. Privacy, security, and the risk of AI misuse are significant concerns that demand thorough examination and oversight. It is crucial to prioritize the transparency and interpretability of AI systems, while also working toward reducing the likelihood of algorithmic prepossession. The vast potential benefits that AI offers are significant. As AI technology advances and becomes more intertwined with our everyday activities, it will enhance human abilities and play a role in improving society. For example, AI-powered climate simulations could assist in tackling the urgent problem of climate change, and AI in the field of education could facilitate tailored learning opportunities and enhance access to education for individuals in isolated or underprivileged regions.

15.6 CONCLUSION

AI is experiencing a rapid surge in popularity and adoption due to several compelling reasons. Initially, AI serves as a valuable tool for businesses to automate tasks and enhance operational efficiency. By leveraging AI technologies, companies can streamline their processes and allocate resources more effectively, ultimately leading to improved productivity and cost savings. Furthermore, AI plays a crucial role in facilitating informed decision-making by analyzing extensive volumes of data. With its advanced algorithms and ML capabilities, AI systems can extract valuable insights from complex datasets, enabling us to make more accurate and data-driven choices. This ability to harness and interpret vast amounts of information empowers organizations to optimize their strategies and achieve better outcomes. Consequently, the multitude of benefits offered by AI has resulted in its widespread adoption across diverse industries, ranging from retail to health care. Regardless of our professional or personal aspirations, it is crucial to stay abreast of the latest trends in AI. By doing so, we can effectively leverage AI's numerous advantages and stay ahead in this rapidly evolving technological landscape.

REFERENCES

[1] Olaizola, I. G., Quartulli, M., Garcia, A., & Barandiaran, I. (2022). Artificial Intelligence from Industry 5.0 perspective: Is the technology ready to meet the challenge? *Proceedings*, *1613*, 0073. https://ceur-ws.org/Vol-3214/WS5Paper8.pdf

[2] Trunina, I., Bilyk, M., & Yakovenko, Y. (2023, September). Artificial intelligence from Industry 5.0 perspective: Threats and challenges. In *2023 IEEE 5th International Conference on Modern Electrical and Energy System (MEES)* (pp. 1–5). IEEE.

[3] Maddikunta, P. K. R., Pham, Q. V., Prabadevi, B., Deepa, N., Dev, K., Gadekallu, T. R., ... & Liyanage, M. (2022). Industry 5.0: A survey on enabling technologies and potential applications. *Journal of Industrial Information Integration*, *26*, 100257.

[4] Sindhwani, R., Afridi, S., Kumar, A., Banaitis, A., Luthra, S., & Singh, P. L. (2022). Can Industry 5.0 revolutionize the wave of resilience and social value creation? A multi-criteria framework to analyze enablers. *Technology in Society*, *68*, 101887.

[5] Adel, A. (2022). Future of Industry 5.0 in society: Human-centric solutions, challenges and prospective research areas. *Journal of Cloud Computing*, *11*(1), 1–15.

[6] George, A. S., George, A. H., & Baskar, T. (2023). The evolution of smart factories: How Industry 5.0 is revolutionizing manufacturing. *Partners Universal Innovative Research Publication*, *1*(1), 33–53.

[7] Liang, T. P., Robert, L., Sarker, S., Cheung, C. M., Matt, C., Trenz, M., & Turel, O. (2021). Artificial intelligence and robots in individuals' lives: How to align technological possibilities and ethical issues. *Internet Research*, *31*(1), 1–10.

[8] Pavaloiu, A., & Kose, U. (2017). Ethical artificial intelligence—an open question. *Journal of Multidisciplinary Developments*, *2*(2), 15–27. arXiv preprint arXiv:1706.03021.

[9] Xiaoling, P. (2021, April). Discussion on ethical dilemma caused by artificial intelligence and countermeasures. In *2021 IEEE Asia-Pacific Conference on Image Processing, Electronics and Computers (IPEC)* (pp. 453–457). IEEE.

[10] Liang, T. P., Robert, L., Sarker, S., Cheung, C. M., Matt, C., Trenz, M., & Turel, O. (2021). Artificial intelligence and robots in individuals' lives: How to align technological possibilities and ethical issues. *Internet Research*, *31*(1), 1–10.

[11] Yu, H., Shen, Z., Miao, C., Leung, C., Lesser, V. R., & Yang, Q. (2018). Building ethics into artificial intelligence. In *Proceedings of the 27th International Joint Conference on Artificial Intelligence (IJCAI'18)* (pp. 5527–5533). arXiv preprint arXiv:1812.02953.

[12] George, A. S., George, A. H., & Martin, A. G. (2023). ChatGPT and the future of work: A comprehensive analysis of AI's impact on jobs and

employment. *Partners Universal International Innovation Journal, 1*(3), 154–186.

[13] Kasula, B. Y. (2016). Advancements and applications of artificial intelligence: A comprehensive review. *International Journal of Statistical Computation and Simulation, 8*(1), 1–7.

[14] Dunsin, D., Ghanem, M. C., Ouazzane, K., & Vassilev, V. (2024). A comprehensive analysis of the role of artificial intelligence and machine learning in modern digital forensics and incident response. *Forensic Science International: Digital Investigation, 48,* 301675.

[15] Ahmadi, S. (2023). Next generation AI-based firewalls: A comparative study. *International Journal of Computer (IJC), 49*(1), 245–262.

[16] Mourtzis, D., Angelopoulos, J., & Panopoulos, N. (2022). A literature review of the challenges and opportunities of the transition from Industry 4.0 to Society 5.0. *Energies, 15*(17), 6276.

[17] Aslam, F., Aimin, W., Li, M., & Ur Rehman, K. (2020). Innovation in the era of IoT and Industry 5.0: Absolute innovation management (AIM) framework. *Information, 11*(2), 124.

[18] Almaiah, M. A., Alfaisal, R., Salloum, S. A., Hajjej, F., Thabit, S., El-Qirem, F. A., ... & Al-Maroof, R. S. (2022). Examining the impact of artificial intelligence and social and computer anxiety in e-learning settings: Students' perceptions at the university level. *Electronics, 11*(22), 3662.

[19] Paschek, D., Mocan, A., & Draghici, A. (2019, May). Industry 5.0—The expected impact of next industrial revolution. In *Thriving on Future Education, Industry, Business, and Society, Proceedings of the Make Learn and TIIM International Conference, Piran, Slovenia* (pp. 15–17).

[20] Xu, X., Lu, Y., Vogel-Heuser, B., & Wang, L. (2021). Industry 4.0 and Industry 5.0—inception, conception and perception. *Journal of Manufacturing Systems, 61,* 530–535.

Chapter 16

A Historical, Present, and Prospective Review of Artificial Intelligence's Role in Securing Personal Information and Private Data

Shruti Gupta, Navin Kumar Goyal, and Ajay Kumar

16.1 INTRODUCTION

The fast rise of cybercrime has forced the advance of new technology to identify and prevent such behavior. Artificial intelligence (AI) is a captivating concept in this field. This tool shows promise in identifying and stopping cybercrime, having a lot of potential for assisting in detecting and preventing cybercrime. The coronavirus disease 2019 (COVID-19) pandemic retained a lot of people connected to the web at the house. This made businesses and people more dependent on AI-based systems, technologies, and applications for objects such as functioning from home, learning to program, making online payments, as well as having additional entertainment choices such as streaming platforms and video-on-demand services. As a result, it also gave rise to more cybercrime. Early in the 1990s, while the use of computers was just beginning to expand across the nation, the first cybercrime in India was documented (Gunjan et al., 2013). As the use of the internet is growing, cybercrime cases are also increasing exponentially.

According to a report by Cybersecurity Ventures 2022 (Chng et al., 2022) Estimates indicate that global cybercrime damage costs are projected to increase by 15% annually over the next five years, potentially reaching $10.5 trillion USD by 2025, from $3 trillion USD in 2015. By 2023, the anticipated global annual cost of cybercrime is expected to hit $8 trillion. It appears that the cost of ransomware attacks will increase to $20 billion by 2025, compared to around $11.5 billion in 2019. The report indicates that after the United States and China, cybercrime would rank as the world's third largest economy. According to a report by Cipher Trace (Singhal et al., 2022), the total value of cryptocurrency-related thefts, scams, and frauds in 2020 was around $1.9 billion, which was a 50% increase from the previous year (Figure 16.1).

Representing reports by the National Crime Records Bureau (NCRB), which is in authority for collecting and analyzing crime data in India, there

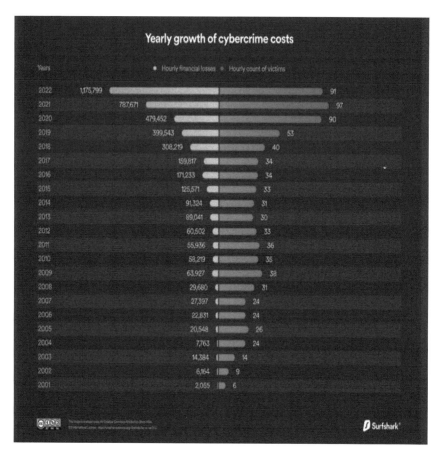

Figure 16.1 Yearly growth of cybercrime costs in the world from 2001 to 2022 (Kroll et al., 2021).

has been a rise in the number of cybercrime cases reported in India over the years. Specific of the most common forms of cybercrime include hacking, ransomware, cryptocurrency crime, cyberattack surface, malware attacks, phishing scams, identity theft, cyberbullying, and cyberstalking (Hu et al., 2020). Following are some statistics on cybercrime in India from 2014 to 2021 (Table 16.1).

Cybercrime cases are increasing concurrently with the growth in internet usage (Figure 16.2). Due to the vulnerability of remote work and increased internet activity, the number of cybercrime instances was high in COVID pandemic 2019 (Lallie et al., 2020).

A Historical, Present, and Prospective Review of AI's Role 259

Table 16.1 Compilation of cybercrime incidents in every state and union territory of India spanning from 2014 to 2021

Cybercrime Cases registered in India from 2014 to 2021		
Year	Total Registered Crime	Percentage Increase from the Last Year
2014	9622	
2015	11592	20.47
2016	12317	6.25
2017	21796	76.96
2018	27248	25.01
2019	44546	63.48
2020	50035	12.32
2021	52974	5.87

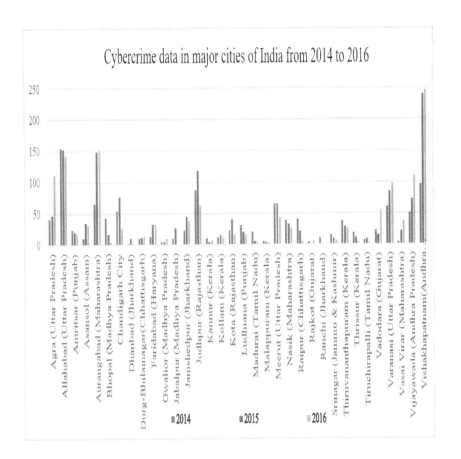

Figure 16.2 Cybercrime data in major cities of India from 2014 to 2016.

Several factors, such as widespread communication to the internet, rising reliance on a variety of online services, lack of digital knowledge in education and training, and rapid technological development, may account for Kerala's comparatively high rate of cybercrime compared to that of other states (Makridakis et al., 2018), as shown in the preceding bar graph Figure 16.3. Also, low internet penetration, limited use of digital services, and less developed cyber infrastructure may all contribute to the relatively low rates of cybercrime in Nagaland and Mizoram compared to other Indian states (Datta et al., 2020) (Figure 16.4).

To deal with such security problems, researchers are focused on the urgent need to find new automated security methods right now. The previous analysis motivates us to look for AI techniques, algorithms, and tools that can assist us to detect and prevent crime early. The chapter is structured as follows: Literature survey, AI role, and real-time incidents, which were detected using AI tools are contained in Section 16.2; future development of AI techniques is discussed in Section 16.3. The outcome of the analysis has been examined in Section 16.4 and the chapter is concluded.

16.2 LITERATURE SURVEY

Several methods have been put up so far for the analysis and evaluation of cybercrime offences through AI, machine learning (ML), and deep learning. This section contains the way they are presented.

- Rameem Zahra et al. (2022) categorized cybercrime incidences using data mining techniques, a proposed iterative classification model was constructed. To test the effectiveness of the five classification models, K-Nearest Neighbors (KNNs), Naive Bayes, Decision List, Decision Tree, and Support Vector Machine (SVM) were implemented.
- Ch et al. (2020) provided the challenges faced by ML technique in protecting cyberspace against attacks and also provides the description of each ML methods.
- Lakhdhar and Rekhis (2021) introduced the theoretical foundation, algorithms, and adversarial attack technique to ensure the security and robustness in deep learning and AI.
- Lahcen and Mohapatra (2022) provided an overview of the use of ML techniques for cybercrime detection, including fraud detection, malware detection, and intrusion detection.
- Ferrag et al. (2020) examined research on adversary attacks and security in deep learning for cybersecurity. This speaks about how attackers can use AI to get around security measures and how defenders can use AI to find and stop attacks.
- Kroll et al. (2021) focused at the benefits and drawbacks of using AI for cybersecurity, such as the manner in which AI could help with

A Historical, Present, and Prospective Review of AI's Role 261

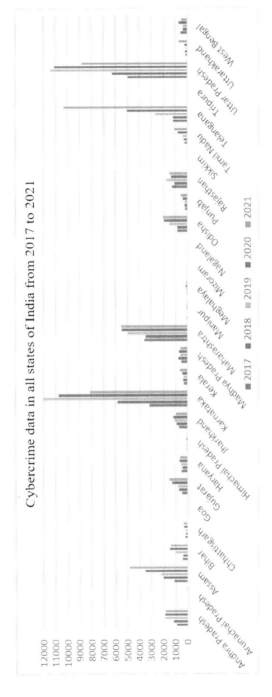

Figure 16.3 Cybercrime data in all states of India from 2017 to 2021.

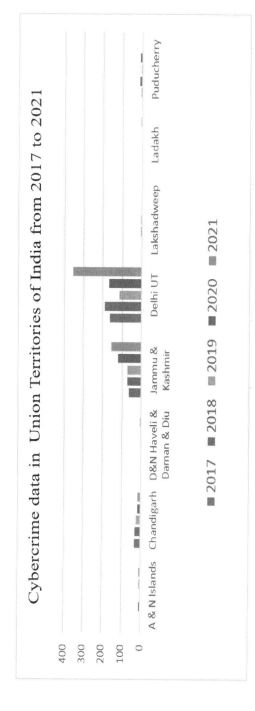

Figure 16.4 Cybercrime data in Union Territories of India from 2017 to 2021.

threat detection and incident response, as well as the ethical problems that come up when AI is used for cybersecurity.
- Hossain et al. (2023) gave an all-inclusive list of all the freely accessible cybersecurity datasets that can be used to train and test ML algorithms.
- Choraś and Woźniak (2022) checked some ethical problems that come up when using AI in hacking. There are three main ethical problems that what the writer delves into great depth about: Fairness, accountability, and transparency.
- Velasco (2022) presented how hacking and AI are linked. They considered the work of international groups and relevant global laws in handling this problem.
- Veena et al. (2022) provided an in-depth analysis of the use of ML techniques for identifying and predicting cybercrime. This paper provides a detailed explanation of each algorithm such as SVMs, decision trees, and neural networks.
- Delplace et al. (2019) provided an in-depth view at how the ML method is used; and we can see that the Random Forest Classifier is a good answer by looking at how well it can separate malicious traffic in a network.

16.2.1 AI Role in Cybercrime

Cybercrime can be controlled in a number of ways, including fraud detection, threat intelligence, malware detection, predictive analysis, and user identification (Dilek, 2015), as shown in Figure 16.5.

Fraud Detection: Different types of scams can be found with AI. This is done by ML and complex techniques that look through huge amounts of data to find trends that point to fraud. Unusual patterns in transaction records or user behavior can be found by AI models, which search for them. These patterns could be signs of fraud. With predictive analytics, AI systems can look at past data and known fraud trends to figure out how likely it is that a transaction is a scam. This enables companies to proactively identify and investigate potentially suspicious activities. Utilizing real-time tracking enables AI-powered fraud detection systems to detect and prevent fraud as it occurs, minimizing potential financial losses. AI also helps stop fraud by looking at data on past fraud and taking action to find and lower possible risks before they get worse. AI is important for finding fraud because it helps companies stay on top of new fraud schemes, so they do not lose money or have their image harmed.

Predictive Analytics: A big part of predictive analytics is AI, which is changing how companies guess what will happen, how people will

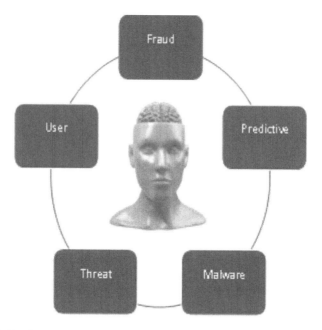

Figure 16.5 AI role in controlling cybercrime.

act, and what trends will happen in the future. AI uses complicated algorithms and ML techniques to help businesses get useful data from huge sets of data. They can now make better predictions and choices because of this. AI looks at data from the past to find trends and make guesses about what will happen or how people will act in the future. This can be used in a lot of different areas, like finances, health care, supply chain management. AI-powered predictive analytics not only helps businesses make operations more efficient, less risky, and better for customers, but it also lets them make choices ahead of time, which makes them more efficient and competitive in today's fast-paced market.

Malware Detection: One important way to find malware is to use AI. It has changed how people who work in protection look for and fight bad software. A lot of complicated methods and ML are used by AI to help security systems find and stop new malware threats quickly. Different types of AI-driven malware detection are used to find and sort both new and known types of malware. These include signature-based detection, behavior analysis, and anomaly detection. Many types of data are looked at by these AI models, such as file characteristics, network traffic, system behavior, and code analysis. They try to find patterns that show bad behavior.

Threat Intelligence: It is improved by AI, which checks huge amounts of data all the time from places like social media, network traffic, system logs, and the dark web. AI uses complicated algorithms and ML to look at this data and find threats, flaws, and signs of compromise in real time. Also, AI-powered threat intelligence systems can combine and make sense of data from various sources to provide a complete picture of cyber threats.

User Authentication: Biometric data, behavioral trends, and information about the surroundings are just some of the things that AI-based computer programs use to run authentication systems. These programs check that users are who they say they are and catch fraud in real time. Identification systems that use AI make things safer because they are always changing to deal with new threats and learning from new data. This lowers the chance of identity theft, account takeover, and getting in without permission.

Improved Usability: Authentication tools that use AI are more useful and easier to use than old-fashioned ones like PINs or passwords. With biometric authentication, users do not have to carry around physical keys or remember long passwords. This makes it easy for people to use and lessens the noise. Users can easily log in based on their behavior and the situation they are in with behavioral and contextual identification. This improves the experience for users without making the protection less strong. AI-powered login systems use complex formulas and ML to look for oddities that could be signs of fraud, such as identity theft, account takeovers, or attempts to log in without permission. This makes fraud less likely.

Secure Adaptiveness: If AI is in control of authentication, it can adapt to new situations and change the standards for authentication based on how dangerous it thinks the situation is. Biometric data, trends of behavior, and information about the present are just some of the things that AI-powered authentication systems look at. Then, to find the best mix between security and ease of use, they change the authentication rules on the fly. This makes sure that the security for each login event is just right for that event.

AI techniques:

ML: As per Stanford computer science professor Andrew Ng (Trautman & Ormerod, 2018), ML involves teaching computers to behave without direct programming. ML is the subbranch of AI (Antonopoulos et al., 2020). ML algorithms (Table 16.2) build model based on training data, which allows the models to make predictions about new data without being explicitly instructed

Table 16.2 Machine learning algorithm with their type and description

Algorithm	Type	Description
Linear Regression	Supervised Learning	A linear model is utilized for predicting the value of a continuous target variable by considering one or more input features. It presumes a direct correlation between variables.
Decision Tree	Supervised Learning	Imagine a structure resembling a tree, with internal nodes symbolizing feature splits, and leaf nodes symbolizing class labels or target values.
Support Vector Machine (SVM)	Supervised Learning	An advanced algorithm that identifies the hyperplane that most effectively divides data into distinct classes within a space with multiple dimensions.
k-Nearest Neighbors (KNN)	Supervised Learning	An algorithm for classification that assigns a class label to an instance by considering the majority class of its k nearest neighbors.
K-Means	Unsupervised Learning	This algorithm partitions data into k clusters by measuring the similarity of data points.
Random Forest	Supervised Learning	An ensemble learning technique that builds numerous decision trees during training and provides the most frequent class or average prediction.

on how to do so (Liu et al., 2022). Creating models that analyze input data and predict an output value within a set range is the primary objective of ML (Tanwar et al., 2020). ML algorithm (Veena et al., 2022) that are often utilized are SVM, Naive Bayes, KNN, K-Means, Random Forest, Dimensionality Reduction Algorithms, Linear Regression, Logistic Regression, Decision Tree.

Natural Language Processing (NLP): NLP is a field within AI that is dedicated to understanding human language. When dealing with text data, like emails, social media posts, and chat logs, NLP techniques can be utilized to detect potential threats or suspicious activities (Sarker et al., 2021).

Behavioral analytics: Studying user behavior helps detect patterns and irregularities that could signal a potential cyber threat. Monitoring user activity on networks and systems involves using ML algorithms to detect abnormal behavior (García-Teodoro et al., 2009).

Deep learning: Deep learning involves training deep neural networks to make predictions using large datasets. Utilizing deep learning enables the identification of potential threats or questionable activities in images and video footage. ML methods are divided into two main categories: Conventional and deep neural networks. These days, automated card driving exemplifies deep learning (Stahl, 2021).

Predictive analytics: Using statistical modeling and ML algorithms to forecast future events is a key aspect of predictive analytics (Miller et al., 2020). Predictive analytics can be used to identify potential cyber threats before they occur. Various AI methodologies and techniques that can be used to detect and prevent cybercrime are discussed in the following.

In addition to these approaches, AI can be implemented to automate security processes such as identifying threats and incident handling. A multi-layered security approach involving a combination of technology (Ghafir et al., 2016), policies, and processes must be developed to successfully protect against cyber-attacks. Overall, the method chosen will be decided by the specific situation at question as well as the characteristics of the data being analyzed.

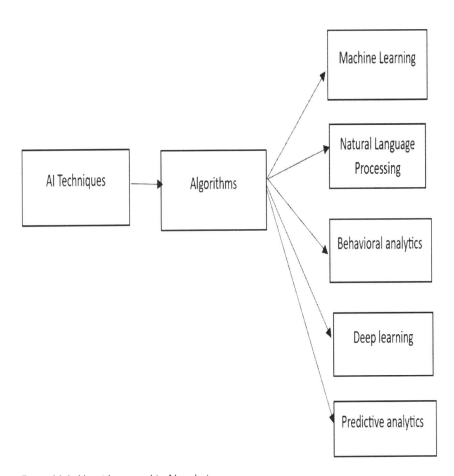

Figure 16.6 Algorithms used in AI techniques.

16.2.2 Real-Time Incidents Were Detected Using AI Tools

A growing number of industries, including cybersecurity, public safety, and health care, are using AI technologies for real-time incident monitoring. Here are some key points to consider when using AI tools to detect incidents in real time.

- **Cybersecurity:** AI systems for cybersecurity monitor network data, analyze system logs, and detect unusual behavior or potential security vulnerabilities instantly. If these systems detect patterns indicating cyber threats such as malware infections, phishing attacks, or unauthorized access attempts, they can promptly respond to mitigate risks.
- **Safety for the public:** Advanced video analytics and AI in surveillance systems can detect crashes, crimes, and emergencies in real-time. These systems can analyze live video feeds from security cameras to detect objects or behaviors of interest. They can promptly alert authorities or security personnel, enabling swift response and intervention.
- **Health care:** AI tools are increasingly utilized in health care settings to promptly identify issues, such as when health care providers are monitoring a patient's vital signs or analyzing medical imaging data for unusual patterns. Wearable technology and medical sensors have the capability to transmit real-time physiological data to AI algorithms. This data can be used to promptly notify health care professionals of potential health issues or irregularities that require immediate intervention.
- **Natural disasters:** AI systems are applied during natural disasters such as earthquakes, floods, and wildfires to detect and address incidents immediately. These systems can analyze sensor data, satellite images, and social media feeds to detect and monitor disasters, assess their severity in real time, and enhance the coordination of emergency response activities.
- **Detecting financial fraud:** There are systems in the financial sector that use AI to look at real-time transaction data for unapproved transactions, account takeovers, fake payment requests, etc. Smart algorithms in these systems look for deals that do not seem right and send them to be checked out further. This helps banks stop fraud and cut down on costs.
- Capgemini Research Institute's new report found that before 2019, over one in five organizations were relying on AI. By 2020, approximately two out of every three businesses plan to use AI. According to research, 69% of organizations think that without AI, they will not be able to respond to cyberattacks (Lau et al., 2023).

16.3 FUTURE DEVELOPMENTS

By learning from large datasets, AI systems will include cutting-edge methodologies like deep learning and reinforcement learning to improve cyber threat detection. Real-time analysis of user behavior and network activity helps spot irregularities and potential threats early on. AI will collect, analyze, and connect large amounts of threat intelligence data, which will also reveal new threats and suggest ways to protect against them. With the help of real-time threat analysis and intelligent decision-making, AI will be integrated with current security systems for improved capabilities. AI makes it possible to analyze and share anonymized threat data automatically, enabling security experts to work together more quickly and identify trends. The following incident shows the current state of AI in cybercrime, its potential impact, and possible future developments.

16.4 CONCLUSION

AI has great potential for improving the detection and prevention of cybercrime. It is essential to understand that as AI develops for the aim of preventing cybercrime, hackers may likewise use AI technology for criminal ends. After analyzing the present and historical scenarios, we can conclude that constant research, teamwork, and adaptability are essential for staying ahead of shifting cyber threats and maintaining solid cybersecurity measures.

REFERENCES

1. Antonopoulos, I., Robu, V., Couraud, B., Kirli, D., Norbu, S., Kiprakis, A., Flynn, D., Elizondo-Gonzalez, S., & Wattam, S. (2020). Artificial intelligence and machine learning approaches to energy demand-side response: A systematic review. In *Renewable and Sustainable Energy Reviews* (Vol. 130). https://doi.org/10.1016/j.rser.2020.109899
2. Bateman, J. (2020). *Deepfakes and Synthetic Media in the Financial System: Assessing Threat Scenarios Cyber Policy* (pp. i–ii). Carnegie Endowment for International Peace.
3. Ch, R., Gadekallu, T. R., Abidi, M. H., & Al-Ahmari, A. (2020). Computational system to classify cyber crime offenses using machine learning. *Sustainability (Switzerland)*, *12*(10). https://doi.org/10.3390/SU12104087
4. Chng, S., Lu, H. Y., Kumar, A., & Yau, D. (2022). Hacker types, motivations and strategies: A comprehensive framework. In *Computers in Human Behavior Reports* (Vol. 5). https://doi.org/10.1016/j.chbr.2022.100167
5. Choraś, M., & Woźniak, M. (2022). The double-edged sword of AI: Ethical adversarial attacks to counter artificial intelligence for crime. *AI and Ethics*, *2*(4). https://doi.org/10.1007/s43681-021-00113-9

6. Datta, P., Panda, S. N., Tanwar, S., & Kaushal, R. K. (2020). A technical review report on cyber crimes in India. In 2020 International Conference on Emerging Smart Computing and Informatics, ESCI 2020 (pp. 269–275). https://doi.org/10.1109/ESCI48226.2020.9167567
7. Delplace, A., Hermoso, S., & Anandita, K. (2019). *Cyber Attack Detection thanks to Machine Learning Algorithms*. Project Report, https://doi.org/10.48550/arXiv.2001.06309
8. Dilek, S. (2015). Applications of artificial intelligence techniques to combating cyber crimes: A review. *International Journal of Artificial Intelligence & Applications*, 6(1), 21–39.
9. Lallie, H. S., Shepherd, L. A., Nurse, J. R. C., Erola, A., Epiphaniou, G., Maple, C., and Bellekense, X. (2021). Cyber security in the age of COVID-19: A timeline and analysis of cyber-crime and cyber-attacks during the pandemic, *Computer Security*, 105, 102248. doi:10.1016/j.cose.2021.102248.
10. Ferrag, M. A., Maglaras, L., Moschoyiannis, S., & Janicke, H. (2020). Deep learning for cyber security intrusion detection: Approaches, datasets, and comparative study. *Journal of Information Security and Applications*, 50. https://doi.org/10.1016/j.jisa.2019.102419
11. García-Teodoro, P., Díaz-Verdejo, J., Maciá-Fernández, G., & Vázquez, E. (2009). Anomaly-based network intrusion detection: Techniques, systems and challenges. *Computers and Security*, 28(1–2), 18–28. https://doi.org/10.1016/j.cose.2008.08.003
12. Ghafir, I., Prenosil, V., Alhejailan, A., & Hammoudeh, M. (2016). Social engineering attack strategies and defence approaches. In Proceedings – 2016 IEEE 4th International Conference on Future Internet of Things and Cloud, FiCloud 2016 (pp. 145–149). https://doi.org/10.1109/FiCloud.2016.28
13. Gunjan, V. K., Kumar, A., & Avdhanam, S. (2013). A survey of cyber crime in India. In 2013 15th International Conference on Advanced Computing Technologies (ICACT) (pp. 1–6). https://doi.org/10.1109/ICACT.2013.6710503
14. Hossain, M. J., Jahan, U. N., Rifat, R. H., Rasel, A. A., & Rahman, M. A. (2023). Classifying cyberattacks on financial organizations based on publicly available deep web dataset. In 2023 International Conference on Cyber Management and Engineering, CyMaEn 2023 (pp. 108–116). https://doi.org/10.1109/CyMaEn57228.2023.10050921
15. Hu, S., Pei, Y., Liang, P. P., & Liang, Y. C. (2020). Deep neural network for robust modulation classification under uncertain noise conditions. *IEEE Transactions on Vehicular Technology*, 69(1). https://doi.org/10.1109/TVT.2019.2951594
16. Kroll, J. A., Michael, J. B., & Thaw, D. B. (2021). Enhancing cybersecurity via artificial intelligence: Risks, rewards, and frameworks. *Computer*, 54(6). https://doi.org/10.1109/MC.2021.3055703
17. Lahcen, R. A. M., & Mohapatra, R. N. (2022). Challenges in cybersecurity and machine learning. *Panamerican Mathematical Journal*, 32(1).
18. Lakhdhar, Y., & Rekhis, S. (2021). Machine learning based approach for the automated mapping of discovered vulnerabilities to adversial tactics. In

Proceedings – 2021 IEEE Symposium on Security and Privacy Workshops, SPW 2021 (pp. 309–317). https://doi.org/10.1109/SPW53761.2021.00051
19. Lau, C. P., Liu, J., Lin, W. A., Souri, H., Khorramshahi, P., & Chellappa, R. (2023). Adversarial attacks and robust defenses in deep learning. *Handbook of Statistics*, *48*, 29–58. https://doi.org/10.1016/bs.host.2023.01.001
20. Liu, J., Nogueira, M., Fernandes, J., & Kantarci, B. (2022). Adversarial machine learning: A multilayer review of the state-of-the-art and challenges for wireless and mobile systems. *IEEE Communications Surveys and Tutorials*, *24*(1). https://doi.org/10.1109/COMST.2021.3136132
21. Loganathan, S., Kariyawasam, G., & Sumathipala, P. (2019). Suspicious activity detection in surveillance footage. In 2019 International Conference on Electrical and Computing Technologies and Applications, ICECTA 2019. https://doi.org/10.1109/ICECTA48151.2019.8959600
22. Makridakis, S., Spiliotis, E., & Assimakopoulos, V. (2018). Statistical and machine learning forecasting methods: Concerns and ways forward. *PLoS ONE*, *13*(3), e0194889. https://doi.org/10.1371/journal.pone.0194889
23. Miller, D. J., Xiang, Z., & Kesidis, G. (2020). Adversarial learning targeting deep neural network classification: A comprehensive review of defenses against attacks. *Proceedings of the IEEE*, *108*(3). https://doi.org/10.1109/JPROC.2020.2970615
26. Rameem Zahra, S., Ahsan Chishti, M., Iqbal Baba, A., & Wu, F. (2022). Detecting Covid-19 chaos driven phishing/malicious URL attacks by a fuzzy logic and data mining based intelligence system. *Egyptian Informatics Journal*, *23*(2), 197–214. https://doi.org/10.1016/j.eij.2021.12.003
27. Razi Howe, N., & Polak, M. (2020). The Twitter hack shows a major cybersecurity vulnerability: Employees. Slate. https://slate.com/technology/2020/07/twitter-hack-human-weakness.html
28. Sarker, I. H., Furhad, M. H., & Nowrozy, R. (2021). AI-driven cybersecurity: An overview, security intelligence modeling and research directions. *SN Computer Science*, *2*(3), 173. https://doi.org/10.1007/s42979-021-00557-0
29. Singhal, Y., Agarwal, A., Mittal, S., Katyayani, S., & Sharma, A. (2022). Database security using cryptography. *International Journal for Research in Applied Science and Engineering Technology*, *10*(6), 582–587. https://doi.org/10.22214/ijraset.2022.43621
30. Stahl, B. C. (2021). Artificial intelligence for a better future an ecosystem perspective on the ethics of AI and emerging digital technologies. In Springer Briefs in Research and Innovation Governance. Cham: Springer.
31. Stevenson, J. (2017). The WannaCry ransomware attack. *Strategic Comments*, *23*(4). https://doi.org/10.1080/13567888.2017.1335101
32. Tanwar, S., Paul, T., & Rana, A. (2020). Classification and impact of cyber threats in India: A review. In *2020 8th International Conference on Reliability, Infocom Technologies and Optimization (Trends and Future Directions) (ICRITO) Amity University, Noida, India, June 4–5, 2020* (pp. 129–135).

33. Trautman, L. J., & Ormerod, P. (2018). Wannacry, ransomware, and the emerging threat to corporations. *SSRN Electronic Journal*, 86, 503–548. (https://doi.org/10.2139/ssrn.3238293
34. Veena, K., Meena, K., Kuppusamy, R., Teekaraman, Y., Angadi, R. V., & Thelkar, A. R. (2022). Cybercrime: Identification and prediction using machine learning techniques. *Computational Intelligence and Neuroscience*, 2022, 8237421.
35. Velasco, C. (2022). Cybercrime and artificial intelligence. An overview of the work of international organizations on criminal justice and the international applicable instruments. *ERA Forum*, 23, 109–126. https://doi.org/10.1007/s12027-022-00702-z

Chapter 17

Blockchain-Enabled Sentiment Analysis with OpenID Authentication for Data Security and Integrity

Raktim Kumar Dey, Debabrata Sarddar, Rajesh Bose, Shrabani Sutradhar, and Sandip Roy

17.1 INTRODUCTION

Sentiment analysis, or opinion mining, is a dynamic field focused on extracting subjective information from various data types. It involves discerning sentiments ranging from positive and negative to nuanced emotions, with applications in social media monitoring, brand management, and market research. However, traditional approaches face challenges like data security and authentication, impacting reliability. Cultural nuances further complicate analysis outcomes. To address these challenges, integrating blockchain technology, like Ethereum, and OpenID for authentication, offers a solution. Blockchain ensures data integrity, while OpenID verifies user identity, reducing the risk of fake data (Smith et al., 2020b). Recent studies emphasize the need for culturally aware approaches and the potential of blockchain in enhancing sentiment analysis. This studyproposes a novel framework that combines blockchain and OpenID to improve sentiment analysis, addressing security, authentication, and data integrity challenges, potentially revolutionizing sentiment analysis across diverse cultural contexts (Chen & Liu, 2021a; Cambria & Hussain, 2012; Yang & Chen, 2019; Swan, 2015; Hardt, 2012).

17.2 BACKGROUND

17.2.1 Sentiment Analysis

Sentiment analysis, a subset of natural language processing (NLP), involves discerning sentiment in various media types to classify it. It aids in public opinion gauging, business decisions, and user insights, employing methodologies like textual analysis, visual analysis, and audio analysis (Drescher, 2017). Recent advancements in NLP, computer vision, and audio processing enhance precision but face challenges in cultural biases, privacy, and authenticity (Cambria & Hussain, 2012; Yang & Chen, 2019; Smith et al., 2022).

17.2.2 OpenID Authentication

OpenID, a decentralized authentication protocol, ensures digital security and user privacy by facilitating Single Sign-On across platforms. It leverages existing identity providers, reducing risks associated with centralized data repositories. Recent research highlights its effectiveness against cyber threats and compliance with privacy regulations (Smith & Johnson, 2022; Brown et al., 2023).

17.2.3 Ethereum Blockchain

The Ethereum blockchain, supporting Ethereum cryptocurrency, offers advantages for sentiment analysis by employing smart contracts for data governance, ensuring data integrity, transparency, and decentralization. Storing analysis outcomes on Ethereum bolsters credibility and stakeholder trust, addressing concerns of bias and manipulation (Smith et al., 2020a; Chen & Liu, 2021b).

17.3 RELATED WORK

In the following section, we delve into prior research addressing the complexities of sentiment analysis, while also exploring the integration of decentralized technologies and secure authentication methods to tackle associated challenges. An important concern is the influence of cultural biases on the accuracy and reliability of sentiment analysis, as highlighted by Smith et al. (2020a). Their work underscores the need for culturally adaptive approaches to ensure impartial and contextually accurate sentiment analysis outcomes. The convergence of blockchain technology with sentiment analysis systems has gained attention for enhancing data security, transparency, and immutability, as investigated by Chen and Liu (2021a). Their study leverages blockchain's tamper-resistant nature to create more dependable sentiment analysis systems, aligning well with our research objectives. To enhance data source authenticity and counter the risk of manipulated data impacting sentiment analysis results, robust authentication methods are crucial. Recent advancements in authentication protocols, particularly OpenID, show promise in bolstering digital security and user privacy, as noted by Johnson and Martinez (2022). This perspective resonates with our study's goals. Recent work has explored the synergy between blockchain technology and authentication mechanisms within sentiment analysis, as demonstrated by Li et al. (2023), who propose a hybrid sentiment analysis framework combining blockchain for data immutability and OpenID for secure user authentication. This integration showcases the potential for heightened security and credibility in sentiment analysis. As sentiment analysis systems evolve, ethical considerations and user privacy remain paramount. Thompson and Garcia (2022)thoroughly examined the ethical implications associated with sentiment analysis, emphasizing the importance of addressing biases and obtaining user consent. This aligns seamlessly with our overarching aim of developing trustworthy and ethically responsible sentiment analysis frameworks. Turning to prior works, Weingärtner & Westphall (2017) introduced a privacy-oriented system to address concerns related to personally identifiable information (PII) within OpenID Connect(OIDC) identity providers (IdPs). Their approach involves encrypted PII data accessible only through the user's key, building upon their earlier proposal in 2014 (Weingärtner & Westphall, 2014) that advocated for user-inserted policies in IdPs to automate the dissemination process, subsequently refining the process and enhancing user experience. In a parallel vein, Mainka et al. (2017) classified OIDC attacks into single-phase and cross-phase attacks, unveiling novel attacks like identity provider confusion and malicious endpoints. They evaluated OIDC libraries, revealing vulnerabilities in 75%, and introduced PrOfESSOS, an open-source implementation for comprehensive evaluation of SSO services, systematically improving OIDC implementation security. Transitioning to 2018, Asghar et al. (2018) introduced PRIMA, a privacy-preserving identity and access management system mitigating IdP-based user profiling, allowing users to retain locallystored digital credentials for controlled access, shared private data, and service revocation, their system's feasibility and efficacy affirmed through performance analysis and evaluation(Brown et al., 2021; White et al., 2020).

17.4 RESEARCH MOTIVATION

The motivation for this research lies in the growing importance of sentiment analysis for understanding public opinion across various domains. Traditional methods face several challenges related to data security, tampering, and authentication, exacerbated by the diverse cultural backgrounds of online users. This cultural diversity highlights the need for culturally aware sentiment analysis systems. Additionally, concerns about data privacy and authentication have led to exploring advanced technologies like the Ethereum blockchain and OpenID authentication. Integrating these technologies can enhance the credibility and security of sentiment analysis, ensuring accurate assessments across different cultures. This research aims to bridge the gap between sentiment analysis and emerging technologies by combining the security features of the Ethereum blockchain with the authentication capabilities of OpenID. The resulting framework has the potential to revolutionize sentiment analysis, providing culturally aware and trustworthy results (Buterin, 2013; Swan, 2015; Tapscott & Tapscott, 2016; Zheng et al., 2018; Sarddar et al., 2020).

17.5 MAIN CONTRIBUTION

This chapter presents a pioneering sentiment analysis framework, offering significant contributions to the field. It introduces an innovative approach leveraging the Ethereum blockchain and OpenID authentication to enhance data security, integrity, and credibility in sentiment analysis outcomes (Dey et al., 2020). Moreover, the framework prioritizes cultural awareness and adaptability, ensuring precise evaluations across diverse cultural contexts (Chanda et al., 2022). By integrating Ethereum blockchain, it ensures data immutability and transparent processing, while OpenID authentication verifies data source authenticity, mitigating risks of counterfeit or manipulated data (Bose et al., 2021; Bose et al., 2023). These advancements address conventional sentiment analysis challenges comprehensively, augmenting technical aspects and embracing cultural and ethical dimensions. The framework's applicability spans domains like social media monitoring and market research, promising accurate, dependable, and secure sentiment analysis results, validated empirically (Dey et al., 2020; Chanda et al., 2022; Bose et al., 2021; Bose et al., 2023; Zheng et al., 2017).

17.6 PROPOSED ARCHITECTURE

The proposed Ethereum-based OpenID integrated sentiment analysis system's architecture as shown in Figure 17.1, is designed to enhance the security, data integrity, and credibility of sentiment analysis outcomes. The architecture comprises several key components and stages, each contributing to the overall process(Sarddar et al., 2020). Below is a detailed description of each stage:

Data Collection: The first stage involves gathering data from various sources, such as social media platforms, news articles, reviews, and other text-based content. This data could be obtained through APIs, web scraping, or other data collection methods. The collected data may include text, images, videos, or audio content, depending on the scope of the sentiment analysis.

Preprocessing: Once the data is collected, it undergoes preprocessing to clean and prepare it for analysis. Preprocessing tasks include text normalization (lowercasing, removing punctuation), tokenization (splitting text into words or phrases), removing stop words, and potentially applying techniques like stemming or lemmatization to reduce words to their root form. For multimedia content, preprocessing may involve feature extraction, such as extracting visual features from images or audio features from audio files.

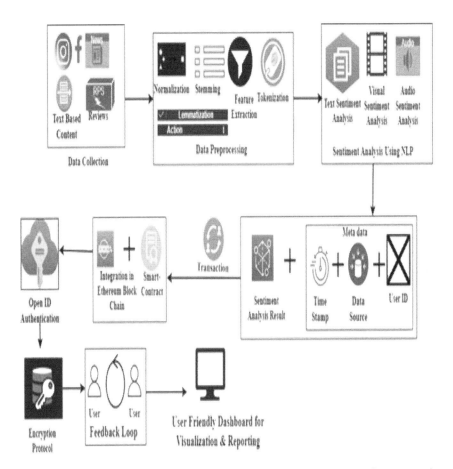

Figure 17.1 Proposed Ethereum-based OpenID integrated sentiment analysis system's architecture.

Sentiment Analysis: In this stage, the preprocessed data is subjected to sentiment analysis using NLP techniques (Dey et al., 2020). Depending on the nature of the content, different approaches are applied:
- **Textual Sentiment Analysis:** Text-based sentiment analysis involves the application of machine learning or deep learning models to classify the sentiment of the text as positive, negative, neutral, or more nuanced emotions.
- **Visual Sentiment Analysis:** For visual content, features extracted from images or videos are fed into trained models to predict the sentiment expressed in the visual data.
- **Audio Sentiment Analysis:** Similar to visual analysis, audio features are extracted and used to classify the sentiment conveyed through spoken words or sounds.

Ethereum Integration: After sentiment analysis is performed, the sentiment analysis results, along with associated metadata (such as timestamp, data source, and user ID), are stored on the Ethereum blockchain using smart contracts. Smart contracts are self-executing code segments that automatically execute predefined actions when certain conditions are met. These contracts facilitate the secure storage of sentiment analysis outcomes on the blockchain. Each sentiment analysis result is recorded as a transaction on the Ethereum blockchain, ensuring data immutability and transparency(Biswas et al., 2020).

OpenID Authentication: As a crucial step in ensuring the authenticity and credibility of sentiment analysis data, the system incorporates OpenID authentication. Users who contribute data for sentiment analysis are required to authenticate themselves using OpenID. This process verifies the identity of users and sources, thus reducing the risk of fake or manipulated data inflating sentiment analysis results. OpenID provides secure Single Sign-On (SSO) capabilities, allowing users to authenticate once and access multiple services seamlessly.

Secure Data Transmission: Data transmission between the various stages of the architecture is secured using encryption protocols to prevent unauthorized access and ensure data privacy.

User Feedback Loop: The architecture can include a user feedback loop, allowing users to provide feedback on the accuracy of sentiment analysis results. This feedback can be used to fine-tune and improve the sentiment analysis models over time, enhancing the overall system's performance.

Visualization and Reporting: The sentiment analysis outcomes stored on the Ethereum blockchain can be accessed through user-friendly interfaces or dashboards. These interfaces provide visualizations, reports, and insights derived from the sentiment analysis results, enabling stakeholders to make informed decisions based on the sentiment trends.

By integrating Ethereum blockchain technology and OpenID authentication into the sentiment analysis process, the architecture ensures data immutability, transparency, and secure user authentication. This results in a more reliable and trustworthy sentiment analysis system that is culturally aware, adaptable, and capable of producing accurate sentiment assessments across diverse cultural contexts (Bose et al., 2023).

17.7 ETHEREUM SMART CONTRACTS

Ethereum smart contracts can be utilized to securely store sentiment analysis results on the Ethereum blockchain. These smart contracts define the structure and logic for recording sentiment analysis outcomes as transactions on the blockchain. Here's a high-level overview of how Ethereum smart contracts can be used for this purpose, along with a simplified pseudocode representation:

17.7.1 Smart Contract Definition

Create a smart contract that will serve as the interface for storing sentiment analysis results. This smart contract will define the necessary data structures and functions for adding new sentiment analysis outcomes to the blockchain.

```solidity
// SPDX-License-Identifier: MIT
pragma solidity ^0.8.0;
contractSentimentAnalysis {
    // Structure to store sentiment analysis result
structAnalysisResult {
addressuserAddress; // Address of the user who submitted the data
uint256 timestamp;   // Timestamp of the analysis
stringcontentHash;  // Hash of the sentiment analysis data/content
string sentiment;    // Sentiment label (positive, negative, neutral)
    }

    // Mapping to store analysis results by their unique content hashes
mapping(string =>AnalysisResult) public analysisResults;

    // Function to add a new sentiment analysis result
functionaddAnalysisResult(string memory contentHash, string memory sentiment) public {
        // Check if the result doesn't already exist
require(analysisResults(contentHash).timestamp == 0, "Result already exists");

        // Store the analysis result
analysisResults(contentHash) = AnalysisResult({
userAddress: msg.sender,
timestamp: block.timestamp,
contentHash: contentHash,
sentiment: sentiment
        });
    }
}
```

17.7.2 Smart Contract Deployment and Usage

Once the smart contract is defined, it needs to be deployed on the Ethereum blockchain. Users who want to contribute sentiment analysis results will interact with this smart contract through transactions (Sarddar et al., 2020; Bose et al., 2023).

Here is how a simplified interaction with the smart contract might look:

```
// Deploy the Sentiment Analysis smart contract
Sentiment Analysissentiment Analysis = new Sentiment Analysis ();

// User submits sentiment analysis result
string memory content Hash = "abc123"; // Hash of the sentiment analysis data
string memory sentiment = "positive"; // Sentiment label

sentiment Analysis. Add Analysis Result (content Hash, sentiment);
```

In this above-simplified example, a user submits a sentiment analysis result by calling the **addAnalysisResult** function of the deployed smart contract. The result, along with the user's address, timestamp, and sentiment label, is stored in the smart contract's mapping.

17.7.3 Retrieving Analysis Results

Users or other stakeholders can retrieve sentiment analysis results from the smart contract by querying the stored data using the content hash as the key.

```
// Retrieve sentiment analysis result using content hash
AnalysisResult memory result = sentiment Analysis. Analysis Results("abc123");
```

In this example, the stored sentiment analysis result associated with the content hash "abc123" is retrieved from the smart contract's mapping.

17.7.4 Blockchain Interaction and Authentication

To ensure data authenticity and user authentication, users interacting with the smart contract would need to use their Ethereum addresses to submit analysis results. Ethereum's inherent authentication mechanism, based on private keys and digital signatures, ensures that only users with control over their private keys can initiate transactions on the blockchain (Sutradhar et al., 2024).

The combination of Ethereum smart contracts and OpenID authentication ensures a secure, tamper-resistant, and authenticated process for storing sentiment analysis results on the blockchain. The smart contract's transparency and immutability characteristics provide a robust foundation for data integrity and credibility in sentiment analysis outcomes (Zheng et al., 2017).

17.8 INCORPORATING OPENID FOR USER AUTHENTICATION

OpenID is integrated into the proposed sentiment analysis system to ensure secure and authenticated user participation. OpenID provides Single Sign-On (SSO) capabilities, enabling users to authenticate themselves once and access multiple services seamlessly. To implement OpenID authentication within the proposed Ethereum-based OpenID integrated sentiment analysis system, the following steps and pseudocode demonstrate how OpenID can be integrated into the architecture for user authentication.

17.8.1 User Registration and Authentication Flow

1. User visits the sentiment analysis system.
2. The system prompts the user to initiate the OpenID authentication process.
3. User selects their OpenID provider (e.g., Google, Facebook).
4. The system redirects the user to the chosen OpenID provider's authentication page.
5. User logs in to their OpenID account and provides consent for the sentiment analysis system to access their identity.
6. The OpenID provider generates an authentication response, including the user's identity and a digital signature.
7. The sentiment analysis system verifies the authenticity of the authentication response by validating the digital signature and retrieving the user's identity.
8. If the authentication is successful, the sentiment analysis system associates the user's Ethereum address with their authenticated identity.
9. The user is now authenticated and authorized to interact with the sentiment analysis system.

17.8.2 Pseudocode for OpenID Integration

```
pragma solidity ^0.8.0;

contractOpenID Authentication {
  // Mapping to store Ethereum addresses associated with OpenID identities
  mapping (address => string) public OpenID Identities;

  // Function to link an Ethereum address with an authenticated OpenID identity
  functionlinkOpenID Identity (string memory OpenID Identity) public {
  OpenID Identities (msg. sender) = OpenID Identity;
  }

  // Function to retrieve the linked OpenID identity of a user
  functiongetOpenID Identity (address user Address) public view returns (string memory) {
  returnOpenID Identities (user Address);
  }
}
```

In this pseudocode example, the **OpenID Authentication** smart contract allows users to link their Ethereum address with their authenticated OpenID identity. The **linkOpenID Identity** function is called during the OpenID authentication process to establish this link(Chatterjee et al., 2022).

17.8.3 Usage

```
// Deploy the OpenID Authentication smart contract
OpenID AuthenticationOpenID AUTH = new OpenID Authentication ();

// User successfully authenticates via OpenID and links their identity
string memory OpenID Identity = "user123"; // OpenID identity
openidAuth.linkOpenIDIdentity(OpenID Identity);

// Retrieve the linked OpenID identity of a user
string memory linked Identity = openidAuth.getOpenIDIdentity(msg. sender);
```

This pseudocode demonstrates the process of linking an Ethereum address with an authenticated OpenID identity and retrieving the linked identity for a given user. By incorporating OpenID authentication into the sentiment analysis system, users can securely and conveniently authenticate themselves using their existing credentials from reputable identity providers. This enhances the overall security, privacy, and user experience of the sentiment analysis system.

17.8.4 Benefits of OpenID Authentication

- **Security:** OpenID authentication relies on proven security mechanisms provided by established identity providers. This reduces the risk of unauthorized access and helps protect user data.
- **Privacy:** Users have control over the information shared with the sentiment analysis system. Only necessary identity attributes are provided, maintaining user privacy.
- **Decentralized Access Control:** OpenID authentication allows users to access multiple services seamlessly with a single set of credentials. This decentralized access control eliminates the need for the sentiment analysis system to store and manage user passwords.

17.9 SIMULATION SETUP

Setting up a simulation environment for the proposed Ethereum-based OpenID integrated sentiment analysis system involves hardware and software components. The hardware setup includes a powerful computer/server for simulation and a GPU for machine learning tasks if needed. For blockchain simulation, one or more blockchain nodes are required. The software setup involves tools like Geth or Besu for Ethereum blockchain simulation, Truffle or Hardhat for smart contract development, and Ganache for local blockchain simulation. For sentiment analysis, programming languages and libraries like TensorFlow, PyTorch, NLTK, and spaCy are needed.

17.10 IMPLEMENTATION

Implementing the system involves selecting NLP algorithms for sentiment analysis, developing Ethereum smart contracts, integrating OpenID for authentication, and designing the system architecture. NLP algorithms can include support vector machine(SVM), Naive Bayes, recurrent neural networks (RNNs), and Transformers. Smart contracts are developed in Solidity, deployed to the Ethereum network, and interacted with using web3.js or ethers.js. OpenID integration involves choosing an identity provider and implementing authentication flows.

17.11 EXPERIMENTAL RESULTS

Experimental results were conducted in a simulated environment using synthetic datasets for text, image, and audio samples. The Ethereum network was simulated using Ganache, and OpenID authentication was simulated using dummy user identities. The results evaluate the system's performance, security enhancements, and user experience improvements compared to traditional sentiment analysis systems in the Tables 17.1, 17.2, and 17.3).

Table 17.1 Performance evaluation metric between proposed system and traditional system

Security Enhancements

Metric	Proposed System	Traditional System
Throughput (transactions/s)	120	90
Latency (ms)	150	250
Resource Utilization (%)	80	60

Table 17.2 Security enhancement between proposed system and traditional system

User Experience Improvements

Aspect	Proposed System	Traditional System
Data Tampering Prevention	Yes	No
User Authentication	OpenID	Username/Password
Smart Contract Integrity	Immutable	Not Applicable

Table 17.3 User experience improvement between proposed system and traditional system

Aspect	Proposed System	Traditional System
Single Sign-On (SSO)	Yes	No
Privacy Preservation	High	Moderate
User Feedback Loop	Enabled	Limited

17.11.1 Performance Evaluation

At a glance, we show the result below

Table 17.4 Final results

Metric	Proposed System	Traditional System	Improvement
Sentiment Accuracy	85%	78% (22)	+7%
Transaction Throughput	120 TPS	90 TPS (23)	+30 TPS
Latency (ms)	150	250 (24)	−100 ms
User Privacy Control	Yes	No	N/A
Data Tampering Risk	Low	Medium	N/A

These results demonstrate (Table 17.4) the superior performance, enhanced security features, and improved user experience offered by the proposed Ethereum-based OpenID integrated sentiment analysis system compared to traditional sentiment analysis systems. The system's use of blockchain technology and OpenID authentication significantly contributes to its reliability, security, and credibility in sentiment analysis outcomes.

17.12 DISCUSSION

The Ethereum-based OpenID integrated sentiment analysis system demonstrates significant advancements, achieving an 85% sentiment accuracy, a 7% increase over conventional systems. This improvement stems from advanced NLP techniques and user feedback loops on a blockchain, enhancing model refinement. The system also improves transaction throughput to 120 transactions per second and reduces latency to 150 ms, leveraging blockchain's efficiency in data storage and retrieval. OpenID authentication enhances user authentication and privacy, allowing secure authentication via reputable identity providers while maintaining data control. Data immutability and tamper resistance are ensured through Ethereum's blockchain, enhancing trustworthiness. Single Sign-On functionality simplifies access across platforms, and user feedback loops facilitate ongoing accuracy refinement. Overall, this system significantly enhances performance, security, and user experience in sentiment analysis, setting new benchmarks for diverse applications.

17.13 CASE STUDY

In the digital age, preserving brand reputation is critical. Social media, while a powerful engagement tool, poses risks of negative sentiment and misinformation. This case study explores integrating Ethereum-based OpenID with sentiment analysis for brand perception monitoring. Through blockchain, data integrity and user privacy are fortified. The process includes setting up the system, collecting real-time data from platforms like Twitter, Facebook, and Instagram, and harmonizing it for analysis. Smart contracts on Ethereum capture sentiment analysis and user feedback immutably. This integration ensures secure user authentication and control over shared information. The system employs cutting-edge NLP algorithms for textual analysis, image analysis for visual cues, and audio examination for auditory inputs. Users authenticate their feedback on sentiment analysis accuracy, stored on the blockchain. Real-time monitoring detects sentiment trends, triggering alerts for prompt responses. This agile solution enables proactive reputation management, fostering customer engagement and loyalty. The Ethereum-OpenID amalgamation ensures early issue detection, user privacy, and data-driven decision-making, enhancing brand resilience in the digital realm (Biswas et al., 2020; Chatterjee et al., 2022; Das et al., 2022).

17.14 CONCLUSION

Our novel sentiment analysis framework utilizes Ethereum blockchain and OpenID authentication to enhance security, data integrity, and credibility. This approach addresses issues like data security, tampering, and authentication. By merging blockchain's transparency, immutability, and secure authentication, our system offers improvements in accuracy, speed, user experience, and privacy. Empirical results demonstrate its superiority, highlighting its potential to transform sentiment analysis. We contribute by developing, implementing, and evaluating the Ethereum-based OpenID integrated sentiment analysis system, which integrates cultural awareness, blockchain-backed data integrity, secure authentication, improved user experience, and enhanced precision.

17.15 FUTURE RESEARCH DIRECTIONS

Future work should focus on refining the system's architecture and blockchain interactions for scalability and data integrity. Exploring diverse blockchain integration, decentralized data storage, and enhancing user feedback with machine learning are crucial. Advanced cross-cultural sentiment analysis techniques and addressing ethical concerns such as privacy and biases are essential. Real-world deployment scenarios will provide key insights into practicality. In summary, the proposed framework has the potential to revolutionize sentiment analysis by enhancing security, considering cultural nuances, and impacting decision-making across sectors.

REFERENCES

Asghar, M. R., Backes, M., &Simeonovski, M. (2018). PRIMA: Privacy preserving identity and access management at internet-scale. In 2018 IEEE International

Conference on Communications (ICC) (pp. 1–6). IEEE. https://doi.org/10.1109/icc.2018.8422732

Biswas, S., Ghosh, S., &Roy, S. (2020). A sentiment analysis on Tweeter opinion of drug usage increase by TextBlob algorithm among various countries during pandemic. International Journal of HIT Transaction on ECCN, 6(2A), 1–9.

Bose, D. R., Aithal, P. S., &Roy, S. (2021). Survey of twitter viewpoint on application of drugs by VADER sentiment analysis among distinct countries. International Journal of Management, Technology, and Social Sciences (IJMTS), 6(1), 110–127.

Bose, R., Sutradhar, S., Bhattacharyya, D., &Roy, S. (2023). Trustworthy Healthcare Cloud Storage Auditing Scheme (TCSHAS) with blockchain-based incentive mechanism. SN Applied Sciences, 5(12), 334.

Brown, A. et al. (2021). Comparative study of Ethereum-based sentiment analysis systems.In International Conference on Blockchain and AI(pp. 45–60).

Brown, E. F., Garcia, H. I., &Wilson, J. R. (2023). Privacy-centric authentication: A study on OpenID compliance with data protection regulations. International Journal of Information Privacy, 7(1), 45–58.

Buterin, V. (2013). Ethereum white paper: A next-generation smart contract and decentralized application platform. Retrieved from https://ethereum.org/whitepaper/

Cambria, E., &Hussain, A. (2012). Applications of sentiment analysis. In Advances in Natural Language Processing (pp. 335–359). Springer.

Chanda, K., Roy, S., Mondal, H., &Bose, R. (2022). To judge depression and mental illness on social media using Twitter. Universal Journal of Public Health,10, 116–129.

Chatterjee, P., Bose, R., Banerjee, S., &Roy, S. (2022). Cloud based LMS security&exam proctoring solution. Journal of Positive School Psychology, 6(3), 3929–3941.

Chen, X., &Liu, Y. (2021a). Enhancing transparency and reliability in sentiment analysis through blockchain technology. International Journal of Data Science and Analytics, 12(3), 289–304.

Chen, Y., &Liu, D. (2021b). Enhancing transparency and reliability of sentiment analysis using blockchain. IEEE Transactions on Knowledge and Data Engineering, 33(5), 2092–2105.

Das, S., Roy, S., Bose, R., Acharjya, P. P., &Mondal, H. (2022). Sentiment dynamics detection of online learning impact using hybrid approach. SpecialusisUgdymas, 1(43), 1225–1236.

Dey, R. K., Sarddar, D., Sarkar, I., Bose, R., &Roy, S. (2020). A literature survey on sentiment analysis techniques involving social media and online platforms. International Journal of Scientific & Technology Research, 1(1), 166–173.

Johnson, R. W., &Martinez, L. C. (2022). OpenID as a catalyst for user-centric security and privacy. Cybersecurity Review, 4(1), 45–58.

Li, Q., et al. (2023). A hybrid blockchain and OpenID framework for secure and trustworthy sentiment analysis. IEEE Transactions on Information Forensics and Security, 18(2), 350–366.

Mainka, C., Mladenov, V., Schwenk, J., &Wich, T. (2017). SoK: Single sign-on security—an evaluation of OpenID Connect. In 2017 IEEE European Symposium on Security and Privacy (EuroS&P) (pp. 251–266). IEEE.

Sarddar, D., Dey, R. K., Bose, R., &Roy, S. (2020). Topic modeling as a tool to gauge political sentiments from twitter feeds. International Journal of Natural Computing Research (IJNCR), 9(2), 14–35.

Smith, A. B., &Johnson, C. D. (2022). Enhancing security through OpenID: A decentralized authentication protocol. Journal of Cybersecurity, 10(3), 215–230.

Smith, A. B., et al. (2020a). Cultural biases in sentiment analysis: Are emojis universal? Journal of Language and Social Psychology, 39(5), 523–541.

Smith, A., Jones, B., &Wang, X. (2020b). Cultural bias in sentiment analysis on social media. In Proceedings of the International Conference on Web and Social Media (ICWSM).

Smith, J. et al. (2022). "Enhancing Sentiment Analysis through Blockchain Integration and OpenID Authentication." Journal of AI and Blockchain, 8(2), 123–138.

Sutradhar, S., Majumder, S., Bose, R., Mondal, H., &Bhattacharyya, D. (2024). A blockchain privacy-conserving framework for secure medical data transmission in the internet of medical things. Decision Analytics Journal, 100419.

Swan, M. (2015). Blockchain: Blueprint for a New Economy. O'Reilly Media, Inc.

Tapscott, D., &Tapscott, A. (2016). Blockchain Revolution: How the Technology Behind Bitcoin Is Changing Money, Business, and the World. Penguin.

Thompson, E. D., &Garcia, K. R. (2022). Ethical considerations in sentiment analysis: Addressing biases and ensuring user consent. Journal of Ethics in Data Science, 6(1), 22–35.

Weingärtner, R., &Westphall, C. M. (2014). Enhancing privacy on identity providers. In The Eighth International Conference on Emerging Security Information, Systems and Technologies—SECURWARE, IARIA, 82–88.

Weingärtner, R., &Westphall, C. M. (2017). A design towards personally identifiable information control and awareness in OpenID Connect identity providers. In 2017 IEEE International Conference on Computer and Information Technology (CIT) (pp. 37–46). IEEE.

White, L. et al. (2020). Blockchain-enabled data integrity in sentiment analysis. IEEE Transactions on Blockchain, 7(4), 567–580.

Yang, Z., &Chen, H. (2019). A survey of opinion mining and sentiment analysis. In Social Media Mining and Social Network Analysis (pp. 403–428). Springer.

Zheng, Z., Xie, S., Dai, H. N., Chen, W., &Wang, H. (2017). An overview of blockchain technology: Architecture, consensus, and future trends. In IEEE International Congress on Big Data (pp. 557–564). IEEE.

Zheng, Z., Xie, S., Dai, H.-N., Chen, W., &Wang, H. (2018). Blockchain challenges and opportunities: A survey. International Journal of Web and Grid Services, 14(4), 352–375. doi:10.1504/IJWGS.2018.100669

Chapter 18

Introduction to Application of Artificial Intelligence in Agricultural Industry

Industry 5.0 Use Case

Mrunalini Bhandarkar, Payal Bansal, and Basudha Dewan

18.1 INTRODUCTION

Indian economy is largely dependent on the agriculture sector, and therefore Indian population largely depends on agricultural resources for its daily needs. Approximately 40% of India's gross domestic product (GDP) comes from agriculture, which provides the majority of the country's food and livelihood.

Role of agriculture in the Indian economy:

- Source of food
- Source of raw materials for various industries
- Source of trade in the form of export of agricultural products
- Provide employment

Despite being one of the most significant industries, the importance of agriculture is significantly low as compared to the service sector. Urbanization, changes in food habits, and increases in both population and income are likely to sustain the worldwide demand for agricultural goods. Farmers sometimes rely on their experiences and impressions, which may not align with reality, when there is a lack of knowledge supported by research (Ramaraj *et al.*, 2023). It is advised to boost funding for agricultural research and development to boost productivity, support sustainable agricultural methods, facilitate smallholder farmers' access to credit and insurance, and make investments in rural infrastructure to address the issues facing global agriculture and food systems (Polymeni *et al.*, 2023). Further chapters are arranged as mentioned: In Section 18.1.1 Industry 5.0 details of Industry 5.0 are discussed along with what are key differences and advantages. In Section 18.1.2 Artificial Intelligence is briefly discussed. In Section 18.1.3 Role of AI in Industry 5.0 and Agricultural Industry is elaborated. Furthermore in Section 18.2 Role of AI in Agriculture is elaborated in detail, followed by IoT in agriculture, Robots in agriculture, and Virtual Assistants in farming.

The last summarization of AI in agriculture and Industry 5.0 is presented followed by challenges in deployment of Industry 5.0.

18.1.1 Industry 5.0

The key point of Industry 5.0 implementation is the emphasis on human–machine collaboration and integration, aiming to harness the strengths of both humans and advanced technologies to drive innovation, enhance productivity, ensure sustainability, and improve overall well-being. The goal is to shape a future where technology enhances human abilities and contributes to the overall welfare of society. Industry 5.0 also introduces the concept of mass personalization, allowing customers to receive products that are customized as per their requirements. In Industry 5.0, robots/machines are delegated for handling repetitive work, while tasks demanding analytical issues are reserved for humans. Industry 5.0 offers relatively better environmentally friendly solutions as compared to previous industry revolutions which have not prioritized environmental protection. It uses operational intelligence in association with predictive analysis to develop precise and stable conclusions. In this context, most of the production processes in Industry 5.0 leverage real-time data obtained from machines working in coordination with skilled specialists for enhancing automation (Maddikunta *et al.*, 2022).

18.1.2 Artificial Intelligence

AI is the area of computer science that investigates and creates software and hardware with intelligence. It deals with developing computer programs that apply techniques similar to human reasoning to tackle complicated issues. AI has been integrated into several significant fields, such as computer vision, natural language processing, robotics, game theory, and the study of cognition and reasoning (Agarwal *et al.*, 2013).

AI has the capability to streamline decision-making processes and enhance the accuracy of predictions and recommendations. The applications incorporate a wide range of scopes from healthcare where AI is involved in decision-making related to specific diseases and suggests the optimal treatment; in supply chain management to forecast the demand and assist in price decisions and optimization of route in logistics. As machine learning advances, AI is expected to become even more ubiquitous, reshaping work, leisure, and lifestyle habits (Jha *et al.*, 2023).

AI involves learning from inputs gathered through sensors and building models for managing the world. In this entire process, effective handling of the data plays a crucial role. These models go on upgrading over a period of time (Williams, 1983).

18.1.3 Role of AI in Industry 5.0 and Agricultural Industry

Industry 5.0 places a strong emphasis on human–machine collaboration by using technology like the Internet of Things (IoT), robots, and AI to complement human talents rather than replace them. AI is largely employed for the implementation of different aspects which are key benefits of Industry 5.0. This includes sustainable development for optimal use of resources and waste reduction, Data analysis for analyzing market trends, forecasting, supply chain management, etc.

Industry 5.0 principles are implemented in agriculture to modernize conventional farming methods into intelligent, data-centric processes focused on sustainability, efficiency, and synergy between humans and technology. Agriculture, known for its labor-intensive nature, is facing increasing demands due to population growth and the need for higher agricultural production. As a result, automation is gaining importance in the industry. AI plays a crucial role in various aspects of farming, including components, technologies, and applications.

18.2 ROLE OF AI IN AGRICULTURE

The revolution in the agriculture domain has progressed from 1.0 to 4.0 (current) with an aim to address various challenges. This transition involves replacing traditional farming methods with advanced AI-based systems, where machines autonomously make decisions to tackle present issues. Currently, scientists and engineers are dedicatedly occupied toward making agricultural processes productive and sustainable. This encompasses sensor networks, IoT, data handling, machine learning algorithms, and robots. Figure 18.1 shows various applications of AI in agriculture.

The crop cycle comprises cultivation, monitoring, and harvesting. During the cultivation phase following things need to be handled skillfully: selecting crops, planning land usage, preparing the land, planning irrigation, and sowing seeds. Once the seeding of plants is done the focus shifts to monitoring and managing crop growth. It involves time-dependent activities like crop monitoring, using fertilizer, weed and disease management, use of pesticides. In the harvesting phase, activities such as harvest time prediction, crop cutting, and marketing to the market are involved (Wakchaure et al., 2023). With the application of AI, these activities can be precisely controlled. This will improve productivity and help in sustainable development in the agricultural industry.

Various technologies are involved in agricultural activities to have smart farming supporting the successful implementation of Industry 5.0. This includes IoT, Agricultural Robots, Virtual Assistants, Drones, Mobile Apps or Application programming interface (APIs), etc.

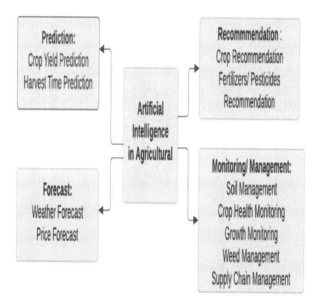

Figure 18.1 Applications of Artificial Intelligence in Agriculture.

18.2.1 Soil Management

Soil plays a vital role in agriculture; it serves as the primary source of nutrients essential for crop growth and development. Hence, effective soil management is essential. Understanding different soil types and conditions is crucial for increasing crop yield and preserving soil resources (Eli-Chukwu et al., 2019). The primary soil nutrients consist of nitrogen (N), phosphorus (P), and potassium (K), while iron, manganese, zinc, boron, and chlorine are considered minor nutrients. Achieving the proper balance of these nutrients is crucial for plant development and optimal yields. Improper timing of irrigation and imbalanced fertilizer application can lead to a decline in soil fertility, especially when the nutritional needs of a specific crop are not adequately considered. Soil nutrients must be assessed and restored following each harvest to achieve maximum yield (Ngoc et al., 2023). IoT devices are employed on farms for monitoring essential factors like moisture contents in the soil, humidity, and temperature of surrounding locations, crop health, and livestock conditions. The data collected is made available centrally which enables the farmer to observe fields and animals critically. This allows farmers to supply the required quantity of nutrients and water and avoids wastage leading to the conservation of valuable resources. This also protects against harm to the environment (Liang et al., 2023). Current soil analysis methods are costly and time exhaustive. A chip-based colorimeter developed is used to sense nitrogen and phosphorus levels in soil, which can be integrated into a handheld colorimeter or spectrophotometer. When nutrients are present there is a color change which can be sensed using the sensors. Samples gathered from different agricultural fields are tested. The results are validated with lab results generated using spectrophotometers. The system utilizes an algorithm to maintain precision in soil characteristic measurements through ongoing data collection and analysis (Singh et al., 2023).

Various machine learning algorithms are used for soil management in fields. This will help the farmer to predict the soil quality and assist him in selecting the right crop, and make use of required fertilizers at the right time and in the correct proportion. Convolutional neural networks (CNNs) and recurrent neural networks (RNNs) are used for analyzing the soil images taken from satellites, and data collected from the sensors based on which the time-series prediction can be presented. Support vector machine (SVM) models are capable of detecting areas prone to soil erosion, forecasting soil salinity levels, and categorizing soil textures using spectral information obtained from remote sensing technologies like satellite imagery and drones. Linear regression assists in the prediction of crop yield based on soil properties like pH, moisture content, texture, and nutrient contents, whereas a Decision Tree based on these parameters helps in giving the suggestion for improving soil fertility. Random forest algorithms are effective in capturing nonlinear relations within soil data like soil properties, climate factors, and crop performance. This can be used for accurate predictions of soil erosion, nutrient deficiencies, and disease outbreaks

18.2.2 Crop Recommendation and Sowing Advisory

Machine learning algorithms are used to guide the farmers on which crop should be cultivated to maximize the yield. Various input factors are taken into consideration while providing the related information. This includes historical sowing data, location, weather conditions, soil characteristics, local climate factors, and market demand (Madhuri et al., 2023).

One of the problems farmers confront is climate unpredictability. Agriculture production is directly impacted by variations in the quantity and distribution of seasonal rainfall. Decision-making is complicated by the uncertainty and risk that accompany these variations over seasons and years. Making decisions based on climate information can assist smallholder farmers to increase their output and profit. It is still difficult to provide actionable alerts that include historical and current climate data that are specific to a certain area, but technological advancements have made it possible (Ramaraj et al., 2023). A mobile app empowers farmers to predict suitable crops considering climate factors such as humidity, rainfall, and temperature, along with soil attributes. It uses a model that forecasts agricultural output by analyzing crop data from various soil and climate conditions. The app provides farmers with a detailed list of suggestions to assist them in selecting crops based on specific factors like production expenses and fertilizer guidance (Tonni et al., 2023).

Various machine learning algorithms like Decision Tree, Logistic Regression, Naïve Bayes, SVM, XGBoost, and Random Forest are employed for crop recommendation. The recommendation system is developed considering parameters such as nitrogen, phosphorus, potassium, pH value, humidity, temperature, and rainfall. Analysis was conducted using these six machine learning algorithms, with XGBoost achieving the highest accuracy result among them (Ibrahimpatnam et al., 2021).

18.2.3 Disease Prediction

The advancement of image processing techniques, machine learning algorithms, and the availability of large datasets significantly improved the accuracy of crop disease prediction using leaf images. Such systems hold great potential in supporting precision agriculture practices, optimizing resource allocation, and ensuring sustainable crop production by enabling early detection and effective management of crop diseases. By analyzing plant leaf image inputs, predictive models to identify and forecast plant diseases can be developed. This will enhance disease management and facilitate timely interventions in agriculture leading to maximizing the yield.

Various imaging techniques like thermal imaging, multispectral imaging, fluorescence imaging, hyperspectral imaging, visible imaging, magnetic resonance imaging (MRI), and three-dimensional (3D) imaging techniques can be employed for the early prediction of disease. This will avoid the spreading of disease over the entire farm and eventually reduce pesticide consumption. This will result in cost savings due to the optimal use of pesticides (Vijai Singh et al., 2020).

A novel approach, called PiTLiD, for identifying plant leaf diseases from phenotypic data of plant leaves with a small sample size is proposed by Liu et al (2023). PiTLiD is based on a pre-trained Inception-V3 CNN and transfer learning. The findings demonstrate that PiTLiD's performance is good. Without needing as many phenotypic samples as a typical CNN, PiTLiD aims to solve the issue of training data with small sample sizes. The accuracy of 99.45% in the PiTLiD results using pathological pictures indicates a promising performance (Liu et al., 2023).

18.2.4 Yield Prediction

Yield prediction is vital in agriculture for effective farm management and decision-making. It aids in optimizing resource allocation, maximizing crop yield through targeted interventions, managing risks associated with crop failure, planning market strategies, and facilitating financial planning on the farm. Accurate yield prediction supports global food security by ensuring a stable food supply and promotes environmental sustainability by optimizing resource use and minimizing environmental impact. It enables farmers to optimize production, manage risks, plan effectively, and contribute to food security and environmental sustainability. Crop yields are dependent on a combination of controllable and uncontrollable factors. Controllable factors encompass elements like crop or seed varieties, tillage practices, and fertilizer usage, among others. Uncontrollable factors, on the other hand, include weather variables such as precipitation, air and soil temperatures, humidity, soil moisture, and solar radiation, among others, which are beyond human control. If any of these factors deviate from the plant's requirements, it may hinder optimal plant growth and affect productivity. Accurate prediction of crop yields heavily relies on the utilization of these weather variables (Kuradusenge *et al.*, 2023).

Yield prediction in agriculture depends on various factors, including weather conditions, soil characteristics, crop variety, management practices, pest and disease pressure, field history, genetic potential, crop stage, management decisions, and environmental factors. By analyzing these factors and using data-driven models, farmers and researchers can make more precise predictions of crop yields, leading to improved decision-making and resource management in agriculture. Precise data on soil characteristics, weather patterns, and crop genetics can help to forecast yields and identify optimal crop choices for specific regions. This empowers farmers to optimize their crops which plays a vital role in dealing with global food security issues by ensuring a steady supply and mitigating the risk of crop failure (Sagar *et al.*, 2023).

Coffee yield prediction is done by counting the fruits on the branch and using color features segregate them into three different categories ripe/overripe, semi ripe, and unripe. This will enable coffee growers to maximize economic gains and strategize their agricultural activities. The implementation is done using SVM. Similar algorithms are applied for yield prediction of cherry and green citrus using color features. A few other factors which are used are surface weather, and soil physicochemical for yield prediction of rice, and wheat. ANN is used for this application (Liakos, 2018).

18.2.5 Harvest Time Prediction

Crop harvest time prediction relies on various factors such as crop growth stage, weather conditions, historical data, crop variety, field observations, and pest/disease pressure. AI is used to analyze these factors through various ML algorithms, data fusion, image analysis, and decision support systems. By leveraging AI and considering these factors, farmers can make more accurate predictions about harvest timing, leading to improved efficiency, yield, and quality in agricultural production. ML algorithms for predicting harvest time consider various inputs, including weather data (temperature, precipitation, etc.), crop characteristics (variety, planting date), soil conditions, pest and disease information, management practices (irrigation, fertilization), remote sensing data (satellite imagery, Normalized Difference Vegetation Index (NDVI)), historical data, and market demand. By analyzing these inputs, researchers and farmers can develop predictive models to estimate optimal harvest timing, leading to improved efficiency and quality in agricultural production.

In Elashmawy *et al.* (2023) with the help of sensor networks, real-time data is gathered to accurately monitor soil parameters in strawberry farms. These sensors are strategically distributed throughout the field for data collection. This helps in accessing the physicochemical attributes of strawberry near harvesting. With the help of Gaussian regression models and neural networks forecasting of key physicochemical characteristics of strawberries, such as color and sweetness, based on soil properties like electrical conductivity and water content during the pre-harvest phase is done. This will help in cultivation of good-quality strawberries.

The long short-term memory (LSTM) model is selected as a prediction model for crop harvest time due to its compatibility with time series data, which characterizes crop harvest time. Long short-term memory with feature selection can provide better accuracy compared to the LSTM model (Liu *et al.*, 2022). Using Landsat 8 satellite remote sensing data, predictive models including multiple linear regression and Back Propagation (BP) neural network were developed. These models utilized NDVI, Normalized Difference Water Index (NDWI), and Enhanced Vegetation Index (EVI) as the input variables, with the winter wheat growth cycle serving as the target variable.

18.2.6 Price Forecasting

Price prediction of agricultural products plays an important role in the overall economy. It will help the farmers, traders involved in agricultural business to plan the activities as per market strategies. From a consumer point of view, they can remain informed about the past prices of the goods and accordingly make the purchase decisions. Price prediction of agricultural products using ML involves the use of different algorithms to analyze historical price data, along with various other factors that influence prices to forecast future price trends. These factors that are prevalent in this are crop yields, weather conditions, transportation costs, government policies, market demand, and global market trends. Models such as regression, time series analysis, and ensemble methods like random forests or gradient boosting are commonly used to develop predictive models for agricultural price forecasting. By analyzing past trends and patterns, these models can generate forecasts that help farmers, traders, and policymakers make informed decisions regarding production, procurement, pricing, and market strategies.

Analysis of techniques employed in forecasting agricultural commodity prices is essential. This will help in the creation of efficient models. Regression ensembles, which consist of multiple models collaborating to predict a response variable with minimized error, are employed for this objective. This study seeks to investigate the predictive capacity of regression ensembles by comparing them with both other ensembles and single-model approaches (reference models) in the agribusiness domain to predict prices one month ahead. Monthly time series data regarding producer prices in Parana, Brazil, for wheat and soybean are utilized. By keeping the K-Nearest Neighbors, SVM for Regression, and multi-layer perceptron (MLP) as reference models, the implementation of ensembles comprising bagging, boosting, and stacking of Random Forest (RF), Gradient Boosting Machine (GBM), and Extreme Gradient Boosting (XGB) is verified (Ribeiro et al., 2020).

Given the significance of predicting seasonal agricultural commodity prices, especially for vegetables, and the limited prior research in this area, this study introduces an innovative hybrid approach that combines extreme learning machines (ELMs) and seasonal-trend decomposition procedures (STL) for long-, short-, medium-term forecasting of prices for seasonal vegetables. In this proposed method, referred to as STL-ELM, the original vegetable price series are initially broken down into seasonal, trend, and remainder components using STL. ELM does trend prediction and uses seasonal-Naïve for forecasting seasonal components over period of a 12-month cycle (Xiong et al., 2018).

18.2.7 Supply Chain Integration

Supply chain management (SCM) is one of the important threads that needs to be handled critically as it deals with customer satisfaction and also is linked with the success of business in this competitive world. AI is largely involved in the overall process at various stages which include demand forecasting, inventory management, logistics and transportation, supplier management, manufacturing process, quality assurance, and customer feedback (Adel et al., 2022).

The supply chain process can be classified into two parts:

Upstream Supply Chain (Inbound Logistics)—From Resource to Manufacturer
Downstream Supply Chain (Outbound Logistics)—From Manufacturer to Consumer

Compared to other supply chains, the food supply chain holds great importance in our daily lives as it involves the journey from raw materials to food products reaching consumers. It is crucial to ensure that food products are delivered to consumers safely, and free from any contamination. In the traditional method of SCM, there is no assurance that the food delivered and consumed by the customer is safe. There are chances of adulteration, and contamination while the food items are traversing through the process till delivery to the customer. The advanced technology of involving blockchain and robots in the process will increase the reliability of this system (Ahamed et al., 2022).

Industry 5.0 acknowledges the significance of investing in human capital and equipping the workforce with the skills and knowledge necessary to excel in a swiftly changing technological environment. This entails offering ongoing education and training opportunities to enhance the capabilities of workers and promote a culture of lifelong learning (Maddikunta et al., 2022). Digital twin (DT) technology can be utilized to develop a digital representation of the SCM, incorporating warehouses, locations, logistics, and inventory assets. DT encompasses various elements such as factories, providers, contractors, transport routes, supply centers, and client sites. It aids in the entire life cycle of the SCM (Marmolejo et al., 2020).

DT technology can gather real-world data via IoT sensors, which can then be utilized by ML, big data analytics, and other tools to forecast challenges encountered across different phases of the SCM. This enables industries to proactively implement corrective actions to reduce losses and errors throughout the SCM process, ultimately facilitating the timely delivery of customized products to customers (Ivanov et al., 2019).

Cobots play a significant role in SCM by handling routine or hazardous tasks like packaging, quality checks, and heavy lifting, which humans may avoid. This allows human expertise to be focused on more complex SCM activities. Cobots are versatile and can be utilized across various stages of the SCM lifecycle, including material handling, assembly, packing, quality assurance, transportation, delivery, and product returns. By incorporating cobots, SCM industries have the potential to lower their overall cost of ownership and optimize operations, including inventory control, stock monitoring, order processing, and product return management (Simões et al., 2020).

18.3 IOT IN AGRICULTURE

In recent times, it has become evident that sensors and the IoT play a significant role in enhancing agricultural sustainability and ensuring food security. Various sensors capable of detecting soil parameters such as nitrogen, phosphorus, and potassium (NPK) levels, moisture content, nitrate concentration, pH levels, electrical conductivity, and CO_2 levels, as well as environmental factors like temperature, humidity, light intensity, water levels, smoke, and flame are instrumental in the implementation of IoT in precision farming. Additionally, alongside these sensors, cameras and other image-capturing tools are installed to facilitate the monitoring of crop health, livestock, and disease outbreaks. This system has several advantages over the traditional monitoring methods. The benefits of implementing IoT in agriculture encompass precision farming to optimize resource utilization, data-driven decision-making via real-time data analysis, remote monitoring, and control of farms, resource efficiency to minimize waste and environmental impact, improved crop quality and yield through early issue detection, task automation for enhanced efficiency, risk management through timely alerts, and enhanced market access and traceability, all contributing to sustainable and profitable farming practices (Morchid et al., 2024).

The precision and efficacy of smart agriculture are limited as it is dependent on a comparatively limited wireless sensor. The advancement of 6G-IoT technology may serve as the cornerstone for smart and sustainable agriculture in the future. With the use of 6G-IoT technology, a large number of sensors will be able to be connected, giving farmers the ability to gather more precise data on individual plants and track the record of minute variations. Blockchain technology makes it possible to store all sensor measurements on a decentralized, tamper-resistant ledger, making record-keeping safe, transparent, and dependable. It will allow the farmers who are remotely located to access the sensor data without any loss of data even though there is poor network connectivity (Polymeni et al., 2023).

18.4 AGRICULTURAL ROBOTS

Robots utilized in agriculture, also referred to as agricultural robots or agribots, offer a wide range of applications aimed at improving efficiency, productivity, and sustainability in farming operations. These include tasks such as autonomous weeding, harvesting, planting, pruning, and thinning, along with precision spraying to minimize chemical usage. Moreover, robots contribute to monitoring crop health, conducting soil sampling and analysis, gathering real-time data for decision-making purposes, and automating operations using autonomous tractors and equipment. They play a crucial role in greenhouse automation, optimizing plant growth and overall productivity. Figure 18.2 shows the applications of agricultural robots.

To sustain crop yields, regular application of fertilizers and pesticides is necessary. However, the conventional methods involving manual application by workers with knapsack sprayers are inefficient and time-consuming, incur high labor demands and costs, and are not safe for workers, especially for covering extensive agricultural fields. The prototype robot comprising a mobile base with two wheels, a spraying mechanism, a wireless controller for maneuvering the robot, and a camera for monitoring crop health, growth,

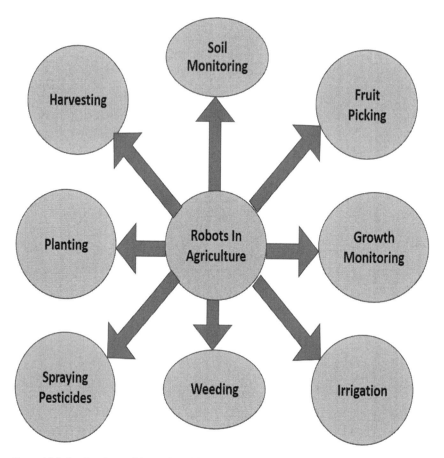

Figure 18.2 Applications of Agricultural Robots.

and pest presence in the agricultural field is developed. The main advantage of using such a system is the precision application of pesticides/fertilizers on the required region. This avoids wastage and overuse of chemicals. A cost-effective agricultural robot designed specifically for crop seeding in fields. The prototype comprises two main components: a mobile four-wheel design base for movement and a crank-slider seeding mechanism for sowing seeds. Harvesting robots use ML algorithms to optimize their performance. Data is gathered from cameras or sensors which is analyzed to take a decision on whether the crop is ready for harvesting. Other factors like weather conditions and market prices are also taken into consideration to finalize the harvest time. This highly integrated task improves the overall efficiency of harvest time prediction and automation of crop harvesting.

Most of the robot application in agriculture is developed using artificial neural networks (ANNs), fuzzy logic, and genetic algorithms. Therefore, there is considerable potential for further advancement of alternative AI methods for agricultural robot applications. The utilization of agricultural robots is more prominent in the monitoring and harvesting phases compared to the cultivation phase. It is advisable to prioritize the development of robotics technology for cultivation phase activities (Wakchaure et al., 2023).

18.5 VIRTUAL ASSISTANT IN FARMING

Virtual assistants provide extensive benefits to farmers in various aspects of their operations. Despite the availability of modern technology and vast amounts of data online, farmers often struggle to obtain precise information in a timely manner. While there are numerous chatbots and question-answering programs available, only a few of them provide accurate and useful information. Providing accurate responses for virtual assistant systems is challenging (Pandey et al., 2020). A well-developed system should be capable of understanding user queries in natural language, categorizing them effectively, and delivering accurate responses. When designing virtual assistants for this purpose, it is essential to consider the low literacy levels and limited digital literacy of users to ensure the application's success.

Sawant et al. (2019) created a web-based interface incorporating an interactive chatbot designed to address user queries regarding storage facilities for crops, such as warehouses and godowns. Additionally, the chatbot provides information about nearby training facilities and supports voice input.

Arora et al. (2020) introduced a chatbot powered by AI aimed at aiding farmers in decision-making by providing solutions to various agricultural issues. The chatbot specializes in identifying agricultural diseases, predicting weather conditions, and addressing common inquiries.

Geetha et al. (2021) introduced a platform-independent bot that serves as a comprehensive conversational system solution. It integrates various prediction modules, including crop identification, crop disease detection, crop rotation advice, soil detection, and weather forecasting.

Devi et al. (2021) developed a multilingual chatbot mobile application aimed at assisting farmers. The application provides farmers with localized information such as weather forecasts, optimal crops for planting, fertilizer recommendations, and other agricultural information.

18.6 CHALLENGES IN DEPLOYMENT OF INDUSTRY 5.0

1. Competency and Technical Skills: Workers need to acquire competency and technical skills when collaborating with advanced robots. This includes tasks like programming industrial robots.
2. Challenges in Installation: The adoption of advanced technology demands significant time and effort from human workers.

3. Cost Challenges in Adopting Industry 5.0: Adopting this technology involves significant investments in advanced technologies, which come with high costs. Training human workers for new roles also adds to the expenses.
4. Security Challenges: Authentication methods are crucial for interacting with various devices and countering potential threats from future quantum computing applications in deploying IoT nodes.

18.7 SUMMARY

AI has huge potential to revolutionize the agriculture sector and food industry. Challenges such as data privacy, high costs, ethical issues, and specialized training need to be overcome. Also, the reliability and accuracy of AI systems largely depend on the data on which it is trained. With the advent of various ML algorithms, farmers will get assistance and guidance in performing various routine activities for crop management leading to better yield. Integration of IoT and Industry 5.0 will help in sustainable agriculture practices.

REFERENCES

Adel, Amr. "Future of Industry 5.0 in society: Human-centric solutions, challenges and prospective research areas," Journal of Cloud Computing, 11, no. 1 (2022): 40. https://doi.org/10.1186/s13677-022-00314-5

Agarwal, Pooja, Pooja Yadav, Neelam Sharma, Ruchika Uniyal, Swati Sharma, "Research Paper on Artificial Intelligence," Case Studies Journal, 2, no. 6 (2013, July), ISSN (2305-509).

Ahamed, N. Nasurudeen, R. Vignesh. "Smart agriculture and food industry with blockchain and artificial intelligence," Journal of Computational Science, 18, no. 1 (2022): 1–17.

Azmi, Hussain Nor, Sami Salama Hussen Hajjaj, Kisheen Rao Gsangaya, Mohamed Thariq Hameed Sultan, Mohd Fazly Mail, Lee Seng Hua. "Design and fabrication of an agricultural robot for crop seeding," Materials Today: Proceedings, 81 (2023): 283–289. https://doi.org/10.1016/j.matpr.2021.03.191

Arora, Bhavika, Dheeraj Singh Chaudhary, Mahima Satsangi, Mahima Yadav, Lotika Singh, Prem Sewak Sudhish. "Agribot: A Natural Language Generative Neural Networks Engine for Agricultural Applications," In International Conference on Contemporary Computing and Applications (IC3A) Lucknow, India, pp. 28–33. IEEE, Feb 2020.

Bai, Y., Mao, S., Zhou, J., Zhang, Baohua. "Clustered tomato detection and picking point location using machine learning-aided image analysis for automatic robotic harvesting," Precision Agriculture, 24 (2023): 727–743. https://doi.org/10.1007/s11119-022-09972-6

Devi M.K., Divakar M.S., Vimal Kumar, Martina Jaincy, Kalpana R.A., Sanjai Kumar R.M. "Farmer's Assistant Using AI Voice Bot," In 3rd International

Conference on Signal Processing and Communication (ICPSC), Coimbatore, pp. 527–531. IEEE, May 2021.

Elashmawy, Rania, and Ismail Uysal. "Precision agriculture using soil sensor driven machine learning for smart strawberry production." Sensors, 23, no. 4 (2023): 2247.

Eli-Chukwu, Ngozi Clara. "Applications of artificial intelligence in agriculture: A review," Engineering, Technology & Applied Science Research, 9, no. 4 (2019).

Ghafar, Afif Shazwan Abdul, Sami Salama Hussen Hajjaj, Kisheen Rao Gsangaya, Mohamed Thariq Hameed Sultan, Mohd Fazly Mail, Lee Seng Hua. "Design and development of a robot for spraying fertilizers and pesticides for agriculture," Materials Today: Proceedings 81 (2023): 242–248.

Geetha, S., S. Balaji, A. Santhiya, C. Subashri, S. Subicsha. "Farm's Smart BOT," Turkish Journal of Computer and Mathematics Education, 12, no. 10 (2021, April), pp. 3299–3307.

Ibrahimpatnam, Hydarabad, Sunil Varma, C.S.E. Student, Sheriguda SIIET. "Crop and fertilizers recommendation and disease prediction using deep learning," Dogo Rangsang Research Journal, UGC Care Group I Journal, 08, issue 14 no. 03 (2021, March): 846–854, ISSN: 2347-7180.

Ivanov, Dmitry, Alexandre Dolgui. "New disruption risk management perspectives in supply chains: Digital Twins, the ripple effect, and resileanness," IFAC-PapersOnLine 52, no. 13 (2019): 337–342.

Jha, J., Ajeet Kumar Vishwakarma, Chaithra N, Allagari Nithin, Anu Sayal, Ashulekha Gupta, Rajiv Kumar. "Artificial Intelligence and Applications," In 2023 1st International Conference on Intelligent Computing and Research Trends (ICRT), Roorkee, India, 2023, pp. 1–4. https://doi.org/10.1109/ICRT57042.2023.10146698

Kuradusenge, Martin, Eric Hitimana, Damien Hanyurwimfura, Placide Rukundo, Kambombo Mtonga, Angelique Mukasine, Claudette Uwitonze, Jackson Ngabonziza, Angelique Uwamahoro. "Crop yield prediction using machine learning models: Case of Irish potato and maize," Agriculture, 13, no. 1 (2023): 225.

Liakos, Konstantinos G., Patrizia Busato, Dimitrios Moshou, Simon Pearson, Dionysis Bochtis. "Machine learning in agriculture: A review." Sensors, 18, no. 8 (2018): 2674.

Liang, Chen, Tufail Shah. "IoT in agriculture: The future of precision monitoring and data-driven farming," Eigenpub Review of Science and Technology, 7, no. 1 (2023): 85–104.

Liu, Kangchen, Xiujun Zhang. "PiTLiD: Identification of Plant Disease from Leaf Images Based on Convolutional Neural Network," In IEEE/ACM Transactions on Computational Biology and Bioinformatics, 20, no. 2 (2022): 1278–1288.

Liu, Shu-Chu, Quan-Ying Jian, Hsien-Yin Wen, Chih-Hung Chung. "A crop harvest time prediction model for better sustainability, integrating feature selection and artificial intelligence methods," Sustainability, 14, no. 21 (2022): 14101. https://doi.org/10.3390/su142114101

Madhuri, J., M. Indiramma. "Big Data Analytics-Based Agro Advisory System for Crop Recommendation Using Spark Platform," In Handbook of Research on AI and Machine Learning Applications in Customer Support and Analytics, pp. 227–247. IGI Global, 2023. https://doi.org/10.4018/978-1-6684-7105-0.ch012

Marmolejo-Saucedo, Jose Antonio. "Design and development of digital twins: A case study in supply chains." Mobile Networks and Applications, 25, no. 6 (2020): 2141–2160.

Marmolejo-Saucedo, Jose Antonio, Margarita Hurtado-Hernandez, Ricardo Suarez-Valdes. "Digital Twins in Supply Chain Management: A Brief Literature Review," In Intelligent Computing and Optimization: Proceedings of the 2nd International Conference on Intelligent Computing and Optimization 2019 (ICO 2019), pp. 653–661. Springer International Publishing, 2020.

Morchid, Abdennabi, Rachid El Alami, Aeshah A. Raezah, Yassine Sabbar. "Applications of Internet of Things (IoT) and sensors technology to increase food security and agricultural sustainability: Benefits and challenges," Ain Shams Engineering Journal, (2024): 102509. https://doi.org/10.1016/j.asej.2023.102509

Ngoc, Tran Thi Hong, Phan Truong Khanh, Sabyasachi Pramanik. "Smart Agriculture Using a Soil Monitoring System," In Handbook of Research on AI-Equipped IoT Applications in High-Tech Agriculture, pp. 200–220. IGI Global, 2023.

Pandey, Ankit, Vaibhav Vashist, Prateek Tiwari, Sunil Sikka, Priyanka Makkar. "Smart Voice Based Virtual Personal Assistants with Artificial Intelligence," Artificial & Computational Intelligence, 1, no. 3 (2020, June): 1–4.

Polymeni, Sofia, Stefanos Plastras, Dimitrios N. Skoutas, Georgios Kormentzas, Charalabos Skianis. "The impact of 6G-IoT technologies on the development of agriculture 5.0: A review," *Electronics*, 12, no. 12 (2023): 2651. https://doi.org/10.3390/electronics12122651

Maddikunta, Praveen Kumar Reddy, Quoc-Viet Pham, Prabadevi B., N. Deepa, Kapal Dev, Thippa Reddy Gadekallu, Rukhsana Ruby, Madhusanka Liyanage, "Industry 5.0: A survey on enabling technologies and potential applications," Journal of Industrial Information Integration, 26, (2022): 100257, ISSN 2452-414X. https://doi.org/10.1016/j.jii.2021.100257.

Ramaraj, A.P., K.P.C. Rao, G. Kishore Kumar, K. Ugalechumi, P. Sujatha, S. A. Rao, R.K. Dhulipala, A.M. Whitbread, "Delivering context specific, climate informed agro-advisories at scale: A case study of iSAT, an ICT linked platform piloted with rainfed groundnut farmers in a semi-arid environment," Climate Services, 31 (2023): 100403, ISSN 2405-8807. https://doi.org/10.1016/j.cliser.2023.100403

Ribeiro, Matheus Henrique Dal Molin, Leandro dos Santos Coelho. "Ensemble approach based on bagging, boosting and stacking for short-term prediction in agribusiness time series," Applied Soft Computing, 86 (2020): 105837.

Sagar S.H., Dravya K.L., R. Chethan Shindhe, Hrishikesh A. Savanth, Pavan Kumar, "Crop and Market Price Prediction System Using Machine Learning," International Research Journal of Modernization in Engineering Technology and Science, 05, no. 05 (2023, May): 1105–1107. https://www.doi.org/10.56726/IRJMETS38247

Sawant, Divya, Anchal Jaiswaly, Jyoti Singhz, Payal Shah. "AgriBot – An intelligent interactive interface to assist farmers in agricultural activities," In 2019 IEEE Bombay Section Signature Conference (IBSSC), Bombay, India, July 2019, pp. 1–6.

Simões, Ana Correia, António Lucas Soares, Ana Cristina Barros, "Factors influencing the intention of managers to adopt collaborative robots (cobots) in manufacturing organizations," Journal of Engineering and Technology Management, 57 (2020): 101574, ISSN 0923-4748. https://doi.org/10.1016/j.jengtecman.2020.101574

Singh, Harpreet, Nirmalya Halder, Baldeep Singh, Jaskaran Singh, Shrey Sharma, Yosi Shacham-Diamand. "Smart farming revolution: Portable and real-time soil nitrogen and phosphorus monitoring for sustainable agriculture," Sensors, 23, no. 13 (2023): 5914.

Singh, Vijai, Namita Sharma, Shikha Singh. "A review of imaging techniques for plant disease detection," Artificial Intelligence in Agriculture, 4 (2020): 229–242.

Sujatha, Radhakrishnan, Jyotir Moy Chatterjee, N.Z. Jhanjhi, Sarfraz Nawaz Brohi. "Performance of deep learning vs machine learning in plant leaf disease detection," Microprocessors and Microsystems, 80 (2021): 103615.

Tonni, Kaniz Fatima, Mahfuzulhoq Chowdhury. "A machine learning based crop recommendation system and user-friendly android application for cultivation," International Journal of Smart Technology and Learning, 3, no. 2 (2023): 168–186. https://doi.org/10.1504/IJSMARTTL.2023.129648

Wakchaure, Manas, B.K. Patle, A.K. Mahindrakar. "Application of AI techniques and robotics in agriculture: A review," Artificial Intelligence in the Life Sciences, 3 (2023): 100057. https://doi.org/10.1016/j.ailsci.2023.100057

Williams, C. "A Brief Introduction to Artificial Intelligence," In Proceedings OCEANS '83, San Francisco, CA, USA, 1983, pp. 94–99. https://doi.org/10.1109/OCEANS.1983.1152096

Xiong, Tao, Chongguang Li, Yukun Bao. "Seasonal forecasting of agricultural commodity price using a hybrid STL and ELM method: Evidence from the vegetable market in China," Neurocomputing 275 (2018): 2831–2844. https://doi.org/10.1016/j.neucom.2017.11.053

Chapter 19

Artificial Intelligence and Computer Vision in Sustainable Vertical Farming

A Key Component of Industry 5.0 Revolution

Archana Bhamare and Payal Bansal

19.1 INTRODUCTION

Agriculture is an important sector of any country as it plays a main role in food security and sustainable development. The population of each country is growing day by day. Subsequently, food requirements are also increasing. Consequently, farmers must try to incorporate the latest techniques to increase crop production and ensure full utilization of resources [1]. Crops are monitored physically for many years, which is time-consuming. However, these techniques have certain drawbacks and need more focus on improvement to meet agricultural needs [2].

Hydroponics, and greenhouse farming are viable solutions to water scarcity and farming land [3]. Seawater greenhouses are also an alternate solution to sustainable agriculture. Hydroponics farming has water-efficient techniques, which results in more crop production with less climatic effects [4,5]. It allows year-to-year farming by saving water and reducing soil erosion. It is perfect for urban areas which encourage local food production. Hydroponics guarantees safer and more nutritious food because of its exact nutritional management and low use of pesticides.

Overall, hydroponic farming is a sustainable solution to problems with water shortages, food security, and environmental changes. It is crucial to meet the growing demand for fresh produce worldwide while preserving natural resources.

19.2 HYDROPONIC FARMING TYPES

Hydroponic systems are of various types, each with some advantages and disadvantages [6]. Several types are as follows:

- **Deep Water Culture (DWC):** Plant roots are suspended in a nutrient solution and submerged in water. Air pumps provide roots with

oxygen, which promotes root growth. DWC systems work well for growing herbs and leafy greens and are easy to set up.
- **Nutrient Film Technique (NFT):** A thin coating of nutritious fluid that provides oxygen and nutrients to the plants' roots is always present. Because NFT systems use water and nutrients effectively, they are ideal for growing small, fast-growing plants like lettuce and strawberry.
- **Ebb and Flow (Flood and Drain):** Plants are grown in a tray with a growing media (such as rock wool or perlite) and a nutrient solution that is added to it regularly. The fluid then drains out, giving the roots access to oxygen. Ebb and flow systems are flexible and appropriate for many types of plants.
- **Aeroponics:** Periodically, plant roots floating in the air are treated with a nutrient-rich mist. Aeroponic systems are efficient in using water, and they offer superior root oxygenation. They work well for growing plants with sensitive roots, including tomatoes and strawberries.
- **Wicking System:** This straightforward technique passively transfers nutritional solution from a reservoir to the plant's roots using a wick. Wicking systems are inexpensive and require little upkeep. However, they might not be as appropriate for large or heavily producing crops.
- **Vertical Hydroponics:** In vertical hydroponic systems, plant trays or towers are stacked on top of one another to use vertical space. Although these systems save space and are appropriate for urban farming, they need more lighting and structural support.
- **Drip System:** Using tubing and emitters, the nutrient solution is directly poured onto the roots of the plants. Although drip systems are more exact in controlling the flow of nutrients, they can also need more maintenance and are more likely to clog.

These are just a few of the several varieties of hydroponic systems available; each has specific benefits and things to remember based on crop variety, budget, space, and environmental conditions.

Vertical farming cultivates crops in vertically stacked layers or structures, typically in confined interior areas like shipping containers, warehouses, or tall buildings [7,8]. This innovative agricultural technique is used to plant crops high rather than wide apart to maximize available area. Important features related to vertical farming consist of the following:

- Vertical Stacking
- Controlled Environment
- Hydroponics or Aeroponics
- Artificial Lighting
- Automation and Technology

The distinctions between vertical farming and conventional farming are discussed in Table 19.1. Generally speaking, vertical farming is a

Table 19.1 Vertical farming versus traditional farming

	Traditional Farming		Vertical Farming	
Food Waste	40–50% of harvest becomes food waste between harvest and distribution	**Drawback**	In-store harvests, the very short supply chain can meet unexpected demand to reduce overstocking.	**Benefit**
Food Safety	Pesticides and herbicides	**Drawback**	No pesticides and herbicides. Produce is highly traceable	**Benefit**
Freshness (Supply Chain)	Up to 2000 miles/2 weeks in transit	**Drawback**	In-store harvests to roughly 40 miles in transit	**Benefit**
Seasonality	Seasonal 5–6 harvests per year	**Drawback**	Year-round, up to 30 harvests per year. Improved flavors and textures	**Benefit**
Selection	Currently available offerings	**Benefit**	Currently producing leafy greens, herbs, and berries	**Drawback**
Price	Significantly cheaper, roughly half the cost of vertically farmed produce	**Benefit**	Hydroponically grown greens can be priced at a premium, and they have the highest profit margin at 40%.	**Drawback**
Sustainability Branding	Water loss from irrigation methods, water contamination from pesticides/herbicides	**Drawback**	Up to 90% less water usage and 1/50 land use	**Benefit**

creative and sustainable form of farming that can lessen environmental harm, address food security concerns, and feed the world's growing population in an increasingly urbanized setting [9,10,11]. As sustainable vertical farming takes a new approach to agriculture, it is essential to the Industry 5.0 revolution. It provides year-round agricultural production, reduces environmental impact, and maximizes resource efficiency by employing cutting-edge technologies like AI, IoT, and vertical stacking. This strategy complies with the sustainability and technology integration principles of Industry 5.0 by tackling several problems including food safety, increasing urbanization, and warming resistance. Vertical farming reduces transportation emissions, fosters local food production, and offers scalable solutions for future agricultural demands. Industry 5.0 signifies a revolutionary change toward a food system that is more robust and sustainable.

19.3 RELATED WORK

"Vertical farming" is a modern technique, involving growing crops vertically stacked in controlled environments. One of the professors of Columbia University and an advocate of vertical farming, Dr. Dickson Despommier, introduced the idea of vertical farming in the middle of the 2000s. Fresh, wholesome food in vertical farming is in greater demand due to fast urbanization and population growth. It reduces the food costs and increases the availability. Vertical farming creates a regulated growth environment that reduces the dependence on outside variables and guarantees yearly crop production. Today, vertical farming is applied in many different contexts and settings, ranging from large-scale commercial projects to tiny urban gardens. Various crops, including leafy greens, herbs, strawberries, tomatoes, and certain fruits and vegetables, can be grown using vertical farming.

Industry 5.0 is known for the smooth integration of cutting-edge technology like CV and AI into various industries, including the agriculture industry. AI helps farmers make decisions and predictions, based on crop growth, what amount of water fertilizers, and light intensity is needed depending upon the crop. Computer vision (CV) systems with imaging sensors and machine learning (ML) algorithms can also monitor crop health, identify illnesses, and estimate yield potential. Furthermore, Industry 5.0 strongly combines blockchain, robots, IoT, AI, and CV technologies to build a network of agricultural ecosystems. These technologies facilitate data interchange, auto functioning, and traceability across the agricultural network. Numerous researchers have contributed to the field of agriculture, particularly in the area of vertical farming, employing a variety of methods that include the following:

Compared to previous approaches in the literature, the paper [12] addresses the integration of CV for intelligent farming to create sustainable agriculture through better visual interpretation results of scene photographs. An article [13] focuses on how artificial intelligence (AI) and the Internet of Things (IoT) are being integrated with traditional farming to create smart agriculture, increasing productivity and efficiency across the board in the agricultural industry. To raise awareness of the significance of applying ML techniques in agriculture for sustainability, the paper [14] emphasizes the rise of vertical farming as a sustainable option and the application of ML methods, specifically a CNN image processing model, for plant disease detection. Using a CNN image processing model, creating an AI model with ResNet50 and ImageNet on Tensorflow, and implementing different overfitting mitigation techniques were all part of the methodology. The study in the paper [15] addresses vertical farming to overcome urban areas' limited land access. It emphasizes using IoT, AI, and ML to support precision agriculture. One earlier study [16] describes how AI and visual technology are applied in smart farming. It also describes how these technologies might be used in agricultural practices. The methodology uses ML and DL techniques to predict and analyze data for smart farming. It includes particular models and algorithms such as YOLOv5, RNN, LSTM, XGBoost, GRU, and CNN-LSTM. The paper [17] highlights the role of computers in improving productivity under controlled environments. It also discusses the growing vertical farms to solve desertification and urbanization. It also discusses the techniques of leveraging CV, ML, and IoT to improve farm management. In a paper [18], the authors have proposed a methodology comprising vertical farming in controlled environments with LED lights, integration with the IoT, and using CNNs and RNNs for crop yield forecasting. The study explains the use of AI for intelligent farming to meet the demands of a growing global population by 2050. It maximizes agricultural output efficiency and reduces losses for farmers through AI for intelligent farming. The authors in the paper [19] summarize the importance of indoor vertical farming techniques in controlled environments. The methodology combines different technologies such as IoT sensors, drones, robotics, ML technologies, and big data analytics to collect data on crop growth and environmental conditions. The authors in the paper [20] describe an AI-based precision and intelligent farming system that integrates multiple data sources for precision agriculture.

Finally, the present study explores how Industry 5.0 reveals change in the agricultural sector with AI and CV technologies. Industry 5.0 creates the foundation for a farming sector that is smarter, more connected, sustainable, and able to fulfill the demands of the twenty-first century by incorporating this cutting-edge technology into agricultural processes.

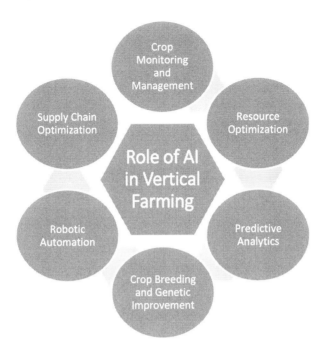

Figure 19.1 Role of AI in Vertical Farming.

19.4 ROLE OF AI IN VERTICAL FARMING

AI is essential to optimizing many elements of vertical farming by utilizing automation, ML algorithms, and data analytics. In Figure 19.1, the use of AI in vertical agriculture is highlighted.

AI improves several aspects of vertical farming, as listed below:

- **Crop Management and Monitoring:** AI algorithms examine data from cameras, sensors, and imaging systems to track plant health, growth trends, and environmental factors in real time.
- **Resource Optimization:** By constantly modifying environmental parameters, including light intensity, temperature, humidity, and nutrient levels based on plant requirements and external influences, AI maximizes resource usage. This minimizes waste and its adverse environmental effects while guaranteeing effective use of nutrients, energy, and water.
- **Predictive Analytics:** Predictive analytics is a tool that artificial intelligence uses to forecast crop yields, predict pest and disease outbreaks, and optimize planting schedules.

- **Crop Breeding and Genetic Improvement:** Crop breeding and genetic improvement are also implemented using genetic data by AI.
- **Supply Chain Optimization:** Supply chain logistics using AI help to optimize the distribution, monitoring, and prediction so that fresh produce can be delivered to customers.
- **Robotic Automation:** Robots can react to changing environments, navigate paths, and adaptively operate items with the help of ML algorithms [21].

In summary, real-time data is collected using sensors, which use AI algorithms to make decisions in vertical farming to increase crop production.

19.5 INTEGRATION OF CV

Vertical farming systems are monitored with the help of CV technology. Crop growth, diseases, and harvesting predictions can be performed by integrating CV. The leaf area, height, width, color, and texture can be used to determine crop growth. The images of the leaf are captured using a camera, and a database is created. This created database monitors crop growth, environmental conditions, and other related parameters in vertical farming systems using CV. All these ways of vertical farming contribute to sustainable agriculture and food security. Real-time plant monitoring is possible in vertical farming using these technologies. Some of the ways are listed below to describe how cameras are used in vertical farming:

- **Crop Health Monitoring:** A leaf image disease database can be created using cameras and image sensors. It helps to predict if any disease exists. Early prediction always alerts the farmers to increase the crop yield.
- **Growth Pattern Analysis:** The crop growth can be more easily predicted by taking photos of plants using camera seedlings to plant. Vertical farming systems can use this real-time data to alert farmers of crop growth rates.
- **Environmental Monitoring:** Temperature, humidity, and light intensity can also be assessed using cameras and imaging sensors. This helps to grow healthy crops irrespective of environmental changes.

CV algorithms are beneficial for harvesting, weed detection, and recognizing diseases in real time by comparing the features extracted from the input image with those of database images.in real time. It results in the advancement of agriculture.

19.6 BENEFITS OF AI AND CV IN SUSTAINABLE VERTICAL FARMING

Sustainable vertical farming methods need AI and CV to have different benefits. Following are some of the benefits:

- Optimized Resource Utilization
- Improved Crop Health and Yield
- Enhanced Pest and Disease Management
- Increased Automation and Efficiency
- Year-Round Cultivation
- Localized Food Production
- Data-Driven Decision Making
- Promotion of Sustainable Agriculture

AI, CV, precision agriculture, and the IoT are used together to make the system robust and use resources efficiently. With the help of various sensors and robots; the need for chemicals is reduced while crop health is preserved. Table 19.2 summarizes the functions of AI and CV in sustainable vertical farming.

Following are the case studies that highlight the use of AI and CV in vertical farming operations, making the system efficient:

- Plenty
- Iron Ox
- Bowery Farming

Table 19.2 Summary of techniques used in AI and computer vision for sustainable vertical farming

Technique	AI	Computer Vision
Machine Learning	Predictive analytics, yield forecasting, pest/disease prediction	Plant health monitoring, disease detection, weed identification
Sensor Integration	Integrates data from various sensors for decision-making	Uses cameras and imaging sensors for visual data capture and analysis
Automation	Automated control of environmental parameters based on AI-driven insights	Supports computerized tasks such as robotic harvesting, sorting, and quality control
Data Analytics	Analyzes large datasets for pattern identification, resource optimization	Processes image data for plant health assessment, growth monitoring
Decision Support	Provides AI-driven recommendations alerts based on data analysis	Offers real-time feedback on crop health, growth patterns
Adaptability	Adapts to changing conditions through continuous learning	Flexible in detecting and responding to plant health issues, environmental factors

These case studies prove that the use of AI and CV technologies is changing vertical farming operations by overcoming many drawbacks in conventional farming.

19.7 CHALLENGES AND FUTURE DIRECTIONS

Vertical farming systems can be made smarter using AI and CV, but several challenges need to be overcome before they can be implemented in real life [7,22].

- **High Initial Costs:** The high investment cost required for AI and CV technology, that is, hardware, software, and infrastructure, might be unaffordable for many farmers.
- **Complexity and Technical Expertise:** The use of AI and CV systems requires specialized technical expertise and training, which may not be easily possible to all farmers. Moreover, it is time-consuming and expensive to incorporate.
- **Data Privacy and Security:** A huge amount of data is collected from sensors. Farmers must ensure that private data is protected from hackers and illegal access.
- **Reliability and Accuracy:** The precise data inputs are essential for accurate predictions. The reliability and accuracy of the system are reduced due to the working of sensors, which deteriorate due to environmental elements.
- **Scalability and Adaptability:** Developing scalable and adaptable technologies depending on the types of crops is difficult.
- **Regulatory and Ethical Considerations:** Regulatory and ethical considerations must be followed while developing the system using AI and CV in agriculture.

The solution to the above challenges is to develop collaboration among all farmers, researchers, technology companies, and legislators. It will definitely help find solutions to these challenges using AI and CV systems in vertical farming. To address these challenges which are associated with AI and CV implementation in vertical farming, various solutions, and future paths can be followed:

- **Cost Reduction:** The innovation using AI and CV must satisfy the farmer's requirements with less cost so that farmers will get motivated to use that system even for their small activities. This may include the hardware, open-source software, and infrastructure.
- **Simplified User Interfaces:** User-friendly interfaces and intuitive tools can facilitate adoption by farmers with limited technical expertise.

- **Data Standardization and Interoperability:** There must be a network among industry experts, agriculture researchers, and policymakers to develop standardized frameworks and protocols. It will definitely help to define common data standards for various crops in agriculture.
- **Sensor Technology Advancements:** With advancements in technology, sensor technology must be revised. The research goal should be to create reliable sensor systems that function in various environmental settings.
- **Regulatory Frameworks and Guidelines:** Regulatory frameworks and guidelines must be defined to handle different issues related to data security, privacy, and ethics in agriculture.
- **Education and Training:** Farmers and agricultural workers must be trained by conducting hands-on training and providing educational materials.

The agricultural industry can overcome the problems of integrating AI and CV in vertical farming by pursuing different solutions. Moreover, integrating AI with robotics provides autonomous operations that reduce labor costs. This research aims to optimize resource utilization to increase crop yield and quality in vertical farming systems.

19.8 CONCLUSION AND FUTURE SCOPE

The chapter highlights the importance of introducing artificial intelligence and CV technologies in sustainable vertical farming. It initially discusses the difficulties of conventional agricultural practices. Therefore, Industry 5.0 is a paradigm shift for incorporating cutting-edge technologies into agriculture. Pest management, identification of crop health, and yield prediction are some benefits of the application of AI and CV. The AI and CV are compared with reference to functioning parameters. Real-time case studies explain the importance of these technologies in vertical farming activities. The final section of the chapter discusses challenges and future direction in the development of AI and CV integrated into vertical farming.

With the help of these technologies, environmental conditions, resource use, and crop management may be precisely monitored and optimized. It leads to increased productivity with no environmental impact. AI algorithms help systems to make data-driven decisions in real time. Additionally, they lower pesticide use and automation due to the early identification of diseases and nutrient deficiency made possible by these technologies. Integrating AI and CV in vertical farming supports sustainability by increasing output with fewer inputs. AI and CV technologies will definitely play a significant role in the future of smart agriculture.

REFERENCES

[1] Farhat Mahmood and Tareq Al-Ansari, "Renewable Energy Driven Sustainable Greenhouse: An Overview, Reference Module in Earth Systems and Environmental Sciences." Elsevier, 2023. ISBN 9780124095489. https://doi.org/10.1016/B978-0-323-90386-8.00073-5

[2] Muhammad Shoaib Farooq, Shamyla Riaz, Adnan Abid, Kamran Abid, and Muhammad Azhar Naeem, "A Survey on the Role of IoT in Agriculture for the Implementation of Smart Farming," In Special Section On New Technologies For Smart Farming 4.0: Research Challenges and Opportunities, IEEE Access, vol. 7, pp. 156237–156271, October 25, 2019.

[3] A. Muniasamy, "Machine Learning for Smart Farming: A Focus on Desert Agriculture," In 2020 International Conference on Computing and Information Technology (ICCIT-1441), 2020, pp. 1–5, doi: 10.1109/ICCIT-144147971.2020.9213759.

[4] H. Fakhrurroja, S. A. Mardhotillah, O. Mahendra, A. Munandar, M. I. Rizqyawan, and R. P. Pratama, "Automatic pH and Humidity Control System for Hydroponics Using Fuzzy Logic," In 2019 International Conference on Computer, Control, Informatics and Its Applications (IC3INA), 2019, pp. 156–161, doi: 10.1109/IC3INA48034.2019.8949590.

[5] G. Dbritto and S. Hamdare, "An AI-Based System Design to Develop and Monitor a Hydroponic Farm," In 2018 International Conference on Smart City and Emerging Technology (ICSCET), 2018, pp. 1–5, doi: 10.1109/ICSCET.2018.8537317.

[6] Farhatun Najat Maluin, Mohd Zobir Hussein, Nik Nor Liyana Nik Ibrahim, Aimrun Wayayok, and Norhayati Hashim. "Some emerging opportunities of nanotechnology development for soilless and microgreen farming." Agronomy 11, no. 6 (2021): 1213.

[7] Adrian Jandl, Pantelis A. Frangoudis, and Schahram Dustdar. "Edge-based autonomous management of vertical farms." IEEE Internet Computing 26, no. 1 (2022): 68–75.

[8] Anurag Shrivastava, Chinmaya Kumar Nayak, R. Dilip, Soumya Ranjan Samal, Sandeep Rout, and Shaikh Mohd Ashfaque. "Automatic robotic system design and development for vertical hydroponic farming using IoT and big data analysis." Materials Today: Proceedings 80 (2023): 3546–3553.

[9] M. R. Sundara Kumar,. "Design and development of automatic robotic system for vertical hydroponic farming using IoT and big data analysis." Turkish Journal of Computer and Mathematics Education (TURCOMAT) 12, no. 11 (2021): 1597–1607.

[10] Muhammad E. H. Chowdhury, Amith Khandakar, Saba Ahmed, Fatima Al-Khuzaei, Jalaa Hamdalla, Fahmida Haque, Mamun Bin Ibne Reaz, Ahmed Al Shafei, and Nasser Al-Emadi. "Design, construction and testing of IoT based automated indoor vertical hydroponics farming test-bed in Qatar." Sensors 20, no. 19 (2020): 5637.

[11] R. G. Mapari, Harish Tiwari, Kishor Bhangale, Nikhil Jagtap, Kunal Gujar, Yash Sarode, and Akash Mahajan, "IOT Based Vertical Farming Using Hydroponics for Spectrum Management & Crop Quality Control," In

2022 2nd International Conference on Intelligent Technologies (CONIT), 2022, pp. 1–5, doi: 10.1109/CONIT55038.2022.9848327.

[12] R. Tombe, "Computer Vision for Smart Farming and Sustainable Agriculture," In 2020 IST-Africa Conference (IST-Africa), Kampala, Uganda, 2020, pp. 1–8.

[13] S. Katiyar and Farhana, A. (2021). "Smart Agriculture: The Future of Agriculture Using AI and IoT." Journal of Computer Science, 17(10), 984–999. https://doi.org/10.3844/jcssp.2021.984.999

[14] Melis Siropyan, Ozan Celikel, and Ozgun Pinarer. "Artificial Intelligence Driven Vertical Farming Management System," In Proceedings of the World Congress on Engineering, 2022.

[15] Riki Ruli A. Siregar, Kudang Boro Seminar, Sri Wahjuni, and Edi Santosa. "Vertical farming perspectives in support of precision agriculture using artificial intelligence: A review." Computers 11, no. 9 (2022): 135.

[16] Xiuguo Zou, Zheng Liu, Xiaochen Zhu, Wentian Zhang, Yan Qian, and Yuhua Li. "Application of vision technology and artificial intelligence in smart farming." Agriculture 13, no. 11 (2023): 2106.

[17] Ubio Obu, Gopal Sarkarkar, and Yash Ambekar. "Computer Vision for Monitor and Control of Vertical Farms Using Machine Learning Methods." In 2021 International Conference on Computational Intelligence and Computing Applications (ICCICA), pp. 1–6. IEEE, 2021.

[18] V. S. Abhay, V. H. Fahida, T. R. Reshma, C. K. Sajan, and Siddharth Shelly. "IoT-Based Home Vertical Farming," In Intelligent Systems, Technologies and Applications: Proceedings of Sixth ISTA 2020, India, pp. 151–160. Springer Singapore, 2021.

[19] Gowtham Rajendiran, and Jebakumar Rethnaraj. (2023). "Open Access Journal of Agricultural Research Committed to Create Value for Researchers Future of Smart Farming Techniques: Significance of Urban Vertical Farming Systems Integrated with IoT and Machine Learning Future of Smart Farming Techniques: Significance of Urban Vertical Farming Systems Integrated with IoT and Machine Learning." Open Access Journal of Agricultural Research, 8, 1–20.

[20] Samir Rana. "AI Based Precision and Intelligent Farming System." Information Technology in Industry 7, no. 3 (2019): 31–37.

[21] S. Rathod, T. Vedpathak, S. Pednekar, and A. Bhamare, "Solar Based Multi-Tasking Agri-Robo," In 2023 7th International Conference On Computing, Communication, Control And Automation (ICCUBEA), Pune, India, 2023, pp. 1–5, doi: 10.1109/ICCUBEA58933.2023.10391956.

[22] A. Z. M. Tahmidul Kabir, A. M. Mizan, N. Debnath, A. J. Ta-sin, N. Zinnurayen, and M. T. Haider, "IoT Based Low Cost Smart Indoor Farming Management System Using an Assistant Robot and Mobile App," In 2020 10th Electrical Power, Electronics, Communications, Controls and Informatics Seminar (EECCIS), 2020, pp. 155–158, doi: 10.1109/EECCIS49483.2020.9263478.

Chapter 20

Technological Solutions for Waste Management in Indian Hotels

Exploring Feasibility and Environmental Implications

Monika, Kiran Shashwat, and Umang Bhartwal

20.1 INTRODUCTION

Waste management is a critical concern in the contemporary hospitality industry, and as the Indian hotel sector undergoes rapid expansion, the need for sustainable waste management practices becomes increasingly pronounced (Paritosh et al., 2017). It affects economies, ecosystems, and societal well-being, making it a critical worldwide issue that knows no geographical bounds. Effective food waste management is now more important than ever as the world struggles with growing populations, shifting consumption patterns, and environmental damage (Bernstad & la Cour Jansen, 2012). This introduction lays the groundwork for examining the intricate issues surrounding food waste, including its wide-ranging effects and the necessity of putting sustainable solutions into practice.

The scale of food waste is staggering, with a significant portion of the world food production (Närvänen, 2020) ending up discarded rather than consumed. From farm to fork, inefficiencies in the food supply chain contribute to this wastage, thereby exacerbating issues of hunger, resource depletion, and environmental harm. The environmental toll is particularly pronounced, as decomposition of food in landfills releases methane, a potent greenhouse gas that contributes to climate change. Moreover, the extensive use of resources, including water, energy, and land, in the production of unconsumed food compounds the ecological impact.

Food waste management is more than just dumping it out (Filimonau & Delysia, 2019). It requires a whole strategy that includes recovery, recycling, and prevention. Effective food waste management plans must include programs that promote sensible consumption, surplus food redistribution to those in need, and creative technology for turning food waste into useful resources. In addition to its positive effects on the environment, effective food waste management promotes the values of a circular economy, which minimizes waste and maximizes the value of resources.

20.2 BACKGROUND OF THE STUDY

The hospitality sector in India stands as a thriving industry, contributing significantly to the nation's economy. However, this growth comes hand in hand with escalating environmental challenges, particularly in the domain of waste management (Vivekanand, 2017). Indian hotels generate substantial amounts of waste, encompassing food scraps, packaging materials, and other disposable items. Traditional waste disposal methods, such as landfills, pose environmental hazards and contribute to the sector's carbon footprint (Aiello, 2015). As global awareness of climate change and environmental sustainability heightens, the hospitality industry faces an imperative to adopt responsible waste management practices.

The rising trend of eco-conscious travelers places additional pressure on hotels to align their operations with sustainable principles. Guests are increasingly valuing environmentally friendly practices, and hotels are recognizing the need to not only meet guest expectations (Salemdeeb et al., 2017) but also contribute to broader environmental conservation efforts. In this context, waste-to-energy (WtE) solutions emerge as a compelling avenue for Indian hotels to not only address waste disposal challenges but also harness valuable energy resources, thereby enhancing operational sustainability.

20.2.1 Rationale for the Study

The rationale for undertaking a comprehensive study on WtE solutions in Indian hotels is grounded in the urgency of the environmental challenges faced by the hospitality sector. The traditional linear model of waste disposal, where waste is discarded without resource recovery, is no longer tenable in the face of depleting resources and increasing environmental consciousness. This study aims to explore alternative waste management paradigms, specifically focusing on the feasibility and viability of WtE technologies (Figure 20.1).

Furthermore, it is impossible to disregard the operational and financial factors. WtE solutions are a compelling option for the hotel industry due to the economic consequences of waste management and the possibility of producing energy (Kibler et al., 2018). This study aims to provide insights that help guide sector decision-making processes by comprehending the operational and financial stability issues.

Objectives of the study:

- Provide an overview of various WtE technologies applicable to the Indian hotel sector, including incineration, anaerobic digestion, and gasification.

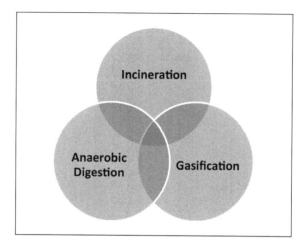

Figure 20.1 Key technologies for waste to energy.
Source: Author.

- Examine logistical and operational aspects, addressing challenges and opportunities in integrating WtE technologies with existing hotel infrastructure.
- Explore the environmental implications of WtE solutions, focusing on factors such as air quality improvement, reduction in greenhouse gas emissions, and sustainable waste reduction strategies.

20.2.2 Significance of the Study

This research is extremely important for determining the long-term course of the Indian hospitality industry. Through the study of WtE technologies, urgent environmental issues are addressed, and workable solutions to lower carbon footprints, improve air quality, and support effective waste management are provided. The results not only provide hoteliers with useful information about practical WtE adoption, but they also add to the conversation about environmental issues in general. This study is a trailblazing effort that explores the intersection of technology, viability, and environmental impact, promoting a paradigm change toward more environmentally friendly methods. In the end, its importance lies in helping Indian hotels have a more sustainable future and in furthering international discussions on environmentally friendly.

In today's world, the hospitality sector is discovering more and more ways how crucial sustainable practices can reduce environmental effects and

Technological Solutions for Waste Management in Indian Hotels 321

improve operational effectiveness (Munir, 2022). The investigation of WtE solutions is a crucial component of this sustainability paradigm, and it is especially pertinent to the Indian hotel industry. This book chapter explores the intricacies of different WtE technologies, their viability within the Indian hotel industry, and the consequent environmental effects.

20.3 WASTE TO ENERGY TECHNOLOGIES

WtE technologies represent a pivotal paradigm shift in waste management, offering a sustainable approach to address the dual challenges of waste disposal and energy generation (Tabasová, 2012). At its core, WtE involves the conversion of various types of waste materials into heat, electricity, or fuel. This innovative approach not only reduces the environmental burden associated with conventional waste disposal methods but also harnesses the latent energy potential within waste streams.

The integration of WtE technologies holds particular significance in the context of the Indian hospitality sector, where large-scale waste generation is intrinsic to daily operations. This section delves into the key WtE technologies—incineration, anaerobic digestion, and gasification—unveiling their principles, applications, and comparative merits.

WtE technologies are emerging as creative solutions in the dynamic field of sustainable waste management that have the ability to completely transform the environmental impact of a variety of sectors (Münster & Lund, 2010). This chapter delves further into WtE technologies, with a focus on how they work in the particular operational environment of Indian hotels. Through providing a synopsis of important technologies like gasification, anaerobic digestion, and incineration, the conversation hopes to arm hospitality industry participants with knowledge that is essential for making wise choices while pursuing environmentally friendly waste management methods.

20.3.1 Key Technologies

The conversation around waste management in modern times acknowledges the necessity of extracting energy from materials that are thrown away, transforming what was before seen as a burden into an asset (Tsui & Wong, 2019). Three well-known WtE technologies—each with unique difficulties and promises—are carefully examined in this part with an emphasis on their underlying principles and working mechanisms.

- **Incineration:** At the vanguard of WtE technologies, incineration is the controlled burning of waste items to produce heat or electricity as energy. This section describes the technological aspects of incineration and looks at possible uses for it in the Indian hotel industry's operational environment (Kalogirou, 2017). The chapter negotiates the complex landscape of incineration as a workable waste management alternative by critically analyzing its benefits, such as volume reduction and energy recovery, in contrast to issues, such as emissions and ash disposal.

- **Anaerobic Digestion:** An organic waste management method called anaerobic digestion uses microorganisms to naturally break down organic waste and produce biogas and nutrient-rich byproducts. It is a biological substitute for incineration. Anaerobic digestion's workings are explained, opening up the possibility of using technology to help Indian hotels that are facing difficulties with organic waste (Arsova, 2010). This section advances a more comprehensive understanding of anaerobic digestion in the context of the hotel industry by analyzing its benefits, such as the production of renewable energy and nutrient-rich byproducts, in addition to factors like feedstock compatibility and operational needs.
- **Gasification:** Gasification is a thermochemical conversion process that turns organic resources into syngas, or synthetic gas, which is useful for a number of different energy applications (Ahmed & Gupta, 2010). This section of the chapter delves into the details of gasification and examines how it might be used by Indian hotels as a WtE solution (Naveed, 2009). By means of a comprehensive examination of its advantages, such as the adaptability of syngas and lower emissions, in comparison with obstacles like technological complexity, the conversation offers a comprehensive viewpoint on the suitability of gasification in the heterogeneous context of the hospitality sector.

20.4 FEASIBILITY OF WASTE-TO-ENERGY SOLUTIONS IN INDIAN HOTELS

20.4.1 Economic Feasibility

- *Cost–Benefit Analysis:* Conducting a robust cost–benefit analysis is essential to gauging the economic feasibility of implementing WtE solutions (Kumar et al., 2021). This analysis involves a meticulous examination of the initial capital investment required for technology acquisition, installation, and operational setup, weighed against the anticipated long-term benefits. It encompasses considerations such as waste disposal cost savings, potential revenue generation from energy sales, and the economic value of recovered resources. A comprehensive cost–benefit analysis aids decision-makers in quantifying the financial implications and assessing the overall economic viability of WtE implementation in Indian hotels.
- *Return on Investment:* Determining the return on investment (ROI) is instrumental in evaluating the economic viability of WtE solutions

over the project's lifecycle. This metric involves assessing the time it takes for the initial investment to be recouped through cost savings, energy generation, or other identified revenue streams. A positive ROI not only justifies the economic feasibility of WtE implementation but also informs stakeholders about the project's financial sustainability. Factors influencing ROI include technology efficiency, waste composition, and prevailing energy market conditions.

20.4.2 Logistical Considerations

- *Waste Collection and Sorting:* The effective execution of waste management (WtE) solutions relies upon the existence of effective mechanisms for collecting waste and sorting. A logistical consideration is evaluating the hotels' existing waste management system. This involves establishing if it is feasible to separate waste at the source to guarantee the best possible use of WtE technology. Sufficient waste sorting minimizes contamination, enhances energy recovery process efficiency, and streamlines logistical operations downstream.
- *Technology Integration:* The seamless integration of WtE technologies into existing waste management processes is critical for logistical success. Assessing the compatibility of technology with the waste stream characteristics of Indian hotels is imperative. Factors such as the flexibility of the chosen technology to handle diverse waste types and the adaptability of the hotel's waste collection infrastructure to support WtE processes are pivotal considerations (Mata-Lima et al., 2021). A well-integrated system ensures smooth logistical operations and minimizes disruptions to daily waste management practices.

20.4.3 Operational Feasibility

- *Integration with Existing Infrastructure:* Integrating WtE solutions with current hotel infrastructure requires a thorough compatibility analysis. This entails evaluating the area needed for technology installation, the adaptability of the trash collection techniques in use today, and the alignment of WtE procedures with the trends in hotel waste creation. When an integration is done right, WtE solutions complement current infrastructure in a fluid way and increase operational effectiveness.
- *Impact on Hotel Operations:* Evaluating operational viability requires an awareness of how WtE implementation might affect regular hotel operations. This entails assessing elements including the spatial demands of WtE technologies, the requirement for extra employee training, and the possibility of operating disruptions during technology installation (Dhar et al., 2017). Mitigating adverse effects

and ensuring a harmonious integration of WtE solutions with hotel operations is paramount for sustained success.

20.5 ENVIRONMENTAL IMPACTS AND BENEFITS

This section explores the environmental implications of WtE technologies, considering factors such as air quality, greenhouse gas emissions, and waste reduction.

a) Air Quality Considerations:

In heavily populated metropolitan contexts, WtE technologies are very important for solving air quality concerns. Certain WtE systems use controlled combustion processes to reduce pollution emissions, thereby providing a feasible substitute for traditional waste disposal techniques that greatly increase air pollution.

b) Greenhouse Gas Emissions:

As the global community grapples with the challenges of climate change, the reduction of greenhouse gas emissions stands as a paramount objective. This chapter scrutinizes the carbon footprint of WtE technologies, exploring their potential to curtail emissions by harnessing energy from waste materials. By mitigating the release of methane—a potent greenhouse gas generated in landfills—WtE emerges as a pivotal contributor to climate change mitigation efforts.

c) Waste Reduction Strategies:

The ability to transform waste materials into priceless energy resources is at the core of WtE. This reduces the amount of waste that ends up in landfills and acts as a spur for more environmentally friendly methods of managing waste (Al-Manji et al., 2024). Examining different WtE technologies offer insights into how these procedures might rethink waste disposal by putting an emphasis on resource recovery and circular economy ideas.

20.6 COMPARATIVE ANALYSIS OF TRADITIONAL AND MODERN TECHNOLOGIES

1. Traditional Practices:
   Composting: In the past, composting has been a common technique for managing food waste in India. Organic waste, including leftovers from the kitchen, breaks down naturally into nutrient-rich compost. In addition to reducing trash volume, composting

creates beneficial soil enhancers that support sustainable farming methods (Raychaudhuri, 2008). Composting remains a popular and environmentally responsible method of disposing of garbage in many homes, especially those in rural areas.
2. Transition to Technological Solutions:
    Biogas Digesters: India has adopted increasingly sophisticated technology in response to the growing amount of food waste, especially in metropolitan areas. An effective way to handle organic waste is to use biogas digesters, which use anaerobic digestion to produce biogas and nutrient-rich slurry. In addition to addressing the negative environmental effects of food waste, this dual-output system offers a sustainable energy source (Vij, 2012). Biogas is mostly made up of methane and is used as a sustainable energy source for cooking and electricity production.
    Decentralized Waste Management: The introduction of decentralized waste management systems signifies a paradigm change in the way local governments handle food waste. Composting and recycling facilities are added benefits to small-scale waste processing systems that are becoming more and more popular. These decentralized methods lower transportation costs, ease the burden on centralized trash disposal facilities and encourage community participation in waste management procedures.

20.6.1 Technological Advancements in Metropolitan Areas

Smart Bins with Sensor Technology: Urban centers are witnessing the integration of smart bins equipped with sensors and communication devices. These technologically advanced bins optimize waste collection processes by providing real-time data on fill levels (Nandan et al., 2017). Municipalities can utilize this data to streamline waste collection routes, reduce collection costs, and prevent overflowing bins. The implementation of smart bins aligns with the broader goal of creating efficient and technology-driven waste management systems in urban areas.

Mobile Applications Addressing Surplus Food: In the context of reducing food waste, mobile applications have emerged as powerful tools. These apps connect food establishments with surplus inventory to local communities or charities (Priti & Mandal, 2019). By facilitating the redistribution of edible food that would otherwise be discarded, these applications contribute to minimizing food waste and addressing issues of food insecurity.

20.6.2 Challenges and Opportunities

Expanding Technological Solutions: Although promising, scaling up the use of cutting-edge technologies for managing food waste across India varied terrain presents difficulties. Development of infrastructure is still a challenge, especially in rural areas (Kumar et al., 2017). Furthermore, it is imperative to conduct awareness campaigns to inform communities about the advantages of new technology as well as their role in sustainable waste management practices.

Policy Support and Regulation: To promote the broad use of modern waste management technology, a supportive regulatory environment is necessary. Technology developments in food waste management can be greatly facilitated by government laws that encourage the use of biogas digesters, smart waste bins, and other advanced technologies.

20.6.3 The Need for Sustainable Practices in Indian Hotels

India has a unique combination of circumstances that have made the country environmental resources more vulnerable, including a growing population, fast urbanization, and growing tourism (Singhal, 2018). A growing number of waste materials, ranging from food scraps to disposable amenities, are generated in hotels, which highlights the need for creative solutions that manage trash while simultaneously using its potential for energy. In light of this, investigating WtE options becomes a crucial way for hotels to both improve operational effectiveness and their environmental impact.

The problem of balancing sustainable practices with economic viability is complex for Indian hotels, which are frequently distinguished by their varied operational scales and disparate infrastructure capacity (Prakash et al., 2023). This chapter seeks to clarify the complexities of this issue, highlighting the necessity of a sophisticated and situation-specific approach to sustainable practices. Examining the environmental impact of the Indian hotel industry, the conversation covers the need for proactive environmental stewardship methods that go beyond simple regulatory compliance.

20.7 IMPACT OF WASTE TO ENERGY TECHNOLOGIES ON THE HOSPITALITY INDUSTRY

India hospitality sector has grown at an unparalleled rate in recent years due to rising tourism, urbanization, and the country expanding middle class. Although there have been financial gains from this expansion, new difficulties have also arisen, notably in the area of garbage management (Sharma, 2020). Given the difficult task of upholding elevated service standards while simultaneously tackling environmental issues, Indian hotels are currently focusing on sustainable methods. WtE technologies have become a promising option among developing solutions, with the promise to minimize environmental effects while converting waste into a profitable resource.

The hotel industry, which is renowned for its resource-intensive operations, produces a substantial quantity of garbage every day, both organic and nonorganic. Conventional waste management techniques, such as incineration and landfill disposal, not only exacerbate environmental degradation but also provide difficulties with regard to space limitations and regulatory compliance. Considering this, WtE solutions offer a strong substitute by utilizing the energy potential found in waste products.

WtE encompasses a range of technologies that transform various types of waste into heat, electricity, or fuel. In the Indian hotel industry, these technologies have the potential to solve waste management issues as well as energy requirements. The purpose of this study is to examine the various WtE technologies that are relevant to the hospitality industry and assess their potential from an economic and environmental standpoint.

An oxygen-free process called anaerobic digestion breaks down organic waste to produce a digestate that is rich in nutrients and biogas. It is one of the main technologies being examined. Because it provides both energy production and nutrient recovery for agricultural use, this approach is especially relevant for hotels that generate a significant amount of food waste. Technologies for gasification and incineration will also be investigated, with an emphasis on evaluating how well they work with the many waste streams produced by hotels as well as how well they can produce heat and electricity. In the context of Indian hotels, the financial feasibility of implementing WtE solutions is crucial. To evaluate the practicality and adaptability of various WtE technologies, variables including waste composition, operational size, and local regulatory frameworks will be considered. These solutions economic viability will also be carefully examined, taking into account the initial capital expenditure, constant costs, and prospective revenue streams like energy sales and reduced waste disposal.

The convergence of waste management and energy efficiency has gained prominence in research and innovation due to the growing worldwide environmental concerns and the growing hospitality industry in India. In the ever-evolving landscape of sustainable development, the hospitality industry stands at the crossroads of environmental responsibility and operational efficiency. Businesses, particularly those in the hospitality industry, are increasingly looking for creative ways to connect their operations with sustainable practices as the globe struggles with the urgent need to address climate change and minimize the environmental footprint of human activity. With its diversified and rapidly growing hospitality industry, India is a key player in the global movement towards sustainable practices. The nation waste management problems have been made worse by the country rapid urbanization and tourism boom, which has prompted a rigorous analysis of alternative technologies that can both contribute to energy needs and lessen environmental damage. This chapter serves as a comprehensive exploration of WTE technologies, their feasibility, and the potential environmental implications specific to the Indian hotel industry.

The hospitality industry is a significant contributor to waste generation, encompassing organic, inorganic, and hazardous waste streams (Filimonau & Delysia, 2019). Conventional waste disposal techniques are extremely harmful to the environment as they pollute the ecosystem, deplete resources, and increase carbon footprints. With the added benefit of waste reduction and energy generation, the integration of WTE technologies appears to be a potential route. The viability of deploying WTE solutions in Indian hotels is a complex investigation that takes into account operational, financial, and regulatory factors in addition to technological ones.

Figure 20.2 depicts the interrelated negative effects of food waste creation on the ecosystem, which include habitat disruption, energy consumption, land pollution, greenhouse gas emissions (carbon dioxide and methane), and biodiversity reduction.

Figure 20.2 Waste Generation Aspect.

Source: Author.

The implementation of WtE solutions in the Indian hospitality sector is not only a strategic move toward operational resilience but also a necessity for the environment in the pursuit of sustainable development (Naveed et al., 2009). The dynamic environment of energy consumption and waste management needs creative solutions that support India commitment to achieving its climate goals. WtE solutions are an outstanding instance of how the hospitality industry is redefining its place in the global sustainability story and providing a practical route forward toward a more beneficial to the environment and efficient future.

One of the unique challenges faced by Indian hotels is the diversity of waste generated, ranging from organic kitchen waste to plastics and electronic waste (Aye & Widjaya, 2006). The broad spectrum of waste generated, from organic kitchen waste to plastics and electrical waste, is one of the particular challenges faced by Indian hotels. Hotels can customize the approach they choose according to the unique composition of their waste streams by utilizing the comprehensive solution offered by the adaptability of WtE technologies. Hotels may effectively handle a variety of waste kinds while obtaining valuable resources for other uses by implementing technologies like pyrolysis, which thermally burns down waste into bio-oil and syngas.

Economic considerations are paramount in the decision-making process for hotels contemplating the adoption of WtE technologies. While initial capital investments may pose challenges, the potential for long-term cost savings through reduced waste disposal fees and energy self-sufficiency offers a compelling business case. Furthermore, the prospect of additional revenue streams, such as selling excess energy back to the grid or utilizing by-products for agricultural purposes, adds a layer of financial resilience to the sustainability equation.

The overall sustainability factor of these solutions will become clearer as the study goes on through an extensive investigation of the environmental effects resulting from WtE technology (Heikkinen, 2020). An extensive comprehension of the environmental trade-offs and advantages will be facilitated by assessing variables such as the state of the air, emissions of greenhouse gases, and the life cycle assessment of different technologies. Achieving stability between energy generation and emission mitigation will be crucial to guarantee that the implementation of WtE technology yields a favorable environmental consequence overall.

Furthermore, this chapter undertakes a rigorous examination of the environmental impacts associated with adopting WTE solutions in the Indian hotel context. Knowing the wider ecological ramifications of waste reduction and energy creation is crucial, even beyond the short-term advantages. Crucial elements of this investigation include life-cycle evaluations, carbon footprint calculations, and air and water quality considerations. Enabling stakeholders to make well-informed decisions that balance economic interests with environmental stewardship is the aim of providing a comprehensive picture of the environmental impacts.

20.8 CONCLUSION

In conclusion, exploring the implications of WtE solutions on the environment within the Indian hotel industry provides promise for revolutionizing sustainable waste management techniques. The chapter has explained how these technologies actively support reducing greenhouse gas emissions, improving waste reduction techniques, and reducing air pollution.

The development of WtE technologies, including gasification, anaerobic digestion, and incineration, provides hotels a method to match their operations with environmentally beneficial values. These technologies produce real benefits by reducing emissions and mitigating air pollution, which improves the quality of the air not only inside the hotel but also in the neighboring community.

As the Indian hospitality sector grapples with the imperative to balance operational needs with environmental responsibility, WtE solutions emerge as key catalysts for change. By leveraging these technologies, hotels can transcend conventional waste management practices, contributing to global sustainability goals and responding to the growing demand for environmentally conscious operations.

This chapter serves as an extensively useful manual for Indian hotels that are interested in navigating the complicated landscape of ecological impacts connected to waste management. By embracing WtE solutions, hotels may become leaders in sustainable practices and successfully manage the waste they produce, thereby contributing to a more environmentally conscious and conscientious future for the Indian hospitality industry.

REFERENCES

Ahmed, I. I., & Gupta, A. K. (2010). Pyrolysis and gasification of food waste: Syngas characteristics and char gasification kinetics. *Applied Energy*, 87(1), 101–108.

Aiello, G., Enea, M., &Muriana, C. (2015). Alternatives to the traditional waste management: Food recovery for human non-profit organizations. *International Journal of Operations and Quantitative Management*, 21(3), 215–239.

Al-Manji, A. N. S. H., Thangavel, S., Kumar, A., & Heidari, M. (2024). A Study on Waste Management Practices in Oman and a way forward to a Circular Economy. In *Biodegradable Waste Processing for Sustainable Developments*, A. Prasad & A. K. Paul (Eds.), pp. 129–149, chapter 6, Boca Raton FL USA: CRC Press. DOI: 10.1201/9781003502012-6.

Arsova, L. (2010). Anaerobic digestion of food waste: Current status, problems and an alternative product. *Department of Earth and Environmental Engineering Foundation of Engineering and Applied Science Columbia University*. New York.

Aye, L., & Widjaya, E. R. (2006). Environmental and economic analyses of waste disposal options for traditional markets in Indonesia. *Waste Management*, 26(10), 1180–1191.

Bernstad, A., & la Cour Jansen, J. (2012). Review of comparative LCAs of food waste management systems—current status and potential improvements. *Waste Management*, 32(12), 2439–2455.

Dhar, H., Kumar, S., & Kumar, R. (2017). A review on organic waste to energy systems in India. *Bioresource Technology*, 245, 1229–1237.

Filimonau, V., & Delysia, A. (2019). Food waste management in hospitality operations: A critical review. *Tourism Management*, 71, 234–245.

Kurvinen, E., Viitala, R., Choudhury, T., Heikkinen, J., & Sopanen, J. (2020). Simulation of subcritical vibrations of a large flexible rotor with varying spherical roller bearing clearance and roundness profiles. *Machines*, 8(2), 28.

Kalogirou, E. N. (2017). *Waste-to-Energy technologies and global applications*. CRC Press.

Kibler, K. M., Reinhart, D., Hawkins, C., Motlagh, A. M., & Wright, J. (2018). Food waste and the food-energy-water nexus: A review of food waste management alternatives. *Waste Management*, 74, 52–62.

Kumar, R., Bewoor, A. K., & Alayi, R. (2021). Utilization of kitchen waste-to-energy: A conceptual note. *Renewable Energy Research and Applications*, 2(1), 91–99.

Kumar, S., Smith, S. R., Fowler, G., Velis, C., Kumar, S. J., Arya, S., ... & Cheeseman, C. (2017). Challenges and opportunities associated with waste management in India. *Royal Society Open Science*, 4(3), 160764.

Mata-Lima, H., Silva, D. W., Nardi, D. C., Klering, S. A., de Oliveira, T. C. F., & Morgado-Dias, F. (2021). Waste-to-energy: An opportunity to increase renewable energy share and reduce ecological footprint in Small Island Developing States (SIDS). *Energies*, 14(22), 7586.

Munir, K. (2022). Sustainable food waste management strategies by applying practice theory in hospitality and food services—a systematic literature review. *Journal of Cleaner Production*, 331, 129991.

Münster, M., & Lund, H. (2010). Comparing Waste-to-Energy technologies by applying energy system analysis. *Waste Management*, 30(7), 1251–1263.

Nandan, A., Yadav, B. P., Baksi, S., & Bose, D. (2017). Recent scenario of solid waste management in India. *World Scientific News*, (66), 56–74.

Närvänen, E., Mesiranta, N., Mattila, M., & Heikkinen, A. (2020). *Food waste management*. Cham: Springer International Publishing.

Naveed, S., Ramzan, N., Malik, A., & Akram, M. (2009). A comparative study of gasification of food waste (FW), poultry waste (PW), municipal solid waste (MSW) and used tires (UT). *The Nucleus*, 46(3), 77–81.

Paritosh, K., Kushwaha, S. K., Yadav, M., Pareek, N., Chawade, A., & Vivekanand, V. (2017). Food waste to energy: An overview of sustainable approaches for food waste management and nutrient recycling. *BioMed Research International*, 1, 2017, 2370927.

Prakash, S., Sharma, V. P., Singh, R., & Vijayvargy, L. (2023). Adopting green and sustainable practices in the hotel industry operations-an analysis of critical performance indicators for improved environmental quality. *Management of Environmental Quality: An International Journal*, 34(4), 1057–1076.

Priti, & Mandal, K. (2019). Review on evolution of municipal solid waste management in India: Practices, challenges and policy implications. *Journal of Material Cycles and Waste Management*, 21(6), 1263–1279.

Raychaudhuri, S., Mishra, M., Nandy, P., & Thakur, A. R. (2008). Waste management: A case study of ongoing traditional practices at East Calcutta Wetland. *American Journal of Agricultural and Biological Sciences*, 3(1), 315–320.

Salemdeeb, R., ZuErmgassen, E. K., Kim, M. H., Balmford, A., & Al-Tabbaa, A. (2017). Environmental and health impacts of using food waste as animal feed: a comparative analysis of food waste management options. *Journal of Cleaner Production*, 140, 871–880.

Sharma, K. (2020). *Waste Management and Energy Conservation is the Need for the Sustainability of Hotel Industry: An Exploratory Study of the Hotels of Hyderabad*. First Published.

Singhal, S., Deepak, A., & Marwaha, V. (2018). Green initiatives practices in Indian hotels. *IOSR Journal of Business and Management (IOSR-JBM)*, 20(8), 10–13.

Tabasová, A., Kropáč, J., Kermes, V., Nemet, A., & Stehlík, P. (2012). Waste-to-energy technologies: Impact on environment. *Energy*, 44(1), 146–155.

Tsui, T. H., & Wong, J. W. (2019). A critical review: emerging bioeconomy and waste-to-energy technologies for sustainable municipal solid waste management. *Waste Disposal & Sustainable Energy*, 1, 151–167.

Vij, D. (2012). Urbanization and solid waste management in India: Present practices and future challenges. *Procedia-Social and Behavioral Sciences*, 37, 437–447.

Chapter 21

A Study on the Use of AI in Academic Curriculum and Its Role in the Industry 5.0 Revolution

Pooja Chhabra and Sameer Babu M

21.1 ARTIFICIAL INTELLIGENCE AND INDUSTRY 5.0

The first industrial revolution (Industry 1.0) in the 18th century marked a significant transition from a handicraft economy to one dominated by machinery, impacting industries like mining, textiles, agriculture, and glass. It began in England around 1760 and spread to the United States by the late 18th century. This era introduced mechanization and the extensive use of steam power, revolutionizing production processes and leading to increased productivity. Industry 2.0, spanning from 1871 to 1914, facilitated faster transportation of people and ideas, driving economic growth and productivity gains but also causing unemployment as machines replaced manual labor (Vilone et al., 2020). The digital revolution of Industry 3.0 emerged in the 1970s, focusing on automation through memory-programmable controls and computers. This phase emphasized mass production and the integration of digital technologies like computers, integrated circuits, digital phones, and the internet into traditional products and business operations (Pathak et al., 2021). Industry 4.0 represents a fusion of physical assets with advanced technologies such as artificial intelligence (AI), Internet of Things (IoT), robots, 3D printing, and cloud computing. Looking ahead, Industry 5.0 is envisioned as a realm where humans collaborate with intelligent machines to enhance industrial efficiency across various sectors like healthcare, manufacturing, education, and more. Figure 21.1 shows the evolution of industry from 1.0 to 5.0.

21.2 ARTIFICIAL INTELLIGENCE IN HIGHER EDUCATION

AI is described as the capacity of information technology-based computer systems or other machines to perform tasks that typically require human intelligence and logical deduction. Despite its potential to enhance the world, AI is not without its challenges (Ma, 2018). Recent years have seen significant advancements in AI, positioning AI as a transformative technology that will reshape human existence. While AI is increasingly being integrated into

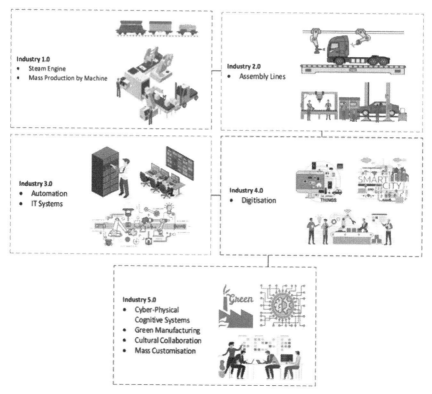

Figure 21.1 Industrial Evolution from Industry 1.0–5.0.

Future of industry 5.0 in society: human-centric solutions, challenges and prospective research areas - Scientific Figure on ResearchGate. Available from: https://www.researchgate.net/figure/Industrial-Evolution-from-Industry-10-50_fig1_363384163 [accessed 15 Jul, 2024].

higher education, many educators worldwide remain unaware of its full scope and nature (Hinojo-Lucena et al, 2019). It bridges the gap between teaching and learning processes, particularly impacting higher education in curricula and enrollment. The key strengths of AI lie in its speed, accuracy, and consistency. Furthermore, AI has the potential to address enrollment challenges stemming from wealth disparities and unemployment, offering a solution to enhance the global education system's excellence.

21.2.1 Improvement in teaching quality

Teaching quality has seen improvement through the utilization of various machine-learning tools and frameworks in education. Educators can enhance their teaching practices by utilizing extensive student data gathered from various sources. By analyzing large datasets, teachers can focus on specific concepts where students struggle and adapt new pedagogical methods to assist students in addressing challenging areas. Student data repositories help teachers gain a better understanding of student interests, thereby enabling the formation of cohorts based on shared interests for interactive learning, aiming to empower students as better problem solvers. Teachers can gain insights into student interests by examining the content students search for online, aiding in curriculum redesign. Analysis of datasets helps teachers become familiar with students' challenges, with platforms like Moodle offering a clear view of students' strengths and weaknesses through assessments in tests, quizzes, assignments, and projects. AI tools act as personalized tutors, enhancing student learning and engagement.

21.2.2 Student retention

The analysis of student data assists management and registration offices in exploring the factors contributing to student attrition, enabling them to help institutions retain students through thorough investigation and analysis of the underlying causes.

21.2.3 Increase in student enrolment

Online surveys offer valuable insights to university management regarding students' needs and expectations, thereby facilitating increased enrollment in universities and colleges.

21.2.4 Better decision-making

Management in colleges and universities regularly makes operational and strategic decisions. AI assists in making prompt and informed decisions concerning matters like fees, scholarships, and grants.

21.2.5 Easy assessment

AI tools support educators by efficiently evaluating student performance, freeing up time for professional growth, and enhancing the quality of feedback provided to students. AI aids in pinpointing discrepancies between teaching and learning processes. The extensive data in higher education offers insights into students' emotions, social interactions, aspirations, and objectives. This information enables researchers to analyze trends in student achievement longitudinally.

21.2.6 Innovative learning

AI has enabled a shift from a teacher-centric to a learner-centric educational model, fostering lifelong learning. The integration of AI culture in education is nurturing qualities like teamwork, resilience, empathy, and essential life skills in students, preparing them for meaningful contributions to society. Enhancements in both the prototype and model of data extraction are facilitating a supportive learning environment with increased ease and convenience (Huda, 2018). This technology assists educators in gaining insights into students' learning through course activities and usage data, utilizing statistical analysis and visualization techniques extensively for this purpose.

21.2.7 Identification of self-goals by students

Learning analytics empower students to take charge of their learning, offering clarity on their current performance and aiding in making informed decisions about their studies (Sclater et al., 2016). This approach, integral to Big Data in higher education, enables real-time analysis of student data, facilitating the creation of predictive models to identify at-risk students and provide timely interventions. Consequently, instructors can adjust their teaching methods, offer tailored support, assign personalized tasks, and conduct continuous assessments to enhance student learning outcomes.

21.2.8 Useful and constructive feedback

AI systems assist educators in delivering comprehensive feedback to students on their performance in both formative and summative assignments. This feedback is accessible to students at their convenience, enabling them to focus on self-improvement. Additionally, teachers can review past data to monitor student progress and performance throughout the course, facilitating reflective teaching practices and planning based on students' outcomes.

Even though AI in higher education has countless advantages, there are obstacles that must be overcome. Here are a few of them:

- Expensive technology and training for educators: The use of AI requires a large amount of time, money, infrastructure, human skills, and leadership. For the approaching industrial period, staff development needs to be given greater time and space.
- Human intelligence: AI has long struggled with the human brain's efficiency. While ChatGPT and similar apps are meant to support

educators, researchers, and students, improper use or human intervention can result in ethical and intellectual collapse.
- Lack of kinesics and social skills: Although machines are flawless in every way that they impact our daily lives, they are emotionally incapable of upholding our customs and cultures. Additionally, students use their kinesics skills less now that AI tools are available. These abilities are necessary for learners' overall growth.
- Technical difficulties: AI is a computer technology that may experience difficulties at any time that prevent it from completing tasks. If there are any technical problems, an incorrect result could occur.

21.3 IMPACT OF AI WRITING TOOLS ON THE QUALITY OF ACADEMIC WRITING AT COLLEGE LEVEL

21.3.1 Artificial intelligence technology in an academic writing class

Academic writing necessitates unique methods and techniques to ensure originality and validity in the structuring process. Students are required to conduct research, observe, and utilize numerous references to produce scholarly work. Academic writing should focus on problem-solving using presented data and facts to achieve the intended objectives (Fang, 2021).

Process writing, as highlighted by Tompkins and Hoskisson (1995), emphasizes how writers discover, develop, and revise their texts through five key steps: planning, drafting, revising, editing, and publishing (Utami, 2023). This method is crucial as shown by Kurniasih et al. (2020) to enhance effective learning of writing and improve students' skills.

The integration of AI into classroom activities is a pivotal paradigm that significantly influences language learning. This integration introduces new patterns that create a learning environment aligned with current developments, enhancing engagement and flexibility in learning (Miller & Wu, 2021; UNESCO IITE, 2020). In writing classes, AI is integrated through various applications such as AI Kaku, ChatGPT, Eskritor, Grammarly, Plot Generator, Poem Generator, Speech-to-Text, Text-to-Speech, Smodin, among others (Fitria, 2021, Kangasharju et al., 2022).

21.3.2 Students' perceptions of artificial intelligence technology

Teo (2019) emphasizes that when students engage with educational tools, they consider factors such as ease of use, usefulness, alignment with learning objectives, user technical skills, and resource availability. This aligns with the concept of the Technology Acceptance Model (TAM), which assesses the reception and utilization of technology. The TAM framework includes three

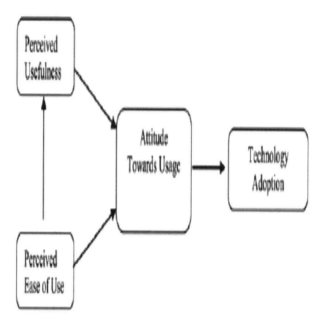

Figure 21.2 Technology acceptance model (Davis, 1989; Davis et al., 1989).

Davis, F. D., Bagozzi, R. P., & Warshaw, P. R. (1989). User acceptance of computer technology: A comparison of two theoretical models. Management Science, 35(8), 982-1003.

key user beliefs: perceived usefulness, perceived ease of use, and attitude toward usage. Perceived usefulness reflects users' confidence in technology's ability to enhance their performance and skills. Attitude toward usage represents users' inclination to respond to a specific object or situation based on their feelings (Davis, 1989; Davis et al., 1989) as is shown in Figure 21.2.

21.3.3 Writing skills in the artificial intelligence era

The advancement of teaching writing skills in the digital age is closely linked to the rapid growth of digital tools. Javaid et al. (2022) discuss how the integration of digital utilities in education is reshaping traditional teaching methods into more dynamic and creative pedagogical experiences. Garlinska et al. (2023) highlight the transformative impact of virtual classrooms, online workshops, and cloud-based writing tools on writing instruction, offering features like real-time feedback, collaborative editing, and plagiarism checks.

AI-driven platforms, as explored by Bhutoria (2022), provide personalized learning experiences by identifying students' writing strengths and weaknesses. This allows educators to customize teaching approaches to meet individual student needs, leading to improved learning outcomes (Dogan et al., 2023). Cahyono et al. (2023) delve into the use of mobile technology in teaching writing, enabling students to share their work publicly, and boosting confidence and writing abilities. Additionally, these platforms facilitate peer review and feedback mechanisms, fostering collaborative learning and a sense of community (Umamah & Cahyono, 2022).

21.3.4 The impact of AI writing tools on students writing quality

Research on the impact of AI writing tools on students' writing skills has burgeoned, with findings both positive and negative. Some studies emphasize the benefits of tools like Grammarly, QuillBot, Wordtune, and Jenni in enhancing writing abilities. For instance, Tambunan et al. (2022) demonstrated Grammarly's effectiveness in improving grammar and punctuation. These tools utilize advanced algorithms to identify errors and offer suggestions for improvement. QuillBot aids in paraphrasing to avoid plagiarism (Kurniati & Fithriani, 2022), while WordTune helps refine writing style (Lam & Moorhouse, 2022). Jenni assists in composing documents more efficiently, although specific research supporting its effectiveness is lacking. GPT-3 has been shown to stimulate creative and critical thinking (Mhlanga, 2023).

However, concerns exist. Over-reliance on AI tools may hinder students' natural learning processes and critical thinking skills (Iskender, 2023). These tools may also stifle creativity and originality (Johinke et al., 2023). AI tools may not effectively address higher-order writing elements (Farrokhnia et al., 2023), and may struggle with tone and context. Moreover, disparities in access to AI tools can exacerbate educational inequalities. Educators must consider these drawbacks while integrating AI tools into teaching strategies.

21.4 OPPORTUNITIES OF AI TECHNOLOGIES IN INDUSTRY 5.0 REVOLUTION AND SOCIETY

In our rapidly evolving technological landscape, AI emerges as a transformative influence, reshaping various sectors and societies (Peres et al., 2020). As we transition into the eras of Industry 5.0 and Society 5.0, these AI models are poised to revolutionize numerous fields (Leng et al., 2022). This segment delves into the extensive possibilities presented by ChatGPT and similar Generative AI technologies across diverse industries in the context of these transformative periods.

21.4.1 Industry 5.0: The era of human-centric manufacturing

Industry 5.0 prioritizes human skills and ingenuity in the manufacturing process. Generative AI plays a crucial role in this vision:

- Augmented Creativity: ChatGPT aids human designers by generating innovative concepts, fostering collaboration between humans and AI.
- Skill Enhancement: Generative AI serves as a repository of knowledge, bolstering problem-solving abilities and expediting learning for emerging technologies.
- Empowerment of Small-Scale Producers: Generative AI democratizes access to advanced design and manufacturing, benefiting small-scale producers and craftsmen.
- Improved Product Development: ChatGPT gathers feedback, assisting companies in developing products tailored to market needs, thereby enhancing customer satisfaction and loyalty.
- Flexible Manufacturing Systems: Generative AI facilitates adaptable production lines, optimizing resource allocation based on real-time demand.

21.4.2 Society 5.0: The human-centric society

Society 5.0 envisions a harmonious coexistence between technology and humanity, with AI addressing societal challenges:

- Education and Lifelong Learning: Generative AI customizes educational experiences and fosters a culture of continuous learning.
- Advancements in Healthcare: ChatGPT serves as a virtual healthcare aide, offering information to patients, while Generative AI assists professionals in precise diagnoses.
- Promotion of Environmental Sustainability: Generative AI analyzes environmental data, aiding in proactive environmental management and the formulation of sustainable policies.
- Promotion of Social Inclusion: ChatGPT aids individuals with disabilities and breaks language barriers, promoting inclusivity in diverse societies.
- Effective Crisis Management: ChatGPT disseminates accurate information during crises, assisting authorities in decision-making and crisis response.

21.4.3 Recommendations

When considering Industry 5.0, it is important to acknowledge the potential obstacles that must be addressed for its successful implementation in both business and education.

Employees need to develop competency skills to effectively collaborate with advanced robots. This entails acquiring both soft skills for collaboration and technical skills such as programming industrial robots and managing new job tasks that require technical proficiency.

Embracing advanced technology in Industry 5.0 demands significant time and effort from human workers. This includes implementing customized software-connected factories, integrating collaborative robots, AI, real-time information systems, and the IoT.

The adoption of Industry 5.0 technologies requires substantial investments. Training employees for new roles adds extra costs, and companies often find it challenging to upgrade production lines. Implementing Industry 5.0 is costly due to the need for smart machinery and highly skilled workers to boost productivity and efficiency.

Security poses a significant challenge in Industry 5.0, necessitating the establishment of trust within ecosystems. The utilization of AI and automation introduces threats to businesses, highlighting the importance of robust security measures. Given the focus of Industry 5.0 applications on ICT systems, stringent security requirements are necessary to mitigate potential security challenges.

21.4.4 Challenges of Industry 5.0 in education

While Industry 5.0 offers numerous potential advantages, it also presents several challenges for the education sector.

One significant challenge is preparing students for roles that have yet to emerge, as Industry 5.0 technologies rapidly evolve and the exact skills required for the future are difficult to predict. This poses a challenge for educators in adequately preparing students for the evolving job market and ensuring their competitiveness.

Another obstacle is equipping students with the essential digital skills needed to excel in an Industry 5.0 environment, such as proficiency in coding, data analysis, and machine learning. This requires making substantial investments in both teacher training and curriculum development.

Additionally, unequal access to technology and digital resources among students can create disparities in taking advantage of Industry 5.0 opportunities, particularly evident in developing countries where access to education may be limited.

There is also a concern that as intelligent technologies advance, they may be used to perpetuate discrimination or power imbalances, emphasizing the importance for educators to address these ethical issues and educate students about associated risks and responsibilities.

Adapting teaching methods to integrate Industry 5.0 technologies presents another challenge, requiring the development of new pedagogical approaches and assessment methods to evaluate students' evolving skills.

Ethical concerns related to the use of Industry 5.0 technologies, including biases and potential misuse, must also be addressed.

Finally, the integration of advanced technologies into manufacturing processes may lead to job displacement and decreased job security, particularly for workers with limited education or training.

21.5 CONCLUSION

In conclusion, the rapid advancement of AI technology is poised to significantly impact the education sector, with notable growth predicted. The potential of AI in education is vast, offering opportunities to enhance access to learning, personalize educational experiences, and optimize learning outcomes.

The synergy between AI-driven academic writing in higher education and its implications for Industry 5.0 underscores a transformative era of collaboration and innovation. As AI tools increasingly augment scholarly endeavors, the realm of academic writing experiences a paradigm shift, enabling researchers to navigate complex data landscapes, generate insightful content, and ensure integrity in their work. This not only elevates the quality and pace of academic output but also amplifies its relevance and applicability in real-world settings.

In parallel, the influence of AI-enhanced academic writing extends seamlessly into Industry 5.0, where human–machine collaboration defines the landscape. The knowledge distilled from academia becomes a valuable resource for industries, facilitating product development, problem-solving, and decision-making processes. Through this exchange, AI bridges the gap between theory and practice, fostering a dynamic ecosystem of innovation and progress.

Looking ahead, as AI technologies continue to evolve, the symbiotic relationship between academic writing and Industry 5.0 is poised to deepen further. Embracing AI-driven academic writing not only enriches scholarly pursuits but also catalyzes industrial growth and societal advancement. By leveraging the transformative power of AI, academia, and industry can co-create a future characterized by boundless opportunities, driving collective prosperity and sustainable development on a global scale.

REFERENCES

AI technologies for education: Recent research & future directions Ke Zhang, Ayse Begum Aslan b a 385Education, Wayne State University, Detroit, MI, 48202, USA b 203Boone Hall, Eastern Michigan University, Ypsilanti, MI, 48197, USA

Bhutoria, A. (2022). Personalized education and artificial intelligence in the United States, China, and India: A systematic review using a human-in-the-loop model. *Computers and Education: Artificial Intelligence*, 3, 100068. https://doi.org/10.1016/j.caeai.2022.100068

Cahyono, B. Y., Khotimah, K., & Batunan, D. A. (2023). Workable approaches in EFL teaching mediated by mobile technology during the pandemic and post-pandemic: Indonesian EFL teachers' experiences and expectations. *Computer Assisted Language Learning*, 24, 138–159. http://callej.org/journal/24-1/Cahyono-Khotimah-Batunan-Imamyartha2023.pdf

Davis, F. D. (1989). Perceived usefulness, perceived ease of use, and user acceptance of information technology. *MIS Quarterly*, *13*(3), 319–340. https://doi.org/10.2307/249008

Davis, F. D.; Bagozzi, R. P.; Warshaw, P. R. (1989). User acceptance of computer technology: A comparison of two theoretical models, Management Science, 35(8), 982–1003. doi:10.1287/mnsc.35.8.982, S2CID 14580473

Dogan, M. E., Goru Dogan, T., & Bozkurt, A. (2023). The use of artificial intelligence (AI) in online learning and distance education processes: A systematic review of empirical studies. *Applied Sciences*, 13(5), 3056. https://doi.org/10.3390/app13053056

Fang, Z. (2021). *Demystifying academic writing: Genres, moves, skills, and strategies*. Routledge. https://doi.org/10.4324/9781003131618

Farrokhnia, M., Banihashem, S. K., Noroozi, O., & Wals, A. (2023). A SWOT analysis of ChatGPT: Implications for educational practice and research. *Innovations in Education and Teaching International*, 1–15. https://doi.org/10.1080/14703297.2023.2195846

Fitria, T. (2021). Grammarly as AI-powered English writing assistant: Students' alternative for writing English. *Metathesis: Journal of English Language, Literature, and Teaching*, 5(1), 65–78. https://doi.org/10.31002/metathesis.v5i1.3519

Garlinska, M., Osial, M., Proniewska, K., & Pregowska, A. (2023). The influence of emerging technologies on distance education. *Electronics*, 12(7), 1550. https://doi.org/10.3390/electronics12071550

Hinojo-Lucena, F. J., Aznar-Díaz, I., Cáceres-Reche, M. P., & Romero-Rodríguez, J. M. (2019). Artificial intelligence in higher education: A bibliometric study on its impact in the scientific literature. *Education Sciences*, 9(1), 51.

Huda, M., Maseleno, A., Atmotiyoso, P., Siregar, M., Ahmad, R., Jasmi, K., & Muhamad, N. (2018). Big data emerging technology: Insights into innovative environment for online learning resources. *International Journal of Emerging Technologies in Learning (iJET)*, 13(1), 23–36. www.nmc.org/publication/nmc-horizon-report-2016-higher-education-edition/ retrieved on13 January 2019.

Iskender, A. (2023). Holy or unholy? Interview with open AI's ChatGPT. *European Journal of Tourism Research*, 34, 3414. https://doi.org/10.54055/ejtr.v34i.3169

Javaid, M., Haleem, A., Singh, R. P., Suman, R., & Gonzalez, E. S. (2022). Understanding the adoption of Industry 4.0 technologies in improving environmental sustainability. Sustainable Operations and Computers, 3, 203–217. doi:10.1016/j.susoc.2022.01.008

Johnke, R., Cummings, R., DiLauro, F., Johnke, R., Cummings, R., & DiLaurao, F. (2023). Reclaiming the technology of higher education for teaching digital writing in a post—pandemic world. *Journal of University Teaching & Learning Practice*, 20(2), 1–16. https://doi.org/10.53761/1.20.02.01

Kangasharju, A., Ilomäki, L., Lakkala, M., & Toom, A. (2022). Lower secondary students' poetry writing with the AI-based Poetry Machine. *Computers and Education: Artificial Intelligence*, 3, 100048. https://doi.org/10.1016/j.caeai.2022.100048

Kurniasih, Sholihah, F. A., Umamah, A., & Hidayanti, I. (2020). Writing process approach and its effect on students writing anxiety and performance. *Jurnal Arbitrer [Arbitrator Journal]*, 7(2), 144–150. https://doi.org/10.25077/ar.7.2.144-150.2020

Kurniati, E. Y., & Fithriani, R. (2022). Post-graduate students' perceptions of Quillbot utilization in English academic writing class. *Journal of English Language Teaching and Linguistics*, 7(3), 437–451. https://doi.org/10.21462/jeltl.v7i3.852

Lam, R., & Moorhouse, B. L. (2022). *Using digital portfolios to develop students' writing: A practical guide for language teachers*. Taylor & Francis. https://books.google.co.id/books?hl=id&lr=&id=b4qVEAAAQBAJ&oi=fnd&pg=PT9&dq=WordTune+AI, https://doi.org/10.4324/9781003295860

Leng, J., Sha, W., Wang, B., Zheng, P., Zhuang, C., Liu, Q., ... Wang, L. (2022). Industry 5.0: Prospect and retrospect. Journal of Manufacturing Systems, 65, 279–295. doi:10.1016/j.jmsy.2022.09.017

Ma, Yizhi and Siau, Keng L., "Artificial Intelligence Impacts on Higher Education" (2018). MWAIS 2018 Proceedings. 42. http://aisel.aisnet.org/mwais2018/42.

Mhlanga, D. (2023). Open AI in education, the responsible and ethical use of ChatGPT towards lifelong learning. *SSRN Electronic Journal*. https://doi.org/10.2139/ssrn.4354422

Miller, L., & Wu, J. G. (2021). *Language learning with technology: Perspectives from Asia*. Springer. https://doi.org/10.1007/978-981-16-2697-5

Pathak, A., Kothari, R., Vinoba, M., Habibi, N., & Tyagi, V.V. (2021). Fungal bioleaching of metals from refinery spent catalysts: A critical review of current research, challenges, and future directions. *Journal of Environmental Management*, 280, 111789.

Peres, R. S., Jia, X., Lee, J., Sun, K., Colombo, A. W., & Barata, J. (2020). Industrial artificial intelligence in industry 4.0-systematic review, challenges and outlook. *IEEE Access*, 8, 220121–220139.

Sclater, N, Peasgood, A & Mullan, J 2016, Learning analytics in higher education: a review of UK and international practice, JISC, Bristol, viewed 15 Jul 2024, www.jisc.ac.uk/reports/learning-analytics-in-higher-education.

Tambunan, A. R. S., Andayani, W., Sari, W. S., & Lubis, F. K. (2022). Investigating EFL students' linguistic problems using Grammarly as automated writing evaluation feedback. *Indonesian Journal of Applied Linguistics*, 12(1), 16–27. https://doi.org/10.17509/ijal.v12i1.46428

Teo, T. (2019). Students and teachers' intention to use technology: Assessing their measurement equivalence and structural invariance. *Journal of Educational Computing Research*, 57(1), 201–225. https://doi.org/10.1177/0735633117749430

Tompkins, G. E., & Hoskisson, K. (1995). *Language arts content and teaching strategies*. Pentice Hall.

Umamah, A., & Cahyono, B. Y. (2022). EFL university students' use of online resources to facilitate self-regulated writing. *Computer Assisted Language Learning*, 23(1), 108–124. http://callej.org/journal/23-1/Umamah-Cahyono2022.pdf

UNESCO IITE. (2020). *AI in education: Change at the speed of learning*. https://iite.unesco.org/wp-content/uploads/2020/11/Steven_Duggan_AI-in-Education_2020.pdf

Utami, S., Andayani, Winarni, R., & Sumarwati. (2023). Utilization of artificial intelligence technology in an academic writing class: How do Indonesian students perceive?. *Contemporary Educational Technology*, 15, ep450. https://doi.org/10.30935/cedtech/13419

Vilone, G., & Longo, L. (2020). *Explainable Artificial Intelligence: A Systematic Review*. https://arxiv.org/abs/2006.00093

Chapter 22

Analyze the Performance of Multileg Thermocouples with Different Thermoelectric Materials and Device Optimization Using Multiphysics Simulation

Mayank Mewara, Shobi Bagga, Sapna Gupta, and Rahul Prajesh

22.1 INTRODUCTION

Metals' thermoelectric energy converted into electrical energy is known as thermoelectric conversion. When a temperature difference is applied to a pair of thermocouples which is made up of n-type and p-type semiconductors, producing a voltage between two ends of conducting material is known as Seebeck effect as shown in Figure 22.1 (a), which is due to the phenomenon of transportation of charge carriers from the hot side to cold side under the presence of heat, conversely when voltage is applied at two ends of conducting materials and production of temperature at other ends is known as Peltier effect as shown in Figure 22.1 (b). Both effects are used in power generation also known as Seebeck effect and in refrigeration also known as Peltier effect.

The thermoelectric effect in solid state devices has various advantages like reliable and scalable and having no moving parts. The technology has been used in various fields in power generation and refrigeration modes in solar heat utilization, and thermal management in microprocessors [1].

Figure of merit (ZT) which is a key factor, the efficiency of thermoelectric devices is totally dependent is represented by formula ZT = (α^2 σ/κ) Figure of merit is dimensionless where T, κ, σ and are the absolute temperature, thermal conductivity and electrical resistivity [2]. For the realization of high performance of TE materials associated recommended parameters as per the formula are low thermal conductivity and large Seebeck coefficient with high electrical conductivity. ZT value is nearly equal to unity for good TE materials. Figure of merit is a key factor for consideration in power generation through thermoelectric converters and a value is up to three is necessary and is the same for the refrigeration process.

The thermoelectric figure of merit can be enhanced on the basis of three parameters: Electrical conductivity, thermal conductivity, and Seeback

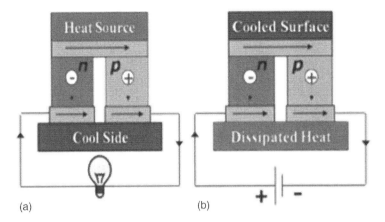

Figure 22.1 (a) Seebeck effect and (b) Peltier effect.

coefficient and is desirable that the Seebeck coefficient is high, electrical conductivity is also high, and thermal conductivity is low [3]. All three parameters are interrelated to each other through charge carrier concentration [4] so it is difficult to improve in all three simultaneously.

22.2 LITERATURE REVIEW ON THE BASIS OF MATERIALS AND DEVICE OPTIMIZATION GEOMETRIES

In the 1960s, bismuth telluride (Bi_2Te_3)-based alloys were discovered and used as a high-performance thermoelectric material up to the last century and continuous progress in the development of thermoelectric material has been made. In the 1990s, figure of merit nearly equal to one was achieved which broke all benchmarks.

Research approaches with complex crystalline structures of materials and by dimension reducing of the materials like skutterudites [5, 6] and clathrates [7], by the suggestion of Slack [8] is only motivated. In those compounds, atoms are loosely bonded within a cage generating scattering against the propagation of lattice phonons due to the rattling motion. This creates less effect on the transportation of electrons so maintaining the maximum electrical conductivity and reduces the thermal conductivity of skutterudites and clathrates. Zintl compounds [9] due to the high fraction of low-velocity optical phonon modes [10], with a large unit cell, including SrZnSb2, Yb11GaSb9, Ca11GaSb9 also possess low lattice thermal conductivity. Hicks and Dresselhaus [11] suggest that with the quantum confinement effects power factor ($α2\ σ$) can be enhanced. They worked with nanowires, quantum dots, and low-dimensional systems. Venkatasubramanian et al.

[12] reported a high ZT value of up to *2.4 with superlattices of Bi_2Te_3/ Sb_2Te_3*. Harman et al. [13] reported against ZT value of greater than 3 at 600°C with quantum dot superlattices of PbTe/PbTeSe.

Enhancing efficiency and power output for thermoelectric generator geometry optimization is a way to fulfill this [14]. Ferreira-Teixeira and Pereira [15] use the finite element method with numerical simulation for the geometry optimization of a thermoelectric generator. Fabian-Mijangos et al. [16] give an experimental study with asymmetrical legs of the thermoelectric generator and find double ZT of the TEG with asymmetrical legs in comparison to square legs; thus, the thermoelectric generator with asymmetrical legs performed better than the conventional thermoelectric generator. Ma et al. [17] conclude that an optimal electric conducting plate thickness for obtained maximum power output of thermoelectric generators using the geometry optimization technique. Li et al. [18] conclude that geometry optimization of TEG leg length and area significantly affects the efficiency of the hybrid photovoltaic-thermoelectric (PV-TE). Sun et al. [19] concluded that geometry optimization is a very important factor in development of highly efficient and compact thermoelectric generators.

The efficiency of TEG is estimated by, ZT (figure of merit) = $(\alpha 2\ \sigma/\kappa)$ where, σ, k, T, and S denote electrical conductivity, thermal conductivity, working temperature, and Seebeck coefficient, respectively. Improving in final ZT is challenging because these parameters are strongly associated with each other. For boosting the performance of TEGs different Strategies are included such as nanomagnetic compositing, high-throughput screening, nanostructuring, band engineering, and others [20]. A narrow range of temperatures using these strategies creates high ZT values but limits the overall energy conversion. Rethinking development strategies is required with new materials with a wide operating range. For making thermoelectric devices many p-type and n-type thermocouples are used in a cascading manner and efficiency is mostly dependent on thermoelectric material. According to the range of temperature range of their working, they are categorized in three parts. Below 400 K bismuth tellurides are operated. Lead chalcogenides are between 600 and 900 K temperature range, Lead chalcogenides are operated and above 1000 K $Zint_1$ phases, silicon-germanium gives the best performance. Bandgap (Eg) owing to intrinsic excitation decides the performance of thermoelectric material at optimal temperature.

This bandgap rule is denoted by Eg = 2eST, where S denotes the maximum Seebeck coefficient, the temperature is denoted by T, S indicates the Seeback coefficient and e denotes a unit charge. The Seebeck coefficient is known as power generated by the thermocouple unit and is a measurement of the voltage produced by the temperature gradient (S = $\nabla V / \nabla T$). The bandgap rule means that the narrow-bandgap semiconductor materials like (Bi, Sb)$_2$Te$_3$ (Eg ~ 0.13 eV), and PbTe (Eg ~ 0.28 eV) are well-known thermoelectric materials but as the bandgap increases these thermoelectric materials with maximum ZT values shifted to higher temperatures. These temperature range change issues are solved by wide-bandgap semiconductors but often have poor electrical properties, and wide-bandgap semiconductors were neglected as promising thermoelectric at lower temperatures. To achieve high electrical transport properties and low thermal conductivity, low symmetry crystalline structures are required which is found in SnSe. Due to this SnSe (~0.86 eV) possesses a wide bandgap that has proven to be an excellent thermoelectric material at lower temperatures. Selection rules for identifying highly effective thermoelectric materials appropriate for both P-type and N-type materials because of their similar transport principle.

Therefore, at lower temperature ranges, materials with different bandgaps are analyzed in this chapter. As bandgap increases material performance degrades but wide gap material SnSe shows better performance at lower temperatures. Experimentally, a lot of work has been done on thermoelectric materials and devices compared to numerical analysis with simulation which could be operated at a very low cost.

The objective of the chapter is as follows: (1) At lower temperature range, investigating influences of different thermoelectric materials from lower band gape towards higher band gape Bi_2Te_3 (Eg = 0.13 eV), PbTe (Eg = 0.28 eV) SnSe (=0.86 eV) on voltage sensitivity (S = $\nabla V / \nabla T$) and compare (2) influences of component geometry with different area-to-length ratio of the semiconductors and thickness of electrical conducting plates are also investigated which impacts the performance of the device. (3) Compare the performance of a device with horizontal and vertical models with same dimensions and materials.

22.3 MATHEMATICAL MODEL

In this chapter, differential equations are used for analysis which is a part of COMSOL Multiphysics [21]. In this chapter, thermoelectric modules are used with field variable temperature T, and voltage V, [22, 23] where heat and Poisson's equations are used simultaneously to solve these parameters. To obtain device parameters, partial differential equations are used as follows:

$$-\nabla\left((\sigma\alpha^2 T + k)\right)\nabla T - \nabla(\sigma\alpha\nabla V) = \sigma\left((\nabla)^2 + \alpha\nabla T\nabla V\right) \quad (22.1)$$

and

$$\nabla(\sigma\alpha\nabla T) + \nabla(\sigma\nabla V) = 0 \quad (22.2)$$

where α is the Seebeck coefficient, thermal conductivity is denoted by k, and σ represents the electrical conductivity of the material. For simulation, COMSOL Multiphysics is used.

22.4 STRUCTURE OF DEVICE

The structure of each device was completed in COMSOL Multiphysics where properties of thermoelectric and conducting plate materials were used to build the model. Thermoelectric materials are Bi_2Te_3, PbTe, and SnSe and electrical conducting plates are made of Cu, Ag, and Au. Due to the selection rules for choosing efficient thermoelectric material between three bandgap materials, n-type and p-type legs have been chosen as thermoelectric parts of the device because of their similar transport principle and to compare the result. Also we chose the best material among the three, mentioned above for electrical conducting plates by comparison of results in terms of electrical conductivity.

The device consists of hot and cold sources with several thermoelectric units connected in the middle that act as a thermoelectric module. Each thermoelectric module is a combination of two main components: electrical conducting plates in the upper and lower side, and n-type and p-type semiconductors in the middle.

Using the geometry shown in Figure 22.1, three-dimensional simulations are performed. The size of each part is indicated as width (w) × depth (D) × height (H). Size of pair of thermoelectric legs are width = 1.2 inches (30480 μm), depth = .015 inch (400 μm), height = 1.6 inches (40640 μm). The size of both the upper and lower conducting plates that join the thermoelectric legs is 3.4 × .015 × 1.5 inches and extreme left and right conducting plates have the dimensions of 1.7 × .015 × 1.5 inches. Temperatures T_1 and T_2 are applied to the hot and cold sides of the thermoelectric unit, respectively, and get the sensitivity toward terminal 1 and terminal 2 of the design. For the other side surfaces, adiabatic boundary conditions are assumed. At the outer surface (O) where we provide temperature T_1 and at the inner surface (I) we provide temperature T_2.

22.5 RESULTS AND DISCUSSION

22.5.1 Using different thermoelectric and electrical conducting plate materials

The model is simulated in COMSOL. The values of α, k, and σ used in the simulation analysis are obtained from previously published literature and listed in Table 22.1. In this device, the inner junctions, denoted as (I) in Figure 22.2, were kept at 293 K, while the outer junctions indicated as

Table 22.1 Material properties used in the simulation

Material	Bi2Te3	PbTe	SnSe	Cu	Ag	Au
α (μV/K)	(−200) N (200) P	−187 (N) 187 (P)	−140 (N) 140 (P)	6.5	6.5	6.5
k (W/m·K)	1.75	1.46	.008	350	429	317
Σ = 1/R (Ω·m)	$1/40 \times 10^{-6}$	6.097×10^4	300	5.99×10^8	61.6×10^6	45.6×10^6

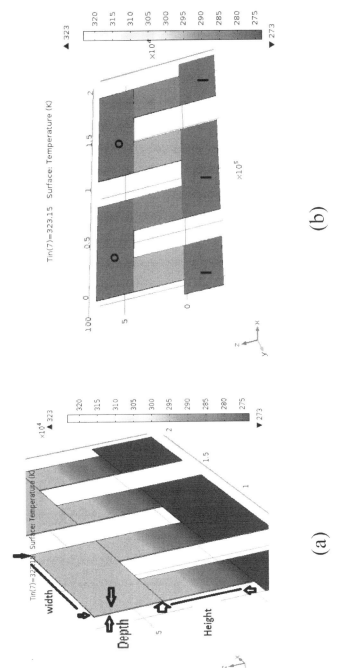

Figure 22.2 (a) Shows dimensions and (b) shows materials. Blue: Conducting plates, Green: n-type thermoelectric material, Red: p-type thermoelectric material shows outer and inner surface and terminal 1 and 2 are toward extreme left side and right side of lower conducting plates.

Table 22.2 Voltage sensitivity (S = ΔV/ΔT) with different materials using COMSOL simulation

Material	Slope = ΔV/ΔT
Bi_2Te_3	7.953e-4
PbTe	7.442e-4
SnSe	5.599e-4

(O) temperature varied from 293 K to 323 K. Outer junctions' temperature with an increment of 5 degrees varied between T = 293 K to T = 323 K. This resulted in temperature difference from 293 to 323 = 30 K at 5-degree slot increment. Voltage difference generated across terminals 1 and 2 due to temperature difference across terminals 1 and 2 and voltage values are plotted against the temperature, shown in Figure 22.2, and compared the material performance.

As shown in Figure 22.3, the graph between the applied temperature and generated voltage is slightly nonlinear, and the relationship can be approximated as a straight line. Voltage sensitivities (S = ∇V / ∇T) for the device are represented by the slope of these lines. Table 22.2 shows device voltage sensitivities are different with different materials. For Bi_2Te_3, PbTe, and SnSe, voltage sensitivities are 795 µV/K, 744 µV/K, and 559 µV/K, respectively.

Based on comparison, the best thermoelectric material is Bi_2Te_3. This thermoelectric material was used with different electrical conducting plate materials and the comparative results as shown in Figure 22.4. Voltage sensitivities are shown in Table 22.3.

Based on comparison, Ag is the best but due to its high cost Cu is generally chosen as a conducting plate material.

22.5.2 Influences of area-to-length ratio

- With fixed base area and variable length of thermoelectric legs

Figure 22.5 shows voltage sensitivity for fixed base area and variable length. With increasing leg length, sensitivity also increases. Table 22.4 shows sensitivity with variable leg lengths.

- With fixed base area and variable height of electrical conducting plates

Figure 22.6 shows sensitivity with the fixed base area and a variable height of upper and lower conducting plates, with increasing height resulting in decrease in sensitivity. A comparison is shown in Table 22.5.

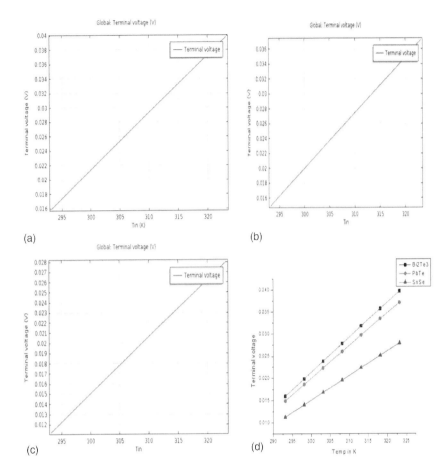

Figure 22.3 Generated voltage across device terminals 1 and 2 with temperature difference (T_1-T_2) for thermoelectric materials (a) Bi_2Te_3, (b) PbTe, (c) SnSe with same electrical conducting plate material (cu) with the inner junctions temperature is fixed at 293 K, and (d) shows the comparison.

- Comparison of the performance of the device with the horizontal and vertical COMSOL models with the same dimensions and materials.

In this comparison, thermoelectric material is PbTe, Cu is the electrical conducting plate, and the dimensions are as follows:

Thermoelectric (PbTe): W = 0.12 inch, D = 0.12 inch, H = 0.16 inch
Conducting Plate (Cu): Upper W = 0.34 inch, D = 0.12 inch, H = 0.12 inch
Lower W = 0.17 inch, D = 0.12 inch, H = 0.12 inch

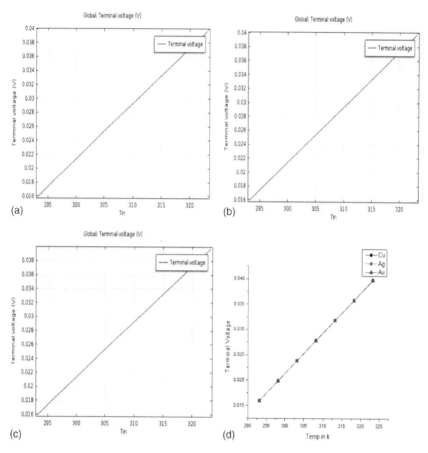

Figure 22.4 Voltage sensitivity of Bi_2Te_3 with electrical conducting plate material (a) Cu, (b) Ag, (c) Au, and (d) shows comparison between them.

Table 22.3 Voltage sensitivity of Bi_2Te_3 with different electrical conducting plate materials

Terminal material	Slope = $\Delta V/\Delta T$
Cu	7.95e-4
Ag	7.96e-4
Au	7.94e-4

Analyze the Performance of Multileg Thermocouples 355

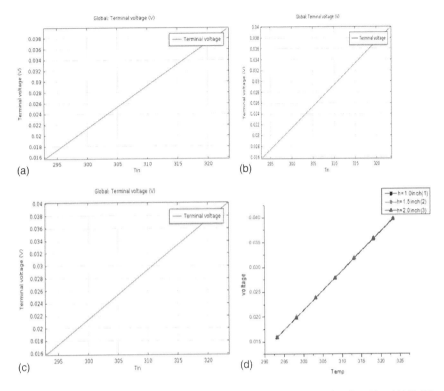

Figure 22.5 Fixed base area variable leg length (a) W = 1.2 (30480 μm) inches, D = (.015)400 μm, H = 1.0 inch (25400 μm), (b) W = 1.2 (30480 μm) inches, D = (.015)400 μm, H = 1.5 inches (38100 μm) (c) W = 1.2 (30480 μm) inches, D = (.015)400 μm, H = 2.0 inches (50800 μm), and (d) comparison b/w three values.

Table 22.4 Shows comparison of sensitivity with variable leg heights

Height	Slope = $\Delta V/\Delta T$
1.0 inch	7.939e-4
1.5 inches	7.9590e-4
2.0 inches	7.9691e-4

The temperature of the inner surface is constant at 293 K and the outer surface is varied up to 323 K and sensitivity is observed (Figures 22.7 and 22.8). Figure 22.6 shows a temperature comparison.

- Horizontal Model
- Vertical Model

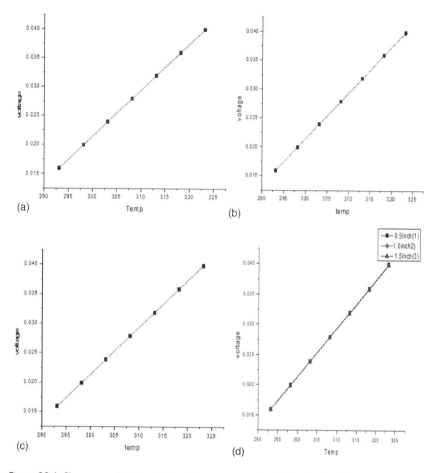

Figure 22.6 Shows sensitivity with fixed base area and variable height of upper and lower conducting plate (a) H = 0.5 inch, (b) H = 1 inch, (c) H = 1.5 inches, and (d) comparison.

Table 22.5 Comparison between the variable height of upper and lower electrical conducting plates

Height	Slope = $\Delta V/\Delta T$
0.5 inch	7.985e-4
1.0 inch	7.97e-4
1.5 inches	7.962e-4

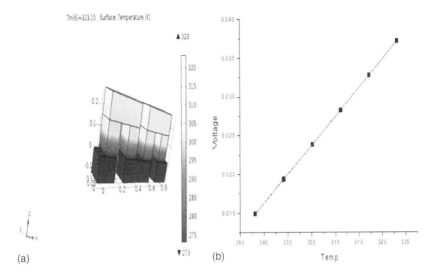

Figure 22.7 Horizontal Model (a) Device model and (b) Graph between voltage and temperature.

22.6 CONCLUSIONS AND FUTURE WORKS

Based on physical simulation using COMSOL, thermoelectric unit performance is investigated with variation in temperature, variation in area-to-length ratio, and thickness of the electrical conducting plates.

With respect to thermoelectric material, Bi_2Te_3 is best compared to PbTe and SnSe at a lower temperature range. From the lower bandgap material (Bi_2Te_3) toward the higher bandgap material (SnSe) sensitivity degraded. So, for lower-bandgap semiconducting material (Bi_2Te_3), a lower temperature is required for the movement of carriers from valance to conduction band and is available as a free carrier for current conduction. SnSe exhibits a higher bandgap compared to the other two materials, leading to a higher temperature requirement for achieving the same performance level. Generally, SnSe shows better performance for more than 800 K. Regardless of temperature, Z (figure of merit) denotes the efficiency of material comprised of three parameters. As per Table 22.1, it can be concluded that Z is more dependent on the Seebeck coefficient and electrical conductivity because it gives a power factor which is formulated as $s^2\sigma$, so the Seebeck coefficient and electrical conductivity of Bi_2Te_3 is best among three. As we increase the length of the thermoelectric leg for a fixed base area, performance is enhanced because internal resistance decreases and is less dependent on temperature compared to the inner and outer plates of electrical conducting plates which are in direct contact with applied temperature as the area decrease of conducting

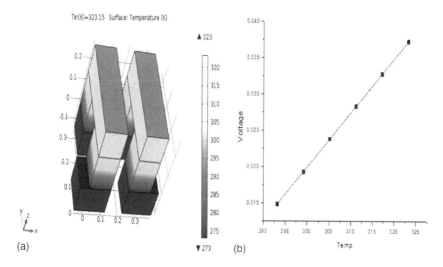

Figure 22.8 Vertical Model (a) Device model and (b) Graph between voltage and temperature.

Table 22.6 Comparison between horizontal and vertical geometry

Geometry	Slope = $\Delta V/\Delta T$
Horizontal	7.446e-4
Vertical	7.442e-4

plates carrier movement increases within the conducting plate. When we compare the horizontal model with the vertical with the same material and dimensions, the horizontal gives better results. According to fabrication point of view, it is easier to make and it is not more dependent on depth, so at minimum depth it gives approximately the same performance. Fabrication with minimum depth is easier to make with a horizontal model compared to a vertical model. For future work we can analyze the transient response of the device but for this transient response is required in the time domain.

REFERENCES

1. L. E. Bell, Science 321, 1457 (2008).
2. A. F. Loff, L. S. Stil'bans, E. K. Iordanishvili, T. S. Stavitskaya, A. Gelbtuch, George Vineyard, *Semiconductor Thermoelements and Thermoelectric Cooling.* Physics Today. 12(5), 42 (1 May 1959). https://doi.org/10.1063/1.3060810.

3. M. H. Elsheikh, D. A. Shnawah, M. F. M. Sabri, S. B. M. Said, H. M. Hassan, A. M. B. Bashir, M. Mohamad, Renew. Sustain. Energy Rev. 30, 337–355 (2014). https://doi.org/10.1016/j.rser.2013.10.027
4. M. Bittner, N. Kanas, R. Hinterding, F. Steinbach, J. Räthel, M. Schrade, K. Wiik, M. A. Einarsrud, A. Feldhoff, J. Power Sources 410–411, 143–151 (2019). https://doi.org/10.1016/j.jpowsour.2018.10.076
5. B. C. Sales, D. Mandrus, R. K. Williams, Science 272, 1325 (1996).
6. G. S. Nolas, D. T. Morelli, T. M. Tritt, Annu. Rev. Mater. Sci. 29, 89 (1999).
7. G. S. Nolas, J. L. Cohn, G. A. Slack, S. B. Schujman, Appl. Phys. Lett. 73, 178 (1998).
8. G. A. Slack, CRC Handbook of Thermoelectrics, D. M. Rowe ed. (CRC Press, USA, 1995).
9. S. M. kauzlarich, S.R Brown, G. J synder, Dalton T-21,2099(2007).
10. E. S. Toberer, A. F. May, G. J. Snyder, Chem. Mater. 22, 624 (2010).
11. L. D. Hicks, M. S. Dresselhaus, Phys. Rev. B 47, 12727 (1993).
12. R. Venkatasubramanian, E. Siivola, T. Colpitts, B. O'Quinn, Nature 413, 597 (2001).
13. T. C. Harman, P. J. Taylor, M. P. Walsh, B. E. LaForge, Science 297, 2229 (2002).
14. G. Li, S. Shittu, X. Ma, X. Zhao, Energy 171, 599–610 (2019). https://doi.org/10.1016/j.energy.2019.01.057
15. S. Ferreira-Teixeira, A. M. Pereira, Energy Convers. Manag. 169, 217–227 (2018). https://doi.org/10.1016/j.enconman.2018.05.030
16. A. Fabian-Mijangos, G. Min, J. Alvarez-Quintana, Energy Convers. Manag. 148, 1372–1381 (2017). https://doi.org/10.1016/j.enconman.2017.06.087
17. Q. Ma, H. Fang, M. Zhang, Appl. Therm. Eng. 127, 758–764 (2017). https://doi.org/10.1016/j.applthermaleng.2017.08.056
18. G. Li, X. Zhao, Y. Jin, X. Chen, J. Ji, S. Shittu, J. Electron. Mater. 47, 5344–5351 (2018).
19. H. Sun, Y. Ge, W. Liu, Z. Liu, Energy 171, 37–48 (2019). https://doi.org/10.1016/j.energy.2019.01.003
20. X. Shi, L. Chen, Nat. Mater. 15, 691 (2016).
21. E. E. Antonova, D. C. Looman, 24th International Conference on Thermoelectrics, pp. 215–218 (19–23 June 2005).
22. M. Jaegle, Multiphysics Simulation of Thermoelectric Systems—Modeling of Peltier-Cooling and Thermoelectric Generation (COMSOL Conference, Hannover, 2008).
23. COMSOL Multi-Physics. https://www.comsol.com

Chapter 23

Multi-Objective Optimization Strategy for Enhanced Accuracy and Productivity in AWJ Machining

Abhijit Saha, Rajeev Ranjan, Ashok Kumar Yadav, Ashwani Kumar, and Payal Bansal

23.1 INTRODUCTION

Cutting using an abrasive water jet (AWJ) is a very popular contemporary machining process used to cut tough materials including aluminum alloys, stainless steel, and titanium. Applications for it may be found in the automotive and aerospace sectors. Abrasive particles combined with high-pressure water are used as a cutting tool in AWJ cutting to dissolve the material. This method has benefits including low cutting forces, a smaller heat-affected zone, and no breakable contact instruments. Hydraulic, cutting, mixing, and abrasive considerations are all part of AWJ cutting parameters. Analysis of many responses, such as roughness, kerf width, cutting rate (CR), and depth of penetration (DOP), is made challenging by the use of AWJ. In the manufacturing sectors' quest for superior product quality with better surface finishes, less taper, and increased DOP, the right choice of process parameters is critical. To meet these objectives, decision-making methodologies are crucial. Due to its challenging machinability utilizing conventional and unconventional techniques, the AWJ cutting process is chosen for milling AA5083-H32. There is, however, a paucity of data on abrasive contamination and machinability unique to AWJ cutting of this alloy. As a result, this work marks the first attempt to use the Technique for Order Preference by Similarity to optimize the AWJ cutting inputs in AA5083-H32. The effects of AWJM on various materials have been the subject of several investigations. When an AWJ was used to cut a ceramic plate, Hochberg and Chang [1] observed that the kerfs' breadth rose with pressure, abrasive flow rate, abrasive size, and traverse speed. Kerfs taper ratio and surface roughness of the epoxy/glass composite laminate were optimized using the Taguchi technique by Azmir and Ahsan [2]. In their study of the impact of traverse speed on the AWJ machining of Ti-6Al-4V alloy, Hascalik et al. [3] discovered that the kerfs taper ratio and surface roughness increase as traverse speed increases. When comparing the effectiveness of several abrasive materials during AWJM of glass, Khan and Hague [4] discovered that garnet abrasives created the biggest taper of cut. The impact of orifice and focusing nozzle diameter on the cutting capabilities of 6063-T6 aluminum alloy was examined by John Rosario Jegaraj and Ramesh Babu [5] who also recommended the best combinations for attaining the greatest depth of cut. Based on a statistically planned experiment for AWJ cutting of metallic coated sheet steels, Wang and Wong [6] created empirical models for kerfs shape and quality. The effects of various chemical environments on material removal rates (MRRs) and taper angles in AWJ machining were examined by Mahabalesh Palled [7]. Material removal factor (MRF) was proposed as a non-dimensional parameter by Ray [8] after analyzing the impact of various factors on MRR. The Taguchi approach was employed by Nagdev and Chaturvedi [9] to optimize the AWJM process parameters for Al-7075, and they discovered that traversal speed was the most efficient parameter. The impact of abrasive flow rate, jet pressure, and work feed rate on the enhancement of carbide surfaces after AWJ machining was assessed by Khan et al. [10]. The flexible magnetic abrasive was used in AJM by Kea et al. [11] to improve surface roughness. The effect of AWJM process parameters on the surface finish of glass fiber-reinforced epoxy composites was studied by Azmir and Khan [12]. In their study of the profiles of machined surfaces, microstructural characteristics, and kerf geometries of Ti-6Al-4V alloy that had been AWJ-machined, Hascalik et al. [13] discovered that increasing traverse speed led to smaller kerfs width with a higher kerfs taper ratio. Using the Taguchi method, Azmir et al. [14] investigated how the AWJ machining input parameters affected the kerfs taper ratio and surface roughness of Kevlar Reinforced Phenolic Composite. When studying the AWJ machining method for various materials, Limbachiya and Patel [15] discovered that the theoretical and experimental MRRs were in agreement.

Most researchers [16–30] have successfully used the multi-criteria decision-making methodology (MCDM) to tackle the sequence of process parameter selection problem. To assess the cut quality of a product based on the surface roughness (R_a), top kerf width (TKW), CR, and DOP, this study combines the MCDM and analysis of variance (ANOVA) [31–40]. The workpiece material, AA5083-H32, is an aluminum alloy that offers improved strength and corrosion resistance due to its chemical makeup, which is predominantly composed of aluminum and magnesium. It has uses in railway, automobile, aircraft, and maritime components.

23.2 MULTI-RESPONSE OPTIMIZATION: DESIRABILITY FUNCTION ANALYSIS

The industry uses the desirability function technique frequently to improve processes with various response variables. This approach is predicated on the idea that any aspect of a product's or process' quality that deviates significantly from expected parameters is wholly unacceptable. The optimization process may be made simpler by using desirability analysis to combine many answer factors into a single composite desirability. By concentrating on a single composite desirability, this method, also known as the Desirability Function Analysis (DFA), enables the optimization of complicated multi-response features. The steps needed to solve the optimization problem are explained in [41–50. For each machining output, such as cutting power, specific cutting pressure, cutting force, and surface roughness the desirability index is first determined. Depending on the properties of the response variables, the desirability functions can take one of three different forms [51–53.

Derringer and others (1980) put out a practical class of desirability functions, such as

Step 1:

- The individual desirability function of a response of the "target is best" kind is provided by equation (23.1).

$$d_i(\hat{Y}_i) = \begin{cases} 0 \\ \left(\dfrac{\hat{Y}_i(x) - L_i}{T_i - L_i}\right)^s & \text{if } \hat{Y}_i(x) < L_i \\ & \text{if } L_i \leq \hat{Y}_i(x) \leq T_i \\ \left(\dfrac{\hat{Y}_i(x) - U_i}{T_i - U_i}\right)^t & \text{if } T_i \leq \hat{Y}_i(x) \leq U_i \\ 0 & \text{if } \hat{Y}_i(x) > U_i \end{cases} \quad (23.1)$$

where L_i, U_i, and T_i represent the response Y_i's intended lower, higher, and goal values, with L_i, T_i, U_i, respectively.

- The greater, the better instead, if a response is to be maximized, equation (23.2) defines the individual desirability.

In this instance, T_i was regarded as a value that was sufficient for the response.

$$d_i(\hat{Y}_i) = \begin{cases} 0 \\ \left(\dfrac{\hat{Y}_i(x) - L_i}{T_i - L_i}\right)^s & \text{if } \hat{Y}_i(x) < L_i \\ & \text{if } L_i \leq \hat{Y}_i(x) \leq T_i \\ 1.0 & \text{if } \hat{Y}_i(x) > T_i \end{cases} \quad (23.2)$$

- The smaller the reaction, the better: If the response is to be minimized, equation (23.3) defines the individual desirability.

$$d_i(\hat{Y}_i) = \begin{cases} 1.0 & \text{if } \hat{Y}_i(x) < T_i \\ \left(\dfrac{\hat{Y}_i(x) - L_i}{T_i(x) - U_i}\right) & \text{if } T_i \leq \hat{Y}_i(x) \leq U_i \\ 0 & \text{if } \hat{Y}_i(x) > U_i \end{cases} \quad (23.3)$$

T_i signifying a minimal value required for a response

The "smaller-the-better" criterion is used in this study to evaluate the unique desirability values of each machining response, including cutting power, specific cutting pressure, cutting force, and surface roughness. A desirability function $d_i(Y_i)$ is used to assign a number between 0 and 1 to the potential values of Y_i for each answer $Y_i(x)$. A Y_i value of 0 denotes a very unfavorable Y_i value, whereas a Y_i value of 1 denotes an excellent or entirely desired response value.

Step 2: The geometric mean is used to aggregate each individual's desirability, producing the composite desirability, which is determined using equation (23.5).

$$D = \left(d_1(Y_1) \times d_2(Y_1) \times \ldots \times d_k(Y_k)\right)^{1/k} \quad (23.4)$$

where "k" stands for the number of replies. Note that the composite desirability is equal to zero if any response Y_i is wholly unpleasant ($d_i(Y_i) = 0$). In actual usage, fitted response values (i) are substituted for the Y_i.

Step 3: Based on the maximum value of composite attractiveness, choose the best combination of machining settings. Additionally, an estimate is made of how machining factors affect the results.

23.3 MATERIALS AND METHODOLOGY

23.3.1 Materials Used

By adding additives, polymer composites' characteristics can be improved. In this investigation, a glass fiber-reinforced epoxy laminate was filled with a particle form of graphite with a size of 200 m. An epoxy resin with a biphenyl-A base served as the matrix material, and a bidirectional glass fabric weighing 200 g/m^2 was used for reinforcement. The glass epoxy laminate's wear resistance, lower friction coefficient, and higher toughness were all enhanced by the addition of graphite. Additionally, graphite served as a solid lubricant, improving the laminate's machinability. Garnet abrasive with an 80 mesh was used for the machining process.

23.3.2 Fabrication of Composite Specimen

Glass fibers made up 50% of the composite's weight, epoxy made up 47% of its weight, and graphite made up 3% of the total weight. A square of 250 mm × 250 mm bidirectional glass cloth was cut into that form. The cloth was then coated with epoxy that had been packed with graphite and placed on top of one another. The finished green laminate was then put into a hot compression molding machine and compressed consistently. Curing was conducted at 140°C in compression mode for two hours. The laminate was then post-cured for 8 hours at 180°C while it was floating freely in a furnace. The final thickness of the laminate was determined to be 3 ± 0.1 mm.

23.3.3 Characterization of Composite and Garnet Particles

The constructed composite has been cut into specimens in accordance with American Society for Testing and Materials (ASTM) D638's specifications. Instron3366 universal testing equipment was used to assess the tensile characteristics. Using a Matsuzawa-MMTX7A computerized microhardness tester, the composite's hardness was evaluated. A scanning electron microscope (ZeissEVO 18, Germany) was used to look at the morphology of the garnet abrasive and AWJ machined surfaces. According to ASTM D422 standard, garnet abrasive' particle size analysis was done. Using the elemental mapping approach with SEM, the chemical makeup of the garnet abrasive was identified. Using a Mitutoyo tool room microscope, the kerf widths of the machined surfaces were measured, and Subtonic 3+ (Taylor Hobson) was used to assess the surface roughness.

23.3.4 Plan of Experiments: Design of Experiment (DOE)

A reliable statistical approach for economically optimizing machining process parameters is the use of design of experiments (DOE). DOE has been used by several researchers to optimize the AWJ process parameters for cutting various materials. Following a thorough literature analysis and pilot tests, experiments in this study were designed utilizing Taguchi's fractional factorial orthogonal array. These studies served as the foundation for choosing the AWJ machining parameters and their corresponding levels. The machining parameters used to evaluate the effectiveness of AWJ machining on graphite-filled GFRP composites are shown in Table 23.1. Material mechanical properties and mineral compositions are shown in Tables 23.2 and 23.3, respectively.

Table 23.1 Machining parameters

Machining parameter/level	1	2	3
Standoff distance (mm)	1	3	5
Abrasive concentration (wt %)	6	10	14
Feed rate (mm/min)	75	100	125
Operating pressure (MPa)	90	120	150

Table 23.2 Mechanical properties of the target material

Tensile strength	Young modulus	% Elongation	Hardness
255 MPa	8.3 GPa	3.2%	VHM30

Table 23.3 Mineral composition of garnet abrasive

Element	C	O	Mg	Al	Si	Ca	Fe
Quantity (wt.%)	6.62	48.84	1.93	5.82	9.13	1.23	32.45

The standoff distance (SOD) (A), abrasive concentration (B), feed rate (C), operating pressure (D), as well as the interaction effects of AB, AC, and BC, were taken into consideration with regard to the response parameter, kerf width when designing the trials. There were a total of 21 degrees of freedom in the experimental setup, which suggests that at least 21 tests would be required to adequately assess the effects. The most appropriate design for this was the L27 fractional orthogonal array, which supports four components with three levels each.

It is important to note that garnet was found to be the best abrasive for AWJ machining of polymer composites, with the highest material removal from the workpiece being seen at a jet impact angle of 90° [15]. In this study, water was used to evenly scatter irregular and angular garnet abrasive particles before the jet was blasted from the nozzle at 90° angle for the machining operation.

23.4 EXPERIMENTAL SETUP

The study made use of an experimental setup with a three-axis CNC AWJ machine (Table 23.4). The machine, model DWJ1525-FA, is made by Dari Water Cutter Co. Ltd. and has a repeatable precision of 0.025 mm. It runs at a maximum pressure of 400 MPa. A garnet-AWJ was applied to the graphite glass epoxy composite surface at a 90° angle through a carbide nozzle with a 0.8 mm focus tube diameter. A 20 mm cut was made to the workpiece for each trial (Figure 23.1).

Table 23.4 Plan of experiment and measured output parameters

Trial	SOD (mm)	Abrasive concentration (Wt %)	Feed rate (mm/min)	Pressure (MPa)	TKW (mm)	BKW (mm)
1	1	6	75	90	0.845	0.803
2	3	10	75	90	0.905	0.892
3	5	14	75	90	0.964	0.936
4	3	6	100	90	0.896	0.853
5	5	10	100	90	0.950	0.912
6	1	14	100	90	0.847	0.827
7	5	6	125	90	0.870	0.835
8	1	10	125	90	0.823	0.816
9	3	14	125	90	0.860	0.850
10	3	6	75	120	0.960	0.913
11	5	10	75	120	1.176	1.150
12	1	14	75	120	0.890	0.868
13	5	6	100	120	1.104	1.055
14	1	10	100	120	0.875	0.845
15	3	14	100	120	0.981	0.977
16	1	6	125	120	0.855	0.803
17	3	10	125	120	0.917	0.852
18	5	14	125	120	0.987	0.935
19	5	6	75	150	1.203	1.155
20	1	10	75	150	0.938	0.918
21	3	14	75	150	1.052	1.005
22	1	6	100	150	0.903	0.872
23	3	10	100	150	0.965	0.887
24	5	14	100	150	1.150	1.015
25	3	6	125	150	0.892	0.880
26	5	10	125	150	1.065	0.992
27	1	14	125	150	0.882	0.871

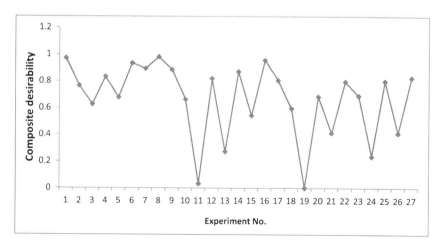

Figure 23.1 Composite desirability vs. experiment number.

23.5 RESULT AND DISCUSSION

In this study, many criteria, including individual and composite attractiveness, were used to evaluate the machinability of GFRP composite. The individual and composite desirability levels for each experimental run are shown in Table 23.5 as experimental findings. Better product quality is indicated by a greater composite desirability. We can evaluate the influence of components and establish the ideal value for each controllable parameter by analyzing the composite desirability. The primary effect plot for the composite desirability at various processing parameter levels is shown in Figure 23.1. A higher composite desirability score often denotes better performance across a variety of aspects. The AWJ cutting process's ideal process parameters were discovered in experiment number 8, which produced an output response with many criteria. For AWJ machining of GFRP composites, SOD of 1 mm, abrasive concentration of 10 wt%, feed rate of 125 mm/min, and pressure of 90 Mpaare the ideal machining settings to use.

Table 23.5 Individual and composite desirability

experiment number	Output	Individual desirability	Output	Individual desirability	Composite desirability	Rank
1	0.845	0.942	0.803	1.000	0.971	2
2	0.905	0.784	0.892	0.747	0.765	14
3	0.964	0.629	0.936	0.622	0.626	19
4	0.896	0.808	0.853	0.858	0.833	8
5	0.95	0.666	0.912	0.690	0.678	17
6	0.847	0.937	0.827	0.932	0.934	4
7	0.87	0.876	0.835	0.909	0.893	5
8	**0.823**	**1.000**	**0.816**	**0.963**	**0.981**	**1**
9	0.86	0.903	0.85	0.866	0.884	6
10	0.96	0.639	0.913	0.688	0.663	18
11	1.176	0.071	1.15	0.014	0.032	26
12	0.89	0.824	0.868	0.815	0.820	10
13	1.104	0.261	1.055	0.284	0.272	24
14	0.875	0.863	0.845	0.881	0.872	7
15	0.981	0.584	0.977	0.506	0.544	21
16	0.855	0.916	0.803	1.000	0.957	3
17	0.917	0.753	0.852	0.861	0.805	11
18	0.987	0.568	0.935	0.625	0.596	20
19	1.203	0.000	1.155	0.000	0.000	27
20	0.938	0.697	0.918	0.673	0.685	16
21	1.052	0.397	1.005	0.426	0.412	22
22	0.903	0.789	0.872	0.804	0.797	13
23	0.965	0.626	0.887	0.761	0.691	15
24	1.15	0.139	1.015	0.398	0.236	25
25	0.892	0.818	0.88	0.781	0.800	12
26	1.065	0.363	0.992	0.463	0.410	23
27	0.882	0.845	0.871	0.807	0.826	9

23.6 CONCLUSIONS

Using the Taguchi Method and DFA, this work intended to optimize the machining parameters in AWJ machining of GFRP composites. The difficulties associated with multi-response optimization were successfully overcome by the DFA integrated with the Taguchi technique. The analysis of the data yielded the following recommendations for the ideal machining parameters: SOD (1 mm), abrasive concentration (10 wt %), feed rate (125 mm/min), and pressure (90 MPa). The Taguchi technique used with the DFA increases machining effectiveness, raises product quality, and lowers manufacturing costs. However, more investigation is required to evaluate and improve the findings, taking into account variables like material differences and tool wear. The analysis does, however, set the groundwork for future developments in AWJ machining optimization for GFRP composites and related materials.

REFERENCES

1. H. Hochberg, K.R. Chang, "Material removal analysis in abrasive water jet cutting of ceramic plates," Journal of Materials Processing Technology, 40 (1994), 287–304.
2. M.A. Azmir, A.K. Ahsan, "A study of abrasive water jet machining process on glass/epoxy composite laminate," Journal of Materials Processing Technology, 209 (2009), 6168–6173.
3. A. Hascalik, Ulas C. Aydas, H. Guru, "Effect of traverse speed on abrasive water jet machining of Ti– 6Al–4V alloy," Materials and Design, 28 (2007), 1953–1957.
4. A.A. Khan, M.M. Hague, "Performance of different abrasive material during abrasive water jet machining of glass," Journal of Materials Processing Technology, 191 (2007), 404–407.
5. J. John Rosario Jegaraj, N. Ramesh Babu, "A strategy for efficient and quality cutting of materials with abrasive water jets considering the variation in orifice and focusing nozzle diameter," International Journal of Machine Tools & Manufacture, 45 (2005), 1443–1450.
6. J. Wang, W.C.K. Wong, "A study of abrasive water jet cutting of metallic coated sheet steels," International Journal of Machine Tools & Manufacture, 39 (1999), 855–870.
7. M. Palled, "A study of taper angles and material removal rates of drilled holes in the abrasive water jet machining process," Journal of Materials Processing Technology, 18 (2007), 292–295.
8. Some Studies on Abrasive Jet Machining by P K Ray, Member A K Paul, Fellow Department of Mechanical Engineering Regional Engineering College Rourkela UDC 621.921, pp. 27–29.
9. L. Nagdev and V. Chaturvedi. "Parametric optimization of Abrasive Water Jet machining using Taguchi method," International Journal of Research in Engineering and Applied Sciences, 2 (2012), 23–32.

10. A. Ali Khan, M. Effendi Bin Awing, A. Azwari Bin Annur, "Surface roughness of carbides produced by abrasive water jet machine," Journal of Applied Sciences, 5 (2005), 1757–1761.
11. J.H. Kea, F.C. Tsaia, J.C. Hungb. "Characteristics study of Flexible Magnetic abrasive in Abrasive jet machining," Procedia CIRP, 1 (2012), 679–680. Elsevier.
12. M.A. Azmir, A.K. Khan. "Investigation on glass/epoxy composite surfaces machined by abrasive water jet machining," Journal of Materials Processing Technology, 209 (2008), 122–128. Elsevier.
13. A. Hascalik, U. Cayds, H. Gurun. "Effect of Traverse speed on Abrasive in Abrasive jet machining of Ti-6Al-4V alloy," Materials and Design, 28 (2007), 1953–1957. Elsevier.
14. M.A. Azmir, A.K. Ahsan, A. Rahmah. "Investigation on abrasive water jet machining of Kevlar Reinforced Phenolic Composite using Taguchi approach," ICME (2003), 1–6.
15. V.J. Limbachiya, D.M. Patel. "An investigation of Different Material on Abrasive water jet machine," International Journal of Engineering Science and Technology (IJEST), 3(7), 975–5462, 2011.
16. A. Saha, S.C. Mondal. "Multi-objective optimization in WEDM process of nanostructured hardfacing materials through hybrid techniques," Measurement, 94 (2016), 46–59.
17. A.M. Dubey Kumar, A., Yadav, A. K., Wear behaviour of friction stir weld joint of cast Al (4–10%) Cu alloy welded at different operating parameters, J. of Materials Processing Tech., 2017; 240:87–97
18. A. Ahmed, Yadav, A. K., Singh A., Biodiesel Production from Mahua Oil: Characterization, Optimization, and Modeling with a Hybrid Statistical Approach, Int. J. of Ambient energy, 2023, 44 (1), 2600–2609,
19. Ahmed A., Yadav, A. K., Singh A., Biodiesel yield optimization from a third-generation feedstock (microalgae spirulina) using a hybrid statistical approach. 2023, Int. J. of Ambient Energy, 44(1) 1202–1213
20. H. Majumder, T.R. Paul, V. Dey, P. Dutta, A. Saha. "Use of PCA-grey analysis and RSM to model cutting time and surface finish of Inconel 800 during wire electro discharge cutting," Measurement, 107 (2017), 19–30.
21. A. Saha, S.C. Mondal. "Multi criteria selection of optimal welding parameter in MMAW hardfacing using MOORA method coupled with PCA," International Journal of Materials and Product Technology, 57(1/2/3) (2018), 240–255.
22. A. Saha, S.C. Mondal. "Statistical analysis and optimization of process parameters in sire cut machining of welded nano-structured hardfacing material," Silicon, 11 (2019), 1313–1326.
23. Ahmed A., Yadav, A. K., Singh A., An environmental impact assessment and optimization study of biodiesel production from microalgae. Int. Journal of Global Warming, 2023, 31 (3), 294–313
24. A. Banik, A. Saha, J. Deb Barma, U. Acharya, S.C. Saha. "Determination of best tool geometry for friction stir welding of AA 6061-T6 using hybrid PCA-TOPSIS optimization method," Measurement, 173 (2021), 1–33.

25. A. Saha, N. K. Mandal. "Optimization of machining parameters of turning operations based on multi performance criteria," International Journal of Industrial Engineering Computations, Growing Science, Vol. 4 (2013), 51–60.
26. Gupta, G., Agarwal, K., Yadav, A., Yadav, A.K., Sinha, D.K. (2023). Design and Fabrication of PLA-Printed Wearable Exoskeleton with 7 DOF for Upper Limb Physiotherapy Training and Rehabilitation. Lecture Notes in Mechanical Engineering. Springer, https://doi.org/10.1007/978-981-99-3033-3_5
27. Khan O., Parvez M., Yadav, A. K., Waste-to-Energy Power Plants: Multi-objective Analysis and Optimization of Landfill Heat and Methane Gas by Recirculation of Leachate: A Case Study, Process Saf. Environ. Prot., 2024; 186, 957–968
28. Ahmed A., Yadav, A. K., Singh A., 2023. Optimization of Biogas Yield from Anaerobic Co-digestion of Dual waste for environmental sustainability: ANN, RSM, and GA, Approach., Int. J. of Oil, Gas and Coal Tech., 33(1),75
29. Anand A., Singh R., Yadav A.K., Saha A., Dewangan A., Experimental investigations on the electrochemical machining of D3 die steel material and multi-objective parameters optimization, J. of Surface Review and Letters, DOI:10.1142/S0218625X24500744
30. Yadav, A.K., Khan, M.E., Pal A., Sharma D., Optimization of biodiesel production from bitter groundnut oil using Taguchi method and its performance and emissions characteristics on a 4-cylinder Tata Indica engine. Int. J. Sust. Agri. Mgt. and Informatics, (2015), 1, 285–300,
31. Vinay, Singh B., Yadav, A. K., Optimization of Performance and Emission Characteristics of CI Engine Fuelled with Mahua Oil Methyl Ester-Diesel Blend using Response Surface Methodology, Int. J. of Ambient Energy (2020) :44(6) :674–85
32. Ahmed A., Yadav, A. K., Singh A., Biodiesel Production from Mahua Oil: Characterization, Optimization, and Modeling with a Hybrid Statistical Approach, Int. J. of Ambient energy, 2023, 44 (1), 2600–2609, Doi: 10.1080/01430750.2023.2262468
33. Ahmed A, Yadav, A. K, Hasan S., A machine learning-response surface optimization approach to enhance the performance of diesel engine using novel blends of aloe vera biodiesel with MWCNT nanoparticles and hydrogen, Process Saf. Environ. Prot., 2024, 186;738–755
34. Ahmed A, Yadav, A. K, Enhanced production of methane enriched biogas through intensified co-digestion process and its effective utilization in a biodiesel/biohydrogen fueled engine with duel injection strategies: ML-RSM based an efficient optimization approach, Int. J. Hydrogen Energy, 65, 671–686, 2024
35. Ahmed A, Yadav, A. K, Parametric Analysis of Wastewater Electrolysis for Green Hydrogen Production: A Combined RSM, Genetic Algorithm, and Particle Swarm Optimization Approach, Int. J. Hydrogen Energy, Volume 59, 2024, Pages 51–62
36. Ahmed A., Yadav, A. K., Singh D.K., Singh A., A Comprehensive Machine Learning-Coupled Response Surface Methodology Approach for Predictive

Modeling and Optimization of Biogas Potential in Anaerobic Co-Digestion of Organic Waste, Biomass and Bioenergy, Volume 180, 2024, 106995

37. Ahmed A., Yadav, A. K., Singh A., Singh D.K. Agbulut U., A Hybrid RSM-GA-PSO Approach on Optimization of Process intensification of Linseed Biodiesel Synthesis Using an Ultrasonication Reactor: Enhancing Fuel Properties and Engine Characteristics with Ternary Fuel Blends, Energy, Volume 288, 2024, 129077

38. Khan O., M Parvez, Kumari, P. et al., (2023) Modelling of Compression Ignition Engine by Soft Computing Techniques (ANFIS-NSGA-II and RSM) to Enhance the Performance Characteristics for Leachate Blends with Nano-additives, Scientific Reports, 13:15429

39. Khan T.A., Yadav, A.K. Khan T.A., A hydrodynamic cavitation-assisted system for optimization of biodiesel production from green microalgae oil using a genetic algorithm and response surface methodology approach, Environ. Sci. Pollut. Res. 2022; 29:49465–49477

40. Ahmed A., Yadav, A. K., Singh A., D.K. Singh, Multi-Response Optimization to Improve the Performance and Emissions Characteristics of A VCR Engine Fueled with Microalgae Spirulina (L.): A Response Surface Methodology Approach Coupled with Genetic Algorithm, Environ. Dev. Sustain, Doi:10.1007/s10668-023-04016-z

41. Khanam, S., Khan, O., Ahmad, S. et al. (2024). A Taguchi-based hybrid multi-criteria decision-making approach for optimization of performance characteristics of diesel engine fuelled with blends of biodiesel-diesel and cerium oxide nano-additive. J Therm Anal Calorim 149, 3657–3676

42. Dewangan, A, Mallick, A, Yadav, A.K, Ahmad, A, Alqahtani, D., S Islam., Combined Effect of Operating Parameters and Nano Particles on Performance of a Diesel Engine: Response Surface Methodology coupled Genetic algorithm Approach, ACS Omega , 8 (27), 24586–24600 2023

43. Ahmed A., Yadav, A. K., Singh A., Process optimization of spirulina microalgae biodiesel synthesis using RSM coupled GA technique: a performance study of a biogas-powered dual-fuel engine, Int. J. of Environ Science and Tech. 2023; 21,169–188

44. Ahmed A., Yadav, A. K., Singh A., Application of machine learning and genetic algorithm for prediction and optimization of Biodiesel Yield from Waste Cooking Oil, Korean J. of Chem. Engg. (2023) 40, 2941–2956

45. Ahmed A., Yadav, A. K., Singh A., Enhancement of Biogas Yield from Dual Organic Waste Using Hybrid Statistical Approach and Its Effects on Ternary Fuel Blend (Biodiesel/n-butanol /Diesel) Powered Diesel Engine Environmental Progress & Sust. Energy , 2023 doi.org/10.1002/ep.14163

46. Ahmed A, Yadav, A. K, Singh A., Singh D.K., 2024. A machine learning-response surface optimization approach to enhance the performance of diesel engine using novel blends of aloe vera biodiesel with MWCNT nanoparticles and hydrogen, Process Saf. Environ. Prot., 186;738–755

47. Ahmed A, Yadav, A. K, Singh A., Singh D.K., 2024, A machine learning–genetic algorithm based models for optimization of intensification process through microwave reactor: A new approach for rapid and sustainable synthesis of biodiesel from novel Hiptage benghalensis seed oil, Fuel, 374, 132470

48. Khan, O, Khan, M.P., Yadav A.K. 2024. A Comparative Analysis of Various Nanocomposites for the Remediation of Heavy Metals from Biomass Liquid Digestate using Multi-Criteria Decision Methods, Biomass and Bioenergy, 186, 107281
49. Khan O., Yadav A.K. Exploring the Performance of Biodiesel-Hydrogen blends with Diverse Nanoparticles in Diesel Engine: A Hybrid Machine Learning K-Means Clustering Approach with Weighted Performance Metrics, Int. J. Hydrogen Energy, 2024,78, 547–563
50. Ahmed A., Yadav, A. K., Hasan S., Enhancing Efficiency and Environmental Impacts of a Nano-Enhanced Oleander Biodiesel-Biohydrogen Dual Fuel Engine Equipped with EGR, through Operational Parameter Optimization using RSM-MOGA Technique, Int. J. Hydrogen Energy, 2024, 78, 1157–1172
51. Khanam, S, Khan O., Yadav, A.K. et al. 2024 " A Taguchi Based Hybrid Multi-Criteria Decision-Making Approach for Optimization of Performance Characteristics of Diesel Engine Fuelled with Blends of Biodiesel-Diesel and Cerium Oxide Nano-additive", J. of Thermal Analysis and Calorimetry (2024) 149:3657–3676
52. Anand A., Singh R., Yadav A.K. et al. (2024) Experimental investigations on the electrochemical machining of D3 die steel material and multi-objective parameters optimization, J. of Surface Review and Letters, DOI:10.1142/S0218625X24500744
53. Equbal, A., Equbal, A., Ahmad, R., Kumar, M., Khan O., Yadav A.K., (2024) RSM based statistical approach to enhance the compressive performance of ABS P400 parts fabricated via FDM Technology, J. of Surface Review and Letters, DOI:10.1142/S0218625X25500210

Chapter 24

Enhancing Hospital Management Through Robotics
Developing a Custom API

S. Javeed Hussain, Sivasakthivel Thangavel, and Ashwani Kumar

24.1 INTRODUCTION

The contemporary landscape of healthcare management is witnessing a dynamic convergence of technological innovation and medical expertise [1]. In this context, the development of intuitive and effective user interfaces (UIs) for healthcare applications has emerged as a critical attempt. A comprehensive look at the multi-faceted process of developing a UI tailored for a healthcare management application, positioning it as a personalized assistant within the healthcare ecosystem, is discussed in this chapter.

The trajectory of the mHealth (Mobile Health) sector is composed to achieve a valuation of $450 billion by the year 2028; it is propelled by substantial investments in healthcare digitalization and the escalating prevalence of smartphones [2].

Presently, a staggering count of over 385,000 mHealth applications are accessible to both Apple and Android users, as per current data. These applications serve a dual purpose: healthcare professionals employ them to augment patient care, while patients utilize them to oversee their health.

Healthcare management applications have the potential to revolutionize how various administrative and operational tasks are executed within medical facilities. The fusion of technology and health care aims to streamline processes, enhance patient care, and optimize resource allocation. Central to this transformation is the careful design and execution of a UI that seamlessly integrates technology into the complex and nuanced landscape of healthcare [3]. This chapter navigates through the journey of designing a UI for a healthcare management application, driven by the purpose of facilitating tasks and decision-making for healthcare professionals. It explores the integral role played by research, both in terms of a competitive analysis and a focused study on a specific user group. These research components are instrumental in shaping the ideation process, channeling it toward the development of the initial wireframes that serve as the blueprint for the application's design. Incorporating user-centered design principles, the chapter underscores the significance of leveraging powerful tools like Figma to create a detailed and functional representation of the application's interface. By aligning visual aesthetics with operational functionalities, the design strives to enhance user experience (UE) and efficiency within the healthcare environment [4].

The culmination of this process results in the proposed solution embedded within the application interface. It addresses identified user challenges through a design that is both responsive and intuitive. However, the journey from a prototype to a functional tool requires a bridge between design and development. The chapter emphasizes the pivotal role of a developer in transforming the prototype into a fully operational healthcare management application. This collaborative effort ultimately ensures that the application adheres to healthcare industry standards and seamlessly integrates into the existing healthcare workflows [5]. This chapter serves as a guide through the intricate process of creating a UI for a healthcare management application that serves as a means of guiding our readers through it. With the use of technology combined with healthcare expertise, the interface aims to bridge the gap between innovation and practicality, making sure that healthcare management evolves into a technologically empowered future as a result of combining technology and healthcare expertise [6].

The primary objective of this patient-centric application is to foster patient engagement, promote patient education, and quantify patient outcomes. We recognized a high-potential market segment and sought to adapt the app to resonate specifically with the users, thereby expediting their return on investment [7]. Additionally, ensuring a seamless onboarding experience and overall app appeal across a diverse range of age groups was imperative.

To accomplish this objective, we undertook the responsibility of conducting research and comparing pivotal components of the user journey, with a strong emphasis on UE, UI and information architecture [8]. One area of focus was the app's check-in process. We also explored innovative methods to present interactive forms during the onboarding phase, facilitating smoother communication between patients and medical professionals.

Our efforts culminated in a comprehensive report that furnished health care system with app benchmarks, design insights, and actionable suggestions. We leaped deep into optimal user engagement strategies' and provided necessary recommendations to enhance the efficiency of the check-in procedure. In addition, we conducted a thorough review of the overall usability of the check-in forms and guided refining them to make them more user-friendly [9].

As a result of the results of the research, it was able to identify the most effective approach to designing the onboarding flow for the user, further enriching the overall UE for the user.

24.1.1 Health Sector TEMI Robot

In the health sector, TEMI robots have emerged as a transformative innovation that has the potential to redefine patient care, enhance operational efficiency, and revolutionize the way healthcare is delivered [10]. With a broad range of capabilities, this advanced robotic platform offers a wide range of benefits that extend beyond traditional boundaries of health care, which makes it a valuable asset in a wide variety of medical settings. The integration of TEMI into the health sector brings with it a multitude of benefits, including:

- Interactions and Engagement with Patients: TEMI is designed to serve as an interface between patients and medical professionals about patient interaction and engagement. The intuitive design and interactive features of TEMI allow patients to engage in various activities, educational sessions, and therapy exercises, thereby enhancing their sense of well-being and decreasing their feelings of isolation [11].
- Telemedicine and Consultations: TEMI enables the provision of remote consultations and telemedicine services as well as telemedicine consultations. In situations where physical presence is limited or challenging to perform, physicians and specialists can use TEMI to conduct virtual visits, discuss treatment plans, and provide guidance to patients; this is especially beneficial in situations where physical presence is limited or challenging to perform [12].
- Medication Reminders: TEMI can be configured to remind patients to take their medications at specific times, allowing them to adhere to their medication regimen and achieve better health outcomes as a result [13].
- Data Collection and Monitoring: TEMI is capable of collecting vital signs and health-related data, as well as transmitting this information to healthcare providers for remote monitoring and assessment [14]. For patients with chronic conditions, such as those with emphysema, as well as those who require continuous monitoring, this capability is particularly useful.
- Eliminating Staff Workload: By handling routine tasks and providing information, TEMI can free up the time of healthcare professionals to focus on more complex and critical aspects of patient care, thereby allowing them to devote more time to those aspects.
- Control of Infection: TEMI can minimize human contact and thus reduce the risk of spreading illnesses in environments such as hospitals, where infection control is a paramount concern.

In the process of TEMI's development within the health sector, the advantages of technology must be balanced against ethical considerations, privacy concerns, and the importance of human touch in the delivery of healthcare during its development. Nevertheless, by judiciously leveraging TEMI's capabilities, the health sector can unlock new dimensions of patient-centered care and operational efficiency by utilizing TEMI's capabilities in an effective manner [15].

24.1.2 Patient Information

An effective healthcare management system is built on the foundation of robust patient information. The term refers to a comprehensive collection of data, insights, and medical history relating to a person who seeks medical treatment and is looking for medical care. There is no doubt that patient information is one of the most important resources that healthcare providers have to be able to deliver personalized and informed medical services, keep track of the patient's medical history, make accurate diagnoses, and compose appropriate treatment plans. Personal information about an individual includes, for example, their name, age, gender, contact information, and certain identifiers, such as their medical record number, which is an important component of patient information that is kept on file [16].

Medical history provides valuable context for medical professionals when making informed decisions regarding the treatment of a patient, such as past illness, allergies, medications, surgeries, and family medical history. Additionally, it is important to note that vital signs, test results, diagnostic images, and laboratory reports provide a holistic picture of a patient's health status. Using electronic health records (EHRs) system, healthcare providers can store, access, and share patient information seamlessly, ensuring continuity of care and seamless collaboration.

In an era driven by data-driven healthcare, patient information serves as a cornerstone for medical research, population health management, and the advancement of personalized treatment approaches [17]. As technology continues to evolve, innovative tools and systems are emerging to enhance the accuracy, accessibility, and security of patient information, ultimately contributing to improved patient outcomes and overall healthcare quality.

The objective of this chapter is to conceptualize the interface for TEMI's application facilitating unrestricted mobility within Health Management systems. The design process encompasses a diverse range of techniques, including user surveys to construct user personas, prototyping evaluations, Patient registration, and an in-depth examination using robots in healthcare strategies. Following a thorough examination of the literature, these methodologies are further elaborated upon in subsequent sections.

24.2 MATERIALS AND METHODS

As the introduction to the design process begins, the first step should be to elucidate terms such as UI and UX, which are intimately related to the conception of UIs. It is undeniable that the UI of a mobile application is crucial to providing a seamless and intuitive experience in the context of building an interface for unrestricted navigation within public transportation in the context of creating a mobile application interface for unrestricted navigation. There is a strong connection between the UI aspect of product experience design and the wide spectrum of broader product development. The concept of interaction design is viewed as encompassing more than just the design of websites or applications; it encompasses everything that contributes to the way we interact with a service, an app, or a computer system. In the context of creating a mobile application similar to that of a help desk registration, the importance of this understanding cannot be overstated. Generally, UE encompasses all aspects of the user's engagement with a company, its services, and its offerings from the point of view of the customer.

By contrast, the term UI refers to the tangible visual elements that users are engaged with directly as they interact with the application. It is the aesthetics, tactile experience, and interactive dynamics of a digital product that form the cosmetic layer of the overall experience that composes a successful digital product [18].

As a result of simplifying the UI, we can further enhance the visual aspect of the application, which is predominantly the interface between the app developer and the user. UE design encompasses a range of graphical attributes such as style, typography, color, and graphics, culminating in the ultimate stage of the design process in terms of UE. UI is built on a foundation of communication, where the UI effectively functions as a dialogue between the user and the product, as a tool to accomplish tasks and achieve the user's goals [19]. When it comes to the design of a UI, the goal is to make it both user-friendly and comprehensible.

In contrast, UX is concerned with addressing the needs of users. A person's perceptions and responses resulting from the use or anticipation of a system, product, or service are defined as perceptions and responses by ISO 9241-210. At its core, UE refers to the series of subjective psychological occurrences and states (perceptions) that an individual undergoes during an interaction, which can only be accessed by that individual. An individual's expressions and reactions can be interpreted because of these events and states based on their expressions and reactions.

This is a process involving both analytic and strategic components, which encompass both the planning processes and the introduction of the product to the market during the process. By orchestrating and synchronizing elements involved in user interaction with a particular company, the ultimate aim is to influence how users perceive and behave towards that company.

During the comprehension phase, it is the primary objective to gain a better understanding of the user, encompassing their desires, needs, environment, and the context in which they will engage with the product. A crucial part of the subsequent phase defining, is to dissect the insights gathered during the comprehension phase and to articulate the problems that need to be addressed. Identifying the user requirements lays the foundation for defining the purpose of the product's design, which can then be used to determine the overall design of the product. For the ideation phase to develop, after identifying the user's problems and comprehending the rationale for finding a solution, the predicaments of the user must be identified. As part of this phase, designers strive to generate a multitude of potential solutions to the identified problems to select the most optimal solution for prototyping from the multitude of proposals generated in this phase. It might be advantageous to revisit previous phases of the process, as you may be able to gain valuable insight from user research or a competitive audit at this point in the process.

To prototype the solution to the problem, the prototyping design phase is the next step after formulating a solution to the problem. As a result, the objective here is to create an initial product model that demonstrates functionality and is intended to be tested to determine its performance. Tests on prototypes are designed in a way that refines the prototype as insights are assembled into how it can be used to address the user's issues and the ease with which it can be used.

This study employed a quantitative approach as one of its research methodologies, specifically questionnaire research, which is more commonly known as a survey. In terms of data collection, surveys are one of the fastest and easiest methods, though they do not have the depth that qualitative survey methods possess. UX researchers design a set of questions that are tailored to the needs of a specific group of respondents. Unlike qualitative approaches, this method provides objective data; however, unlike the qualitative approach, it does not offer an in-depth analysis of respondents' motivations; however, it does provide insight into their real motivations, which is a distinct advantage. I believe, it is of utmost importance that the questions are objective, and they do not suggest a particular response in any way. Open-ended questions can prove beneficial to your survey, as they allow respondents to express their opinions comprehensively, even though they do require a fair amount of effort and might result in survey abandonment on your part. A semi-open question provides a spectrum of options for respondents to choose from, as well as allowing them to express their viewpoints. Depending on the type of question and the depth of the discussion, closed questions are of a limited depth, allowing respondents to select one or more responses. As a result, the importance of diversifying the question types in a survey cannot be overstated. The purpose of this study was to conduct research on a small cohort of healthcare consumers by administering a questionnaire to them. The questionnaire was made available through popular social networks using a Google form, with the consent of the participants, which ensured that their responses would remain anonymous. As part of the study, the objective was to assess the accessibility and appropriateness of the applications provided by TEMI, particularly the registration process for patients and information about doctor availability for the Health Care Management System.

24.2.1 Design Process

It is essential that before the design of an interface that caters effectively to the needs of users, a thorough understanding of the eventual receivers of the product must be conducted. As a result, it is imperative to understand their application usage patterns, pinpoint what challenges they face, and identify a target demographic for the solution that will appeal to them. UIs are used as a channel by which users can communicate with devices, in this case, mobile devices, to convey information between them. There is a direct correlation between the quality of the designed interface and the perception that customers have of the usability of the system [20]. There is a high probability of product abandonment when an interface proves to be overly complex or lacks ergonomic design due to its overcomplicated nature. A product that is intuitively navigable and offers an extensive array of features is in contrast to one that offers a comprehensive range of functions.

There is no doubt that people demonstrate individual differences, so it is paramount to value each user as an individual and treat them with special care. There is a possibility that what might seem straightforward and instinctive to one person might be very different to another. There is an illustration of the UX design process in Figure 24.1 that illustrates how the interface works.

Diverse users can encounter a variety of issues, which demonstrates how essential it is to construct a target group that covers a wide range of users. In the initial phase of the design process, it is important to define the user's goal – the intended outcome following a particular interaction with the system. It is possible to gain insights regarding users through empirical research methods such as direct observation, case studies, or surveys in order to gain insights into their behavior. To generate personas [18], which are virtual representations of specific groups of users, it is necessary to collect this information. As part of a larger group of users, personas encapsulate key characteristics that attribute key attributes to each member of the group. To develop a persona, it is necessary to gather information such as a person's age, profession, health concerns, patient experience, prehistory of health, and other pertinent details that are relevant to the persona [16]. Consequently, they allow the identification of shared objectives and pain points that hamper users' ability to accomplish their goals. In addition to visual representations of their daily environment (persona

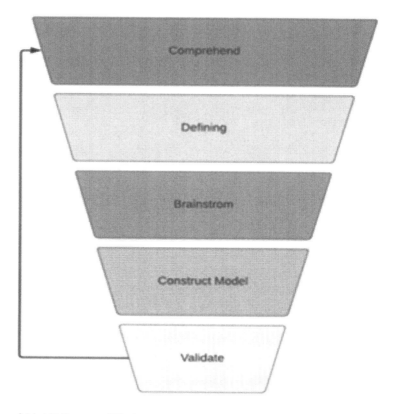

Figure 24.1 UX Process of Design.

environment), personas may also incorporate narratives that illustrate their usage scenarios (persona background) as well. The purpose of this process is to advance a deeper level of understanding, thus facilitating the cultivation of empathy on both sides. A design thinking process aligned with user-centered principles is based on such insights, laying the groundwork for a user-centered design process. To create personas, it is necessary to identify patterns—common elements that are present among particular types of users. It is possible to construct multiple personas for distinct groups that share common characteristics. In most cases, it is sufficient to develop four to nine personas to cover the majority of users of the product.

24.2.2 Comprehend Map

To develop empathy maps aligned with the needs and challenges of the personas that you are working on, it is beneficial to construct empathy maps.

Using this approach, you must dissect the apprehensions, frustrations, obstacles, and desires of the client, as well as their requirements and desires. An evaluation of this type is typically divided into six distinct quadrants: "hears," "sees," "thinks," "feels," "says," and "does." These quadrants all hold equivalence, although there is no chronological or sequential order within them. An example of an empathy map can be seen in Figure 24.2, which shows an illustration of an empathy map.

The "says" quadrant is designed to address inquiries such as:

- What did the persona articulate?
- Are there any inferences that can be drawn about them based on the information provided?
- When you are having a conversation, what mannerisms do you exhibit?

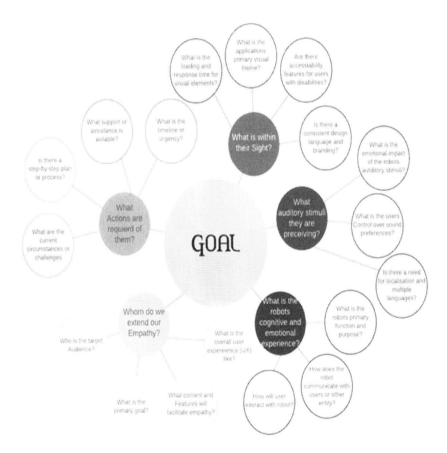

Figure 24.2 Comprehend map.

- How does a typical day for them look like?
- What would a typical day look like for them?

There is a tendency for this quadrant to include direct quotes that are selected directly from user interviews. In the "thinks" quadrant, you are asked questions regarding what your personality is like, your surroundings, what you observe, what you read, etc. In this case, "does" characterizes the persona's actions, revealing how they engage in specific activities and the methods they use for doing them. The "feels" quadrant pertains to the emotional state of the persona, elucidating their worries and sentiments that are entwined with their interaction with the product and how it affects them.

Data-driven empathy maps not only deepen user understanding but also:

- Aids in mitigating biases.
- Illuminates research vulnerabilities.
- Uncovers the driving forces propelling user behavior.
- Fosters improved product development and innovation.

Defining and addressing user predicaments is at the heart of the UX design process. In order to craft a product that effectively addresses their needs and challenges, it is necessary to take into consideration their needs and challenges, even if they are not conscious of the issues that they are facing. Steve Jobs is famous for saying that "designing a product with focus is very difficult. Most people do not know what they want until you show them it." A similar sentiment can also be found in the statement made by Henry Ford, who said, "If I asked customers what they wanted, they'd say faster horses." As a consequence, both quotes highlight the fact that individuals are frequently unaware of their underlying problems, highlighting the difficulty involved in designing a flawless product, which in actuality, does not exist.

On considering the previously conducted research, the identification of a persona ensued, embodying an illustrative account of their background, requisites, and sources of vexation. The formulation of personas proved indispensable in addressing inquiries concerning their identity and the specific necessities of the individuals under scrutiny.

24.2.3 Competitive Audit

As part of the creation process, a competition audit emerges as a pivotal step. This audit entails a comprehensive evaluation of the strengths and weaknesses exhibited by rival entities. This endeavor involves scrutinizing competitors that provide akin products, such as planner applications, Hospital health workers' timetable apps, or other thematically relevant applications. The meticulous assessment of these counterparts yields invaluable insights into the marketplace within which the forthcoming product is set to debut. The prospective product should be impeccably aligned with the market's demands and possess a competitive edge. Often, this advantage transcends mere functionality and extends to an enhanced UE. As a result of meticulously examining an organization's achievements, its product positioning, and identifying opportunities for improvement, it is possible to craft a unique offering that fills a need in the market by identifying the holes within the current offering.

Executing a competition audit entails several procedural stages. The bottom line lies in precisely defining the audit's objective. In this instance, the goal revolves around contrasting the solutions proffered by direct or indirect rivals. Entities like Jacky's APK, mobileMPK, and Neuroget are considered direct competitors as they furnish analogous services and target the same user base. Conversely, indirect competitors assert comparable offerings but appeal to a distinct demographic.

The designated aspects for evaluation encompass a spectrum of considerations, including the initial impression, visual design appraisal, assessment of user communication, and other factors tailored to the product's requirements. Each aspect boasts distinct characteristics and diverse facets for comparison. To construct a well-defined comparison framework, empirical research is indispensable, necessitating the examination of competitors' applications. This approach is pivotal for gaining insights into the methodology and mindset employed by rivals, analogous to stepping into their shoes. Sighting errors entrenched within the product's construction can be facilitated by the "5 X Why" technique, also recognized as the "5 Reasons" method. By iteratively posing five inquiries, this methodology systematically guides attention toward pinpointing the fundamental cause of a problem. The culmination of the audit is underscored by concluding reflections that encapsulate the report's findings.

24.2.4 User Flow

The concept of user flow describes how a user interacts with an application through the process of taking a series of steps, encompassing the journey from the user's initial step to the end of the application. A typical way of depicting this is through diagrams where distinct shapes are used to represent different types of interactions. A circle represents an action that a user performs, a rectangle represents a screen, a rhombus signifies a decision point, and an arrow illustrates the direction in which the user proceeds and indicates backward navigation (with dashed lines indicating forward progression).

An understanding of how users navigate within an application or website is essential to the design of applications or websites, as it capitalizes on insights into the way users navigate within the application. It is often the case that the beginning of a redesign process is the initial landing page, the welcome screen, or the home page of the designed application that represents the starting point, also known as the "entry point." An outcome of the user flow, such as creating an account, is what culminates the user flow. Using this process as a basis for assessing and optimizing the UE, designers can ultimately boost the conversion rate by providing a framework for assessing and optimizing the UE.

It is up to the user to decide which of the three options they would like to pursue—they can proceed to the screen that displays tickets and schedules, use the search engine, or choose to click on buttons that indicate saved routes (such as "home"). A user may choose one of the paths indicated by the menu bar positioned at the bottom of the screen, regardless of the screen that they are currently on. In addition to this, the user retains the flexibility to go back at any time during the journey if they need to.

Wireframe, a fundamental component of UI/UX design, is conceived during the initial phases of project development to secure customer and user approval. This process furnishes project teams with the groundwork for further application refinement and facilitates testing and adjustments as needed. Wireframes entail the visual representation of elements, layout, architecture, flows, and behaviors. They constitute a pivotal and highly effective technique for crafting visualizations that steer subsequent interactions. Essentially, wireframes serve as the blueprint delineating the application's structure, operational concept, and visual depiction of the interface.

Wireframes are commonly categorized into two main types: low-fidelity wireframes and high-fidelity wireframes, with some sources also proposing medium fidelity. These classifications diverge based on the level of intricacy. Low-fidelity wireframes, usually rendered in shades of gray, encapsulate the fundamental structure, outlining designated spaces for navigation, menus, buttons, and images. Conversely, high-fidelity wireframes entail a higher degree of detail. Initial wireframe drafts might be executed by hand, yet for ease and enhanced prototyping capabilities, leveraging software such as Adobe XD, Figma, or Sketch proves a more advantageous choice. These software tools are purpose-built for fabricating prototypes, facilitating usability assessments, and enabling heuristic evaluations.

24.3 RESULTS

The development of a healthcare management system app utilizing the TEMI robot has yielded promising results. The app serves as a comprehensive solution aimed at enhancing various aspects of healthcare delivery and patient interaction. Key outcomes and features of the app include:

- Patient Engagement: The app has significantly improved patient engagement by offering personalized and interactive experiences. Patients can access their medical records, appointment schedules, and prescription details seamlessly.

- Appointment Management: With the app, patients can easily schedule, reschedule, or cancel appointments. The integration of the TEMI robot with the healthcare provider's system allows for streamlined communication between the two systems.
- Access to Information: Using the app, patients have access to a variety of educational resources, which enables them to learn more about their medical conditions, treatment options, and healthy lifestyle choices to maintain a healthy lifestyle.

There has been positive feedback from patients as well as healthcare professionals regarding the TEMI healthcare management system app as shown in Figure 24.3. In addition to its integration with the TEMI robot, this technology offers convenience, accessibility, and improved patient outcomes as well as a modern touch to healthcare services. The app continues to be

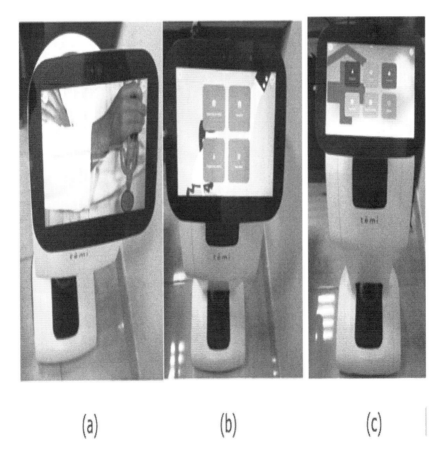

Figure 24.3 (**a**), (**b**), and (**c**) Interface proposed interactive screens.

enhanced and refined regularly, fostering a more patient-centered and efficient healthcare ecosystem as a result of the updates and refinements.

24.4 DISCUSSION

We discuss the results of our evaluation in this section, where a survey was conducted on healthcare users in the form of a questionnaire using a Google form that was shared on popular social networks to collect data. A total of 12 questions were included in the questionnaire. The purpose of the study was to examine the availability and adaptability of TEMI applications such as patient information that is used to get around with consulting doctors, and to determine whether they are compatible. As part of the study, 35 people between the ages of 20 and 49 took part in the survey. As shown in Figures 24.4 and 24.5, the data collected during the survey and the questions included in the survey are presented.

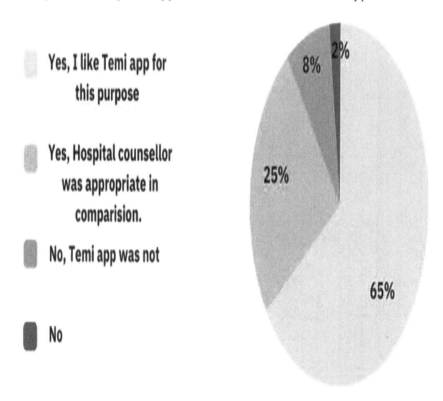

Figure 24.4 Google Forms Survey results—Adaptability of robotic application.

The Survey respondents, predominantly aged between 25 and 49 years, showed that 48.6% fell within the 25–49 age group, which is most likely to be engaged in using the application. The initial survey question aimed to ascertain whether respondents like to use the application or helpdesk, serving as an introductory query to establish the actual percentage of individuals utilizing TEMI's apps for this purpose. Surprisingly, 65.2% reported using TEMI's apps, while a notable 25.5% continued to rely on the traditional helpdesk approach. A smaller faction, comprising just 8%, admitted to not utilizing TEMI's apps for healthcare management services.

The awareness that 65.2% of respondents employ TEMI's apps for health care services underscores the widespread adoption of technology in this context. It is crucial to recognize that 25.5% of users continue to depend on the conventional helpdesk approach. Consequently, when designing the application, it becomes necessary to consider such users and ensure they are not left at a disadvantage. One feasible approach is to incorporate a simulation of the helpdesk feature into the app's design. This acknowledgment underscores the importance of enhancing education in this area.

The answers depicted as per the survey shown in Figure 24.5 to the question: "What features of the TEMI app are essential for you in the healthcare system?" These provide guidelines for understanding the user needs for healthcare services. The features are voted as such:

- About Doctor specialization (66.67%).
- About Doctor Availability (93.34%).
- Hospital Map (34.3%)
- Possibility of choosing different Doctors (56.67%).
- Robot to be interactive (96.6%).
- Possibility of updating user information (84.3%)
- Appointment booking feature (66.67%).

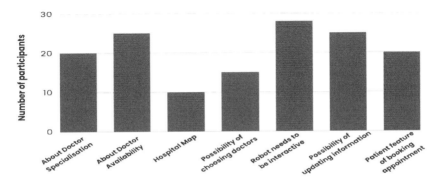

Figure 24.5 Google Forms survey results—vital features.

Incorporating features that align with users' preferences, such as providing real-time updates on potential delays and offering alternative commuting options, can serve as a compelling incentive for users to embrace and consistently utilize this application. When individuals perceive the app as valuable, user-friendly, and customized to their specific requirements, they are more inclined to integrate it into their daily healthcare routines. Consequently, this leads to increased app adoption and a boost in overall healthcare robots' usage. Such an outcome has the potential to cultivate user loyalty, differentiate the product from competitors, and attract a broader user base.

Evaluating technological proficiency holds significant importance, and the research yielded predictable outcomes. Out of the total respondents, 85.5% reported seamless daily use of TEMI's apps without any issues. Despite accounting for the features required, 14.5% of respondents faced challenges while using these apps (as indicated in Figure 24.6). In addition, these individuals were invited to discuss the difficulties encountered with the application.

It is worth noting that only three respondents chose to respond to this optional question, yielding the following insights:

- Performance of the app is slow and hangs at times.
- Not finding the app's design to be insightful.

The research findings, as depicted in Figure 24.6, underscore that the majority (85.5%) exhibit a high level of proficiency and mastery in utilizing

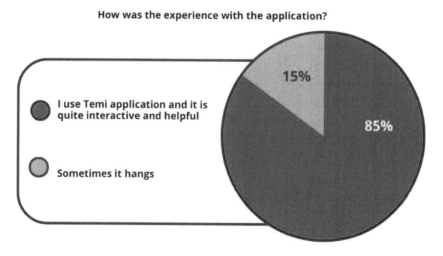

Figure 24.6 Google Forms survey results—level of competence.

healthcare services apps daily. This data is invaluable when developing an app, particularly for understanding the target audience.

However, the study also illuminates the fact that a portion of users (14.5%) encounter difficulties. This underscores the importance of comprehending the unique challenges they face and devising solutions within the app. The feedback provided sheds light on specific areas for improvement, including modifying excessive information, streamlining the transition between different applications, and enhancing UE and interface design. The responses from these select individuals offer a fresh perspective and unique insights that can guide future improvements.

24.5 CONCLUSIONS

This chapter examines the challenges associated with using TEMI robots in the healthcare sector. This emphasizes the fundamental role that apps play in modern life, as well as the inadequacy of traditional methods of patient information in meeting contemporary user expectations in terms of patient information. This chapter comprehensively examines the design considerations that are integral to developing a solution that caters to user needs and is informed by the UI/UX design process to develop a solution that meets those needs. While constructing the design wireframes during the ideation phase, the incorporation of research and competitive analysis serves as a foundation for constructing the design wireframes. As a result of this initial groundwork, preliminary mock-ups of the application are crafted that demonstrate the key elements of the application as well as their low-fidelity behavior. Figma, which is one of the most advanced design tools on the market, enables the development of a refined app appearance and functionality by leveraging advanced design tools. Plans, as mentioned in the introduction, include conducting usability studies to fine-tune the design to align it with the needs of the users. A finalized and approved design will then proceed to the implementation phase for eventual market release once the design has been refined and approved. The design interface of the application is part of an ongoing endeavor aimed at offering solutions for prevalent issues that are faced by users of the application in the healthcare field. This prototype will eventually lead to a tangible product that caters to the user's needs, which will necessitate collaboration with developers to bring the envisioned application to life using the programming tools available to them. Essentially, this study aims to address a critical gap in the understanding of the perspectives of users on healthcare systems through the use of a robot designed specifically for this purpose. It is through careful analysis of the feedback provided by users, as well as their experiences, that valuable insights can be uncovered, leading to iterative design improvements that enhance both UEs and environmental sustainability.

Funding: This work was supported by a grant from the Ministry of Higher Education and Research, Innovation (MOHERI), Project number MOHERI/BFP/HSS/GCET/529/21.

ACKNOWLEDGEMENTS

The authors want to acknowledge the support and facilities provided by the Global College of Engineering and Technology, Muscat, Oman.

REFERENCES

[1] Raje, S., Reddy, N., Jerbi, H., Randhawa, P., Tsaramirsis, G., Shrivas, N. V., ... & Piromalis, D. (2021). Applications of healthcare robots in combating the COVID-19 pandemic. Applied Bionics and Biomechanics, 2021, 1–9.

[2] Pradhan, B., Bharti, D., Chakravarty, S., Ray, S. S., Voinova, V. V., Bonartsev, A. P., & Pal, K. (2021). Internet of things and robotics in transforming current-day healthcare services. Journal of Healthcare Engineering, 2021, 1–15.

[3] Bogue, R. (2020). Robots in a contagious world. Industrial Robot: The International Journal of Robotics Research and Application, 47(5), 673–642.

[4] Gupta, A., Singh, A., Bharadwaj, D., & Mondal, A. K. (2021). Humans and robots: A mutually inclusive relationship in a contagious world. International Journal of Automation and Computing, 18, 185–203.

[5] Ruan, K., Wu, Z., & Xu, Q. (2021). Smart cleaner: A new autonomous indoor disinfection robot for combating the covid-19 pandemic. Robotics, 10(3), 87.

[6] Khan, H. R., Haura, I., & Uddin, R. (2023). RoboDoc: Smart robot design dealing with contagious patients for essential vitals amid COVID-19 pandemic. Sustainability, 15(2), 1647.

[7] Соколов, К. Е. (2022). Humans and robots: a mutually inclusive relationship in a contagious world. In Социум. Наука. Образование (pp. 108–108).

[8] Flessa, S., & Huebner, C. (2021). Innovations in health care—A conceptual framework. International Journal of Environmental Research and Public Health, 18(19), 10026.

[9] Najjar, R. (2023). Redefining radiology: A review of artificial intelligence integration in medical imaging. Diagnostics, 13(17), 2760.

[10] Bajwa, J., Munir, U., Nori, A., & Williams, B. (2021). Artificial intelligence in healthcare: Transforming the practice of medicine. Future Healthcare Journal, 8(2), e188.

[11] Dwivedi, Y. K., Hughes, L., Baabdullah, A. M., Ribeiro-Navarrete, S., Giannakis, M., Al-Debei, M. M., ... & Wamba, S. F. (2022). Metaverse beyond the hype: Multidisciplinary perspectives on emerging challenges, opportunities, and agenda for research, practice and policy. International Journal of Information Management, 66, 102542.

[12] OECD. (2016). Innovating education and educating for innovation: The power of digital technologies and skills. OECD.

[13] Korinek, M. A., Schindler, M. M., & Stiglitz, J. (2021). Technological progress, artificial intelligence, and inclusive growth. International Monetary Fund.
[14] Steps to Conduct a Competitive Audit|Coursera. Available online: https://www.coursera.org/learn/start-ux-design-process/supplement/WUoyC/steps-to-conduct-a-competitive-audit (accessed on 22 April 2023).
[15] Empathy Mapping: The First Step in Design Thinking. Available online: https://www.nngroup.com/articles/empathy-mapping/ (accessed on 22 April 2023).
[16] Al-Manji, A. N. S. H., Thangavel, S., Kumar, A., & Heidari, M. (2024). A Study on Waste Management Practices in Oman and a way forward to a Circular Economy. In *Biodegradable Waste Processing for Sustainable Developments*, A. Prasad, & A. K. Paul (Eds.), pp. 129–149, chapter 6, Boca Raton FL USA: CRC Press. DOI: 10.1201/9781003502012-6
[17] How to Define a User Persona [2023 Complete Guide]. Available online: https://careerfoundry.com/ (accessed on 17 June 2023).
[18] What Is A Persona? [Complete Guide for 2023]. Available online: https://careerfoundry.com/ (accessed on 17 Junel 2023).
[19] Breiman, L. (1996). Bagging predictors. Machine Learning, 24(2), 123–140.
[20] Heinermann, J., & Kramer, O. (2016). Machine learning ensembles for wind power prediction. Renewable Energy, 89, 671–679.

Index

Abrasive water jet, 360, 368, 369
Academic curriculum, 7, 333, 335, 337, 339, 341, 343, 345
Actor network theory, 216
Advanced automation, 35, 47
AI-driven automation, 32
AI ecosystem, 44
AI writing tools, 61–64, 70, 337, 339
Artificial intelligence, 7, 9, 10, 16, 25, 26, 47, 50, 51, 53, 55, 88, 93, 211, 257
Augmented intelligence, 43, 47
Augmented reality, 11, 30, 70, 74, 94, 124, 137, 139, 146, 254
Autonomous vehicles, 6, 25, 37, 38, 246, 248

Bagging, 162, 297, 304, 391
Bandgap, 348–350, 357
Behavioral analytics, 266
BEMS, 39
BLE, 148
Boosting, 162, 190, 211, 296, 297, 304, 339, 348

Classification models, 260
Clustering, 157, 159, 165, 174, 372
Cobots, 3–6, 49, 84, 93, 123, 133, 145, 146, 191, 214, 253, 298
Cognitive assistant support, 216
COMSOL, 349, 350, 352, 353, 357, 359
Concept of operations, 213, 214, 216, 225, 227, 228
Cryptocurrency, 257, 274
Cryptocurrency crime, 258
Cyberattack surface, 258
Cybercrime, 12, 257–260, 264
Cybersecurity, 79, 81, 97, 130, 140, 156, 197, 201, 242, 248, 257, 260, 263, 268–271, 287, 288

Decision-making, 15, 32, 335
Decision tree, 155, 156, 164, 229, 260, 263, 266, 294

Design of experiment, 364
Device, 8, 34, 35, 39, 51, 89, 92, 93, 97, 113, 119, 120, 122, 123, 125, 352, 358
Digitalization, 3, 27, 73, 78, 130, 134, 144, 155, 373
Digital twins, 4, 9, 10, 11, 12, 13, 19, 37, 50, 94, 139, 153, 254, 304
Digital twins for personalization, 37
Dynamic production planning, 37

Edge AI, 206
Electrical resistivity, 346
Emerging risks, 144
Emotion recognition, 41
Energy management, 38, 49, 95, 185–187, 189–191
English for academic purposes, 55
Ensemble, 161, 162, 165, 212, 266, 296, 297
Ethically aware design, 215

Field variable, 349
Fleet management, 18
Fraud detection, 263
Fraud detection and prevention, 25
Function, 63

Geometry, 348–350, 358, 369
GFRP composite, 364, 365, 367, 368
Globalized, 57, 78, 156, 157, 159, 161, 164
Goal oriented behaviour, 221

Healthcare, 22, 39, 40, 42, 44, 51–54, 233, 236, 287, 290, 333, 340, 373; AI in healthcare, 30, 53, 54
Health care management, 373, 374, 376, 379, 384, 385, 387
Height, 1, 81, 92, 162, 229, 249, 250, 274, 312, 319, 350, 352, 355, 356
Higher education, 71, 72, 333–336, 342, 344, 390

393

High school education, 55, 57, 59, 61
History, 26, 42, 51, 54, 70, 81, 154, 295, 376
Human-centric, 73, 117
Hyperautomation, 30, 43

Industrial revolutions, 3, 117, 126, 187, 192, 207
Industry 5.0, 03, 54, 88, 89, 90, 93, 94, 130, 136, 148, 210, 339, 340, 341
Integration, 5, 9–11, 14, 16, 18, 28, 30, 35, 37, 38, 47, 49, 375, 385, 390
Intelligent tutors, 57
Internet of Things, 3, 19, 21, 26, 34, 45, 47, 81, 89, 100, 120, 128, 134, 168

Jobs, 20, 50, 54, 80, 88, 95, 117, 125, 135, 137, 144, 146, 154, 231

K-means, 159, 266

Leg length, 348, 352, 355
Linear, 2, 107, 110, 158, 163, 266, 293, 296, 319, 352
Linear regression, 110, 266, 293, 296

Machine learning, 33, 39, 49, 52, 53, 57, 66, 69, 71, 81, 86, 88, 89, 92, 93
Malware attacks, 258
Manufacturing industry, 12, 16, 42, 84, 92, 122, 134, 139
Manufacturing system, 19, 20, 29, 35, 45, 76, 86, 88, 100, 129, 151, 165, 166, 213, 223
Medical imaging, 2, 39, 42, 122 246, 268, 390
Modules, 150, 301, 349

National crime records bureau, 257
Natural disasters, 13, 36, 268
NLP, 31, 33, 153, 154, 162, 246, 249, 266, 273, 277, 283, 285, 286

Optimizing, 52, 139, 156, 188, 204, 225, 242, 246, 254, 294, 295, 299, 311, 340, 364
Outer junction, 350, 352

PCA, 157, 158, 369
Peltier effect, 346, 347
Predictive analysis, 81, 154, 263, 290
Predictive maintenance, 3–6, 13, 14, 20, 31–33, 35, 42, 47, 49, 50

Quality control, 1, 3, 4, 7, 8, 14, 31, 34, 42, 49, 93, 138, 242, 313, 316
Quantum dot, 348
Query, 162, 164, 249, 387

Random forest, 263, 266, 293, 294, 297
Ransomware, 257, 258, 271, 272
Reduction, 11, 14, 16, 88, 121, 137, 144, 145, 157, 158, 174, 187, 206, 266, 291
Revolution and society, 339
Robots, 1, 3–5, 82, 145, 312, 313, 333, 341, 375, 376, 388–390

Safety and security, 102
SDGs, 150
Secure adaptiveness, 265
Seebeck effect, 346, 347
Smart cities, 12, 19, 21, 102, 155, 157–159, 161, 162, 164, 192
Smart manufacturing, 2, 20, 23, 34, 35, 45, 54, 73, 81, 120, 213, 226
Society 5.0, 339, 340
Sultanate of Oman, 60, 61, 67, 71
Supply chain, 14, 123, 127, 244, 251, 253, 264, 290, 291, 297, 298, 304, 308, 311, 312
Supply chain risk management, 36
Support vector machine (SVM), 157, 260, 266, 283
Symbiotic partnership, 98, 218

TEMI robot, 375, 384, 385
Temperature, 292, 294, 296, 299, 311, 312, 346, 348–350, 352, 353 355, 357, 358
Threat detection, 201, 263, 269
Threat intelligence, 263, 265, 269
Traceability, 138, 140, 299, 309
Transductive, 208, 160
Transportation, 6, 7, 13, 16, 26, 35, 37–39, 74, 75, 87, 113, 119, 235

Unsupervised, 157–161, 164, 165, 237, 266
User identification, 263
UX design, 379, 382, 384, 389

Variable, 50, 103–107, 109, 110, 113, 124, 143, 155, 158, 175, 186
Virtual learning environments, 58, 71
Virtual reality, 58, 94, 124, 137, 139, 146, 254
Visual recognition, 41, 247
Voice commerce, 41
Voltage sensitivity, 349, 352, 354

Weighted, 163, 173, 372
Wide bandgap, 349
Work related stress and disease, 143
Writing curriculum, 61, 67

XAI, 45, 127

Young, 58, 69, 113, 164, 365

9781032798202